Culinary SCHOOLS

WHERE THE ART OF COOKING BECOMES A CAREER

Peterson's
Princeton, New Jersey

About Peterson's

Peterson's is the country's largest educational information/communications company, providing the academic, consumer, and professional communities with books, software, and online services in support of lifelong education access and career choice. Well-known references include Peterson's annual guides to private schools, summer programs, colleges and universities, graduate and professional programs, financial aid, international study, adult learning, and career guidance. Peterson's Web site at petersons.com is the only comprehensive—and most heavily traveled—education resource on the Internet. The site carries all of Peterson's fully searchable major databases and includes financial aid sources, test-prep help, job postings, direct inquiry and application features, and specially created Virtual Campuses for every accredited academic institution and summer program in the U.S. and Canada that offers in-depth narratives, announcements, and multimedia features.

Visit Peterson's Education Center on the Internet
(World Wide Web) at www.petersons.com

Editorial inquiries concerning this book should be addressed to the editor at Peterson's, P.O. Box 2123, Princeton, New Jersey 08543-2123.

ISBN-1-560-79-943-9

Executive Editor: Robert Crepeau
Senior Book Editor: Arthur Stickney
Production Editor: Karen Rae
Xyvision Product Manager: Gary Rozmierski
Research Director: George Geib
Research Analysts: Dan Karleen; Rick Sears
Designer: Cynthia Boone
Programmer: Alex Lin

Printed in the United States of America

10 9 8 7 6 5 4 3 2 1

CONTENTS

CHOOSING A COOKING SCHOOL

ANDREW DORNENBURG and
KAREN PAGE

No matter how strong a chef's inspirations, they are not enough to give rise to greatness. They must be carefully honed and refined through directed effort. The palate, which allowed a chef to first learn what he or she found most enjoyable, must be trained to discern subtleties in flavors and flavor combinations, and to critique as well as taste.

Similarly, basic cooking techniques must be mastered, with speed and efficiency developed over repeated efforts, in order to be able to create desired effects. This is what leads chefs into professional kitchens and, increasingly, into professional cooking schools.

The Role of Education in a Cooking Career

Cooking is a profession which places extraordinary emphasis on continuous learning, and today a chef's first formal education often takes place in a cooking school classroom. There are hundreds of cooking schools in both the United States and abroad, offering opportunities to learn about specialties ranging from vegetarian to confectionery to microwave cooking. (In fact, *Peterson's Culinary Schools* lists more than 340 programs.)

While many of the country's leading chefs reached the top of the profession without the benefit of a cooking school degree, an overwhelming majority of the chefs recommend cooking school as the most expeditious start for an aspiring chef today. Cooking school offers an opportunity to gain exposure in a concentrated period of time to an immense amount of information, from cooking tech-

Andrew Dornenburg and Karen Page are coauthors of the 1996 James Beard Award-winning book Becoming a Chef: With Recipes and Reflections from America's Leading Chefs *(New York: Van Nostrand Reinhold, 1995, © by the authors), from which this article is excerpted with permission. Their other books include* Culinary Artistry *(Van Nostrand Reinhold, 1996), which has won national media acclaim, and* The Food Intelligentsia *(Van Nostrand Reinhold, forthcoming in 1998). Mr. Dornenburg has worked in the kitchens at top restaurants in New York and Boston, including Arcadia, Biba, The East Coast Grill, Judson Grill, and March. Ms. Page has been an industry consultant on culinary trends, marketing strategy, and new product development.*

niques (knife skills, sauté, grill) to theory (nutrition, sanitation), to international/regional cuisines (French, Italian, Asian).

A cooking school diploma can also be an important credential in opening doors and demonstrating commitment to the field. "I only hire cooking school graduates," says Patrick O'Connell, who himself doesn't hold a cooking school degree. "If I had to do it again at this point in history, I would probably go to culinary school." Alfred Portale is even more adamant. "I think that if you can go to cooking school, you should. I feel very strongly about it," he says. "It immediately legitimizes you as a professional, and exposes you to a broad base of information, even though not much of it is practical. It certainly puts you at a greater advantage than someone who's self-taught or learns going up through the ranks."

While attending cooking school full-time represents a certain trade-off in terms of the opportunity cost of foregoing a full-time income while at school, the vast majority of chefs interviewed see it as an investment well made. In fact, the cooking school naysayers have little criticism for the cooking schools themselves; they reserve it for the popular misconception that merely attending cooking school can create a chef, which they believe often misleads people without a real passion for food and cooking into the profession. "Cooking schools do an important job," says Anne Rosenzweig, "but the final results depend a lot on the students." Victor Gielisse concurs: "Cooking school gives aspiring chefs a tremendous foundation. But school alone cannot give you a passion for food. It's impossible. Not even the best teacher can do that."

Given the abundance of reputable cooking schools and programs, there is likely to be an option to suit everyone's specific budget, time frame, and other needs. From four-year bachelor degree programs, to certificate programs which can be completed in a few brief months, to one-session cooking demonstrations by culinary experts, there are numerous opportunities to learn about cooking in a classroom. With the hundreds of cooking schools available, it is up to the prospective students to research various options to determine which offers the best fit.

CHOOSING THE *RIGHT* COOKING SCHOOL

The decision as to whether and where to attend is highly personal and dependent on many factors: How long a program best suits your needs? Is location a factor for you? What is your budget for school? Are you looking to attend full-time or part-time? During the day, or in the evening? While it is not our intention to recommend specific schools, we hope that the following tips on evaluating cooking schools will help guide your eventual decision.

- Consider taking the time to ask some chefs at restaurants in your area to recommend cooking schools to you.
- Read all of the literature available from each of the schools you're considering. This alone should address some of your most basic concerns, such as the range of course offerings and whether the program costs are within your reach.
- Plan to spend a day at each of the schools you're seriously considering.
- Ask a lot of questions of administrators, instructors, and even students.

Find out whether they offer (or require) an externship program or an international study program. Ask what kind of placement assistance they provide to graduating students and alumni. Find out where some of their recent graduates are working and what the breadth of the school's network is. Who are some of their most successful alumni? Are typical alumni working in the kinds of places you'd like to work in someday? Spend time talking with other students, and decide if they're the type of people with whom you'd enjoy learning and spending time. You'll be happiest at a school that offers a "good fit."

Certain schools attract a national (and even international) student body. Other schools tend to attract a greater proportion of their students from their regions, and local chef alumni may show some preference to graduates of their alma maters over graduates of other programs. Still other schools tend to draw more closely from their immediate vicinities.

Schools also differ in their orientation to particular types of cuisine—everything from the haute cuisine of France to low-fat vegetarian cooking.

Programs vary in length and emphasis. Four-year programs often offer courses in management, marketing, communications, nutrition, and financial management in addition to hands-on cooking. Two-year programs augment course work with intensive kitchen experience.

For those unable or unwilling to dedicate two to four years to their culinary education, another option is a short-term program. Upon completing a course and passing practical and written examinations demonstrating mastery of those courses, a diploma is awarded. Students can extend their training by participating in an extern program, which places students in the professional area of their choice for an additional period.

Don't think that if you can't attend school full-time you can't receive a fine education. "If you don't have enough money,

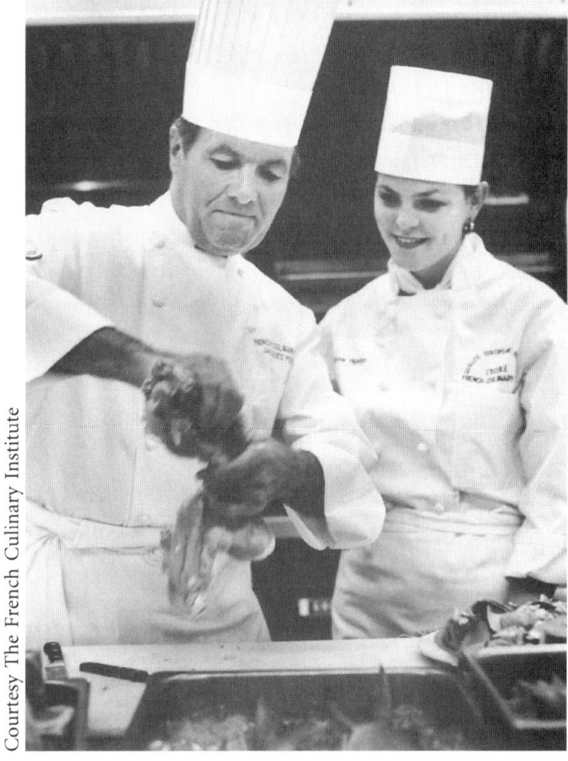
Courtesy The French Culinary Institute

you can still attend part-time," emphasizes Dieter Schorner. Schorner, an experienced chef, has taught classes in the evening at New York City Technical College. "I found some of the most dedicated teachers at this little college," he says. There are more than three hundred vocational schools and community colleges offering programs in culinary arts listed in this guide.

Admissions Requirements and Processes

If you're aiming for a competitive program, getting in presents the next hurdle. Some of the top schools aren't able to admit everyone who applies. At other, less competitive programs, merely submitting an application and application fee is basically all it takes. Given the increasing demand for trained cooks and chefs, many cooking schools have been expanding their programs to be able to accommodate more students.

The admissions process, depending on the school, may be simple and straightforward or relatively more involved. The Culinary Institute of America, for example, requires a completed application form, an application fee, and a high school transcript, as well as transcripts from any postsecondary studies and at least two letters of reference from employers, food service instructors, or Culinary Institute alumni. The application form itself asks whether applicants have traveled extensively in or outside the United States, attended seminars/lectures on the food industry, and read books and/or magazines about the food service industry. In addition, applicants must submit an essay explaining why they wish to enter the food service field, what research on the industry they have done, details of their food service background, and why they're interested in attending The Culinary Institute of America.

The New England Culinary Institute requires applicants to submit a written personal statement on their background and experiences, why they have decided to seek a career in the culinary arts, and their reasons for applying to the New England Culinary Institute. Also required are three letters of recommendation; copies of high school, vocational school, or college transcripts; and the application form and an application fee.

While some local cooking schools will have less rigorous application processes, most schools will require you to apply well in advance of your desired date of attendance, so plan ahead.

Before Attending Cooking School

Some schools see prior work experience as a plus but not a prerequisite. Many cooking schools, however, require work experience in a kitchen as a condition of admission.

Indeed, getting some real-world exposure to a kitchen is a good idea whether the school admissions process requires it or not. Not everyone who's attracted by the perceived "glamour" of the profession finds that the reality of the work is a good fit. Is cooking something you could come to love as a profession? It's best for all concerned that a reality check be taken sooner rather than later.

Larry Forgione says, "I think people really ought to think about whether this is what they really want to do before they jump into it with both feet. I might suggest stepping in with one foot—working weekends at a place, hanging out at a restaurant to understand what the restaurant business is all about." While the glamorous image of restaurant business might help to attract people to the profession, only those who have actually worked in a kitchen know the intense effort involved.

In addition to practical experience, other preparation can provide an edge at cooking school. If you can't work full-time in a kitchen, consider getting some part-time experience working for a caterer, gourmet store, or as a waiter or waitress. Spend as much time as you can reading about food and cooking, if you don't already. In addition, restaurant experience before cooking school offers an opportunity to determine what area a student might want to specialize in—for example, cooking on the line or pastry.

You may, in addition, have prior academic or work experience in a field unrelated to cooking. Prior experience can be a great help in preparing you for the rigors of a tough cooking school program and difficult first jobs. You are likely to have already developed study habits and writing skills. Also, food is so basic that few fields don't offer some degree of overlap—from art to science to history. Alfred Portale's work in jewelry design certainly hasn't hurt his reputation for artistic presentation, and Barbara Tropp's academic training has helped her to succeed as both a chef and an author. If you have already invested time and energy in another field, you are likely to ask good questions and have a clear idea of what you are looking to get out of your education.

AT COOKING SCHOOL

While at cooking school, it's important to make the most of your opportunity. Use classroom time as an opportunity to ask questions and learn as much as possible. Take advantage of extracurricular learning. Volunteer to assist professors and visiting chefs with special events, and make an effort to get to know them. Aside from the knowledge they can pass along, these interactions get you noticed and may spawn leads on future jobs.

Some chefs found that the most important lessons they learned came outside the classroom. "I spent a lot of time in the library," says Gary Danko.

Danko advises, "Be serious when you go to cooking school. Make sure you have some books to draw from, and that you spend some time and really pick the instructors' brains. Ask them every question you can. Spend extra time cooking with instructors. Volunteer for things—because that's where the real learning comes from, putting in the extra time."

Many chefs emphasize how much their in-class learning was complemented by their outside work in a kitchen while attending school and suggest that students maximize their learning by working while attending school. Allen Susser says, "I think you need to see the real side of the ideal things you're cooking in the school kitchen." Mary Sue Milliken goes so far as to recommend that students work full-time while they attend

Courtesy The French Culinary Institute

cooking school: "If the schedule is too rigorous, this is probably not a good career choice."

Working while attending school can serve many purposes. It can reinforce what you are learning in the classroom, and point out what you are not. Jasper White advises working during cooking school "because then you can start applying things every day, and retain a lot more." It can give you the extra practice on your knife and other skills, and allow you to pursue other interests in food, such as ethnic cuisine or catering, that may not be covered. White loved his work at the Waccabuc Country Club in Westchester County, NY: "We did buffets on Sundays; I got to use a lot from school that I wouldn't have had a chance to use otherwise."

Any experience is what you make of it—you don't have to work at a four-star restaurant in order to learn. Susan Feniger worked at a fish market in Poughkeepsie, NY, while attending The Culinary Institute of

America. A former coworker of mine worked at the sandwich counter at a wine and cheese store while attending cooking school, in order to educate himself about cheeses. I know another cooking school student who chose to work in catering instead of a restaurant because of the flexibility in scheduling, which allowed her to learn and practice her culinary skills without the intense pressure of daily service at a restaurant. She was also able to see a wide variety of food at the catering company—from 200 wontons to an entire theme dinner centered around the early settlers in Boston. There is a price to pay for all of this—less sleep, not to mention less energy and time for other school activities. While the burden of such a workload might give pause to some, leading chefs seemed to be in agreement with Nietzsche's principle: "That which does not kill me makes me stronger."

Externships

An externship is a period of time spent working in a restaurant for the purpose of gaining practical on-the-job experience and may be paid or nonpaid. Some schools mandate serving an externship as a requirement for graduation, while others offer it as an option to students. Certain schools even offer a chance to work abroad or at leading resorts. It's important to research the options available through the school you're interested in and, if there's a particular restaurant you'd like to serve an externship with, to take an active role in doing all you can to pursue it. Find out which faculty or administration members have the strongest ties to the restaurant and speak with them, write a letter to the chef, or possibly offer to work for free until the time your official externship commences.

Every experience is different. Some restaurants have students cook on the line, while others allow externs to do only basic prep work. André Soltner requires that The Culinary Institute of America students he takes into his kitchen at Lutèce (NYC) commit to a minimum one-year externship, which maximizes both the student's learning and his restaurant's ability to benefit from it. If an externship is offered or required, make the most of it. Work in the best kitchen you can get hired into. If you make a good impression, you may find yourself with an offer for permanent employment; at worst, you'll end up with a good reference for future jobs. At the New England Culinary Institute, 70 percent of second-year internships turn into first jobs.

One extern I know was passed over for a permanent pastry position due to his inexperience, yet kept coming in on his own time to keep his skills sharp while looking for a job at another restaurant. When the pastry chef left unexpectedly, he was hired into the position. Another extern received an offer to join a restaurant as garde-manger because of her excellent work habits and attitude (including coming in early, staying late, working fast, and taking direction well). While she had no experience, her work habits spoke volumes about her potential and she was given a chance to prove herself.

AFTER GRADUATION

Chefs emphasize the importance of viewing graduation from cooking school as merely a starting point in one's education. André Soltner puts it this way: "I think cooking schools give students the basics they need. But they are not accomplished chefs. They are just coming out of school. A doctor, after his four years, goes to a hospital not as the chief surgeon but as an intern. We have to look at cooking school graduates as what they are." Schools and parents were attributed with feeding students' expectations. "Parents send their kids for two years at The Culinary Institute of America and then think they are André Soltner or Paul Bocuse, " Soltner notes. "But they are not."

Cooking school graduates might find themselves tempted with offers to become full-fledged chefs upon graduation—welcomed by students, at least in part, due to the sometimes substantial debt incurred through financing cooking school. However, leading chefs speak discouragingly of the notion of accepting a job as a chef too soon. In her speech to a graduating class at The Culinary Institute of America, Debra Ponzek used the opportunity not to pump up graduates' hopes and expectations but to implore them: "Don't take a chef's job!" She explained, "It's hard to go back, once you realize there are things you didn't learn. Many people want to make the jump [to a chef's position] too quickly." Patrick Clark agrees. "Don't look for glory right away. When you get there, it's harder."

Some chefs advise that after graduation you should work with your "idols," in order to continue your education. Upon graduating from The Culinary Institute of America, Alfred Portale answered an ad and was selected to work at the food shop that Michel Guérard was opening at Bloomingdale's. He recalls: "Here I'm just out of school, and I'm standing in a kitchen with Michel Guérard, the Troisgros brothers—all these huge French guys, all my idols. It was thrilling. I learned all the butchering and the charcuterie, the poaching and the smoking, and the stuffing and the sausage making, and all that kind of stuff that young cooks dream about. After putting in a year with these guys, they invited me to France. So I spent a year working first at Troisgros and then with Guérard. I had a car and toured France, spending the last six weeks in Paris, going out every day and every night, going to every bookstore, every cooking store, just learning and submerging myself in everything."

After graduating from The Culinary Institute of America, Gary Danko began a dogged cross-country pursuit to track down Madeleine Kamman, to persuade her to let him study with her. While Danko says she expressed reservations about working with a newly minted culinary school graduate, he set out to change her mind. "I pulled up in my car the first day of class with all these local products that I'd been working with—goat cheese, guinea hens, ducks, geese, lamb, you name it. She saw that I was very serious about cooking and she sort of took me under her arm."

CONTINUING EDUCATION

Just because you graduate from cooking school doesn't mean the learning process ends. In this profession, it should never end.

Allen Susser says, "Going to school is only one of the first steps to growth and development in understanding what to do in a kitchen, and what to do with cuisine."

While I was cooking during the lunch shift at Biba, I made it a point to attend Boston University's Seminars in the Culinary Arts in the evening, where I was able to take classes with local chefs, such as Jody Adams and Gordon Hamersley, and visiting luminaries like Julia Child, Lorenza de'Medici, Julie Sahni, and Anne Willan. The Boston Public Library sponsored a Cooks in Print series, which, in lecture format, offered wonderful opportunities to learn about food from leading chefs like Jasper White. The Schlesinger Library at Radcliffe College also offers panels of leading culinary experts on various issues, from food safety to customer service, which are open to the public. To find similar kinds of programs in your local areas, check with adult education and continuing education programs, with your local library, or with local chapters of national associations such as the International Association of Wine and Food.

Many cooking schools also sponsor continuing education programs for working chefs, where one can spend anywhere from a day to a month learning more about cooking.

Nancy Silverton took a break from her career to attend Lenôtre (the namesake school of noted French pastry chef Gaston Lenôtre), in France, to study pastry. "I was working at Michael's, which at the time was considered one of the top restaurants in Los Angeles, and my desserts were very well regarded. I went to Lenôtre and thought I might end up teaching them a few things." Silverton was surprised and humbled to find herself in classes with pastry chefs, some of whom owned their own pastry shops and who had been working in the field for 30 to 40 years. Silverton admits, "That's when I first came to realize that you never learn it all."

If you do decide to pursue cooking school, as recommended by the majority of chefs we interviewed, the most important point to keep in mind is to stay focused on what you're hoping to get out of it. Work for at least a year, in order to confirm your interest, before making such an important investment in your career. This will give you a more realistic view of the profession and also make you more focused once you're at school. If an externship is offered, take advantage of it. Work at the best restaurant you can get into, and learn and absorb everything you can. And, once you graduate, beware the paradox of "commencement": you're not a master chef yet—you've just taken the first important step in beginning to acquire an important base of knowledge on which to build. As Jimmy Schmidt says, "Remember: a building is only as strong as its foundation. If you don't have a strong foundation, you can never erect a skyscraper."

PAYING FOR YOUR CULINARY EDUCATION

MADGE GRISWOLD

Culinary training can be expensive—nearly as expensive as attending a fine private university and sometimes more expensive than attending a fine public university. Few prospective students can simply pay the bills from their own savings or depend on their parents' generosity to fund their studies. Fortunately, help is available through loans and other financial aid programs and also in the form of scholarships.

Finding and Applying for Scholarships

Once you have decided which schools are most appealing to you, contact the financial aid department at each to determine exactly what kinds of assistance are available. Some schools even have work-study arrangements that allow students to work part-time and study as well.

Financial aid departments at culinary schools administer financial aid programs of various kinds and also provide basic advice about securing educational loans. You should consider this advice to be an integral part of your overall career planning. Some loans take many years to pay off, for example. Also, students may not travel abroad while holding certain kinds of federal assistance.

You may be eligible for grant assistance if financial need is proved. A grant is an outright award of money, whereas a loan must be paid back. Financial aid officers are more than happy to counsel you about such options.

Madge Griswold is an author and culinary researcher, Executive Chairman of the International Association of Culinary Professionals Foundation, founding Chairman of The American Institute of Wine & Food's Baja Arizona chapter, and member of the James Beard Foundation. Among her most recent publications are two sections of Culinaria: America Food, *Cologne: Könemann.*

Financial Assistance at a Glance

Loans—available to persons who qualify for them. Loans, unlike grants and scholarships, must be paid back once you graduate. Ask the financial aid advisers at the schools you are interested in for details or ask about student loans at your local bank.

Work-Study Programs—programs that allow students to work while studying. Ask the financial aid advisers at the schools you are interested in whether such programs exist there.

Grants—outright awards based on financial need. Ask the financial aid advisers at the schools you are interested in for details.

Scholarships—awards for culinary study based on talent or potential for excellence in the culinary field. Scholarships are awarded both by schools and organizations related to the culinary field. If you are interested in applying for culinary scholarships, ask about them when discussing options with financial aid officers. In addition, you should contact the appropriate scholarship-awarding organizations listed at the end of this section.

Scholarships, unlike grants based solely on financial need, are awards based on talent and potential. Financial need may or may not be considered when a scholarship is awarded. Requirements for scholarship candidates are established by the donors of specific scholarships. Scholarships are awarded by culinary schools themselves and also by a number of professional culinary associations. A serious applicant in need of substantial financial assistance should explore all of these avenues.

When you plan for your culinary education, you'll want to find the best mix of loans, work-study programs, grants, and scholarships. In many cases, loans for education can be used to support a student's whole educational experience, including tuition, room and board, books, tools, and transportation. Grants and scholarships, on the other hand, may be limited to specific uses. And scholarships might only be applied to tuition.

If finances play a large part in planning your culinary education, consider programs at some of the less prominent schools described in this guide. Many schools have excellent reputations locally and offer generous financial aid packages and scholarship assistance. In fact, you may get more personal attention throughout your education at such places. And remember, culinary education is only part of preparing for a career in the culinary field. The rest is *you*—your knowledge, your talents, your creativity, and your overall work experience.

Applying for Scholarships

Almost all culinary schools award scholarships to truly promising students, so remember to ask for information about each schools' requirements. In addition, many organizations associated with the culinary field award scholarships.

You might want to begin addressing the scholarship application process even before you send in your admission application because the scholarship process can take longer than the admission process. Write or call the organizations listed below if they give scholarships appropriate to your educational needs.

Each organization that provides scholarship aid has its own criteria for making awards, its own application process, and its own time frame. Many of these organizations evaluate applications only once a year. Some do it two or three times a year. Some organizations offer scholarships only for specialized kinds of study or to assist specific kinds of persons. Others administer a broad range of scholarships.

Scholarships awarded by organizations (other than schools) are of two kinds:

- Awards for specific schools. These awards are usually for tuition credit, although occasionally some aid is given for room and board, books, uniforms, or tools.
- Awards of a specific cash value that can be used at a variety of institutions. Awards like these are usually designated by the donors to be applied only against tuition. They cannot be used to help pay for room and board, books, tools, uniforms, or getting to and from school. Cash-value awards generally are paid by the awarding organization directly to the chosen institution. Rarely, if ever, is money paid directly to a student to use as he or she wishes.

Since decisions about scholarship aid are based on the promise of achievement in the culinary field and not just financial need, it's a good idea to think through what kind of impression you want to make on the person or committee who will be evaluating your application. Here are some pointers given by an experienced scholarship committee judge:

- Fill the application out neatly. Type it if possible; neatness makes a good impression. Remember, you will be asked to perform your culinary work neatly. A neatly prepared application gives an indication to readers that you can do that.
- Fill in all the information requested. You could be disqualified for not supplying requested information.
- Check your spelling carefully. If you are unsure of how to spell a word, look it up in a dictionary. Spelling errors detract from the message you are trying to communicate.
- Submit all materials requested to accompany the application.
- Be sure that all materials requested to be sent separately *are* sent separately. Frequently letters of recommendation are requested separately to give some privacy to the referees and to ensure that they actually are the authors of the letters.
- List all work experience in the culinary field. Culinary work often demands long hours and considerable physical and mental effort. The fact that you have worked in the culinary field and understand the demands of your chosen career is important to judges. If you have done volunteer work in the culinary field, be sure to list that as well.
- Throughout the evaluation process, be prepared to explain your goals in life and your plans for the next few years. Lofty ambitions may be lauded but unrealistic plans are not.
- If you are asked to write an essay, write it yourself. This should be obvious; after all it's cheating if someone else writes your essay. A reviewer who suspects that you have not written your own essay may disqualify your entire application. Don't try to impress reviewers with flowery language or French culinary terms unless you actually have worked in a French kitchen and need to describe a station or task in French because of it. Either of these will be seen as an affectation. Don't list impossible or outrageous goals. If you are 18, it's unlikely that in five years you will be the executive chef in a prominent hotel or own your own restaurant. Remember that the readers of your application are food professionals who are well aware of how long it takes to achieve a

position in the field or how much it costs to start a restaurant. If you are having difficulty setting time frames, get advice from a teacher or mentor.

- When you write an essay, be original. If you have had unique experiences that have influenced you to become involved in the culinary profession, by all means include them, but don't say that ever since you were a little child you have wanted to go into the culinary field. It may be true, but it's trite. Tell the reader what is unique and special about you, why you deserve the scholarship, and what it will enable you to do with your life.

The Scholarship Interview

At least one of the organizations listed here conducts telephone interviews with award finalists. Actually, these interviews can be fun because you may find that you and your interviewer have many experiences and ideas in common. Your interviewer will probably put you at ease quickly. Your interviewer is interested in learning how well you speak and how well you present yourself. Conversation with you will convey to the reviewer how committed you are to your culinary goals and something about yourself other than your culinary side. You might be asked "what do you do when you are not cooking?" Your interviewer will also try to make sure you really understand just what the scholarship can do for you and what it cannot do. Often interviewers act as advisers to candidates, pointing out opportunities they may have overlooked. An interview can be an excellent opportunity for you to present what is unique about you and why you deserve to be given a specific award. Look forward to this opportunity.

Where to Find Scholarships

In addition to the schools themselves, a number of organizations award scholarships for culinary education. Read this list carefully before requesting information. Some organizations give scholarships only for very specific purposes. Others give scholarships only for management training or for graduate work.

American Culinary Federation (ACF)

This long-established association of professional cooks has a membership of more than 25,000 and chapters in many cities. In addition to its apprenticeship program, which provides an excellent alternate approach to culinary training, the ACF awards some scholarships on the national level. Many chefs are members of the ACF. Contacts at the local level will be able to provide information about any scholarships awarded by a local chapter. For information about a local chapter, contact the ACF at 800-624-9458. Information about scholarships at the national level may obtained from this number or by writing to:

ACF
P.O. Box 3466
St. Augustine, FL 32085

American Dietetic Association (ADA)

This organization is comprised of 69,000 persons, 75 percent of whom are registered dietitians. It only awards scholarships for registered dietitians working toward master's degrees. If you think you are eligible, contact them at:

American Dietetic Association
216 W. Jackson Blvd.
Chicago, IL 60606-6995
Phone: 312-899-0400
Fax: 312-899-4845
URL: http://www.eatright.org.

American Institute of Wine & Food (AIWF)

A nonprofit organization created to promote appreciation of wine and food and encourage scholarly education in gastronomy, this 10,000-member organization has 30 chapters in cities inside and outside the United States. Many chapters of the AIWF give scholarships for culinary education. Some are administered by the individual chapters; others are administered through the facilities of the International Association of Culinary Professionals Foundation scholarship committee. For more information, see the information about the International Association of Culinary Professionals Foundation, below. You can reach AIWF at

The American Institute of Wine & Food
1550 Bryant St., Suite 700
San Francisco, CA 94103
Phone: 415-255-3000 or 800-274-2493

Confrerie de la Chaine des Rotisseurs

A long-established organization for promoting appreciation of fine food and wine, this society has chapters in nearly 150 cities and generates approximately $150,000 in scholarships each year. Scholarships are established directly with culinary schools. Interested candidates should ask financial aid officers at specific schools about these awards.

Council on Hotel, Restaurant and Institutional Education (CHRIE)

This umbrella trade and professional organization offers no scholarships of its own but instead produces *A Guide to College Programs in Hospitality & Tourism* that lists some available scholarships. Contact them at:

CHRIE
1200 17th St., NW
Washington, DC 20036-3097
Phone: 202-331-5990
Fax: 202-785-2511

Careers through Culinary Arts Programs (CCAP)

A school-to-work program, established in a number of major metropolitan areas, CCAP integrates culinary training at the high school level with work and business experience. Students in these programs are eligible for nearly $1,000,000 in scholarships annually. If you are a high school student, ask your guidance counselor if there is a CCAP program in your area and how you might participate.

Les Dames d'Escoffier

This association has chapters in many major cities. One of its major purposes is the creation and awarding of scholarships to assist women with culinary training. These scholarships usually apply to a specific geographical area surrounding the chapter's home city. Some of these are awarded directly by the chapters. Others are administered by the scholarship committee of the International Association of Culinary Professionals Foundation (IACPF). Women who are interested in applying for such scholarships should watch their local newspapers for announcements of scholarship competitions or write to the IACPF at the address given below.

Educational Foundation of the NRA (National Restaurant Association)

An educational organization that produces a variety of courses, video training sessions, seminars, and other educational opportunities for persons in the hospitality industry, this group also offers scholarships for management training.

Educational Foundation of the NRA
250 S. Wacker Dr., Suite 1400
Chicago, IL 60606
Phone: 312-715-1010

International Association of Culinary Professionals Foundation (IACPF)

This charitable and educational affiliate of the International Association of Culinary Professionals has, as one of its functions, the administration of over 50 scholarships that provide either tuition-credit assistance at specific institutions or financial assistance that can be applied to a variety of institutions. In 1997, awards totaled more than $220,000. The IACP Foundation scholarship committee awards scholarships on an annual cycle, with an application deadline of December 1 for scholarships beginning the following July 1. Interested applicants should consult the IACPF office, since deadlines sometimes change. Contact them at:

IACP Foundation
304 W. Liberty St., Suite 201
Louisville, KY 40202
Phone: 502-587-7953 or 800-288-4227
Fax: 502-589-3602

International Association of Women Chefs and Restaurateurs (IAWCR)

An association specifically designed to promote the education and advancement of women in the culinary professional and to promote the industry overall, the IAWCR awards three scholarships for women annually. All awards are for locations in California. One of these is for undergraduate study; the other two are for continuing education. For further information, contact them at:

IAWCR
110 Sutter St., Suite 305
San Francisco, CA 94104
Phone: 415-362-7336
Fax: 415-362-7335
E-mail: iawcr@well.com

International Food Service Executives Association (IFSEA)

A long-established educational and community service association, this group provides some scholarships of its own and also provides information about scholarships offered by other organizations.

IFSEA
100 S. State Rd. 7, Suite 103
Margate, FL 33068
Phone: 954-977-0767
Fax: 954-977-0874

International Foodservice Editorial Council (IFEC)

Dedicated to the improvement of media communications quality in the food field, this small organization of food service magazine editors and public relations executives awards between four and six scholarships annually to persons seeking careers that combine food service and communications. Contact Carol Metz, IFEC Executive Director, at:

IFEC
P.O. Box 491
Hyde Park, NY 12538
Phone: 914-452-4345
Fax 914-452-0532

James Beard Foundation, Inc.

This prominent organization of food professionals, devoted to the ideals and principles of legendary American cook and writer James Beard, awards a number of substantial scholarships each year. The application deadline is March 1 and awards are made in early May.

James Beard Foundation, Inc.
167 W. 12th St.
New York, NY 10011
Phone: 212-675-4984 or 800-362-3273

Roundtable for Women in Foodservice

As association established to promote and encourage women in all areas of food service, this organization is in the process of establishing a scholarship program.

Ms. Debbie Hicks, Executive Director
Roundtable for Women in Foodservice
1372 La Colina Dr. #B
Tustin, CA 92780
Phone: 714-838-2749
Fax: 714-838-2750

CHARTING A SUCCESSFUL CULINARY CAREER

BARBARA SIMS-BELL

You love to cook; garlic essence smells better to you than an expensive French perfume; your friends and family say you should have a restaurant (well, maybe not your family); and right now you're seriously studying the choice of the best culinary training you can afford to allow you to live your dream. But it is never too early to contemplate the future, to think about opportunities that will come along after culinary training.

What Else Can You Do with a Cooking Degree (besides open up a restaurant?)

Chef, caterer, pastry cook, and restaurant cook are merely the most familiar four; there are hundreds of jobs in the food industry. You may want to consider preparing for positions in management as executive chef, or sales as catering director, or administration in food and beverage management. Maybe you'll want to explore developing specialty products—a line of sauces, dressings, or convenience foods, for example—for retail or wholesale markets. There are also teaching opportunities in professional cooking schools (possibly even the one you choose to attend). Others set out to become a restaurant consultant to entrepreneurs who want to start a restaurant or improve the one they own. Still another option is food writing and editing for magazines and books devoted to food and cooking.

Barbara Sims-Bell is the founder of the Santa Barbara Cooking School and the author of two award-winning cooking books, Career Opportunities in the Food and Beverage Industry, *New York: Facts on File, 1994, and* FoodWork: Jobs in the Food Industry and How to Get Them, *Santa Barbara, Calif.: Advocacy Press, 1993.*

For any of these career directions, you'll find the best and the broadest preparation in an accredited school program. You will come out with a certificate or a degree, and forever after when you are asked "where did you get your training?" you can refer to an accepted and respected credential in professional cooking. This training will provide you with a lifelong basis for understanding quality raw ingredients, creating balance and pleasure in combined flavors, and presenting a beautiful plate to the diner.

Yes, you keep learning, but culinary school gives you a base of knowledge to test and compare to new trends, new ingredients, and your own creativity.

WHERE CAN I GO FROM HERE?

Most culinary students, when they start their training, believe they have found the work they want to do for the rest of their lives, and many are right. Some are surprised when they find so much routine and boredom and repetitive tasks. You haven't seen appetizers until you've assembled 3,000 identical stuffed puffs for a hotel reception. House salad? You'll clean and prep cases of the same greens and garnishes day after day. And the signature white chocolate mousse and meringue dacquoise layers you always wanted to perfect? You'll be preparing untold orders for it every evening. You have to love it.

Rick Webb

Rick Webb is Director of an alternative vocational culinary program in the El Paso High School District. He is training teenagers in professional kitchen skills to equip them for jobs in restaurants and institutional kitchens, and he motivates the more ambitious students to go on to culinary academies and other restaurant and hospitality industry school programs.

Rick got hooked on cooking as a young child in his mother's kitchen and got an after-school job cooking at Pizza Hut when he was only 15. His manager saw him as a promising youth and sponsored him for Pizza Hut management training. After high school graduation, his first full-time restaurant job was at the El Paso Westin Hotel and again he had a boss who encouraged him, this time to go on to culinary school.

Following his professional training at Scottsdale Culinary Institute, Rick was rehired, just where he wanted to be, at the El Paso Westin, working again with the chef who had been his mentor. It was the "rush" of cooking on the line that Rick found so exciting; he speaks happily of being "slammed" with orders. He lent some of his minimal free time to El Paso school district cooking classrooms to guest teach and encourage teens to consider a career as professional chefs. His life was fully satisfying: a chance to improve and advance in his job and a solid marriage at home. Then something happened.

When his son was born, he began to resent the industry hours. Sixty or more hours a week didn't leave him much time to be with his infant son. Rick's priorities did a complete flip-flop. He says, "I wanted to prepare myself for him." Luck just happens. The high school district wanted a chef/director to initiate a vocational school program in restaurant skills, and they came to Rick to seek his input and offer him the job. The time he can spend with his child is precious, and now it's shielded by a job that accommodates his home life. Now, he says, it's the rush he feels when his students succeed that gives him a satisfying thrill in the work he's doing.

If managerial positions are more to your liking, you'll need skills in addition to cooking. Managers create the working environment for the staff, often developing a sixth sense to recognize problems before they rupture. They are the motivational force that drives the staff. They must understand finance and business reports and their implications. They must have highly sensitive character judgment and the ability to manage people from hiring to mentoring and to firing.

If your interests take you into catering and sales, think about these skills: You'll need to be able to research a product and explore your market. You'll need to really enjoy being with people. You'll need to draw on strong self-esteem to take "no" and not take it personally. You'll need internal discipline to keep the work flowing. You'll need communication skills to persuade people that your product is best. And you'll need to be strongly motivated to make a sale.

Where Do I Fit?

To choose a career path that seems right for you, you'll need to define your own personality profile, whether it gives you the skills you need if you want to move higher or take a detour and move sideways. Or do you need to add some skills that you haven't yet developed in yourself?

One approach is to see a qualified career counselor for an evaluation of your strengths and weaknesses. Even if you reject or overrule the findings, you may gain an understanding of yourself that you didn't have before. Career testing extracts from us an inventory of our preferences.

Professional career counselors have the training, experience, and credentials to help you explore some of the possible choices that tempt you. They use finely tuned tests such as Myers-Briggs, Holland, and Strong Interest. Then they interpret the test reports to give you additional guidance, either to follow your obvious bent or to stretch yourself into other areas with training and exploration. As in everything, there are quacks and there are bona fide wizards. The best course is to check the credentials of anyone you're considering with the National Career Development Association in Worthington, Ohio. (see page 23).

Whether you seek outside career guidance or not, you should do some soul searching on your own. Take stock of who you are. What are your best skills? Break them down into culinary, service, finance, research, communication, and management. Some of the categories will be longer than others; that tells you where you've placed your learning emphasis and where you'll have to work a bit harder if you want to beef up a meager one. Think about your lifestyle and workplace values. Is independence something you seek, or do routine and stability matter more? Are you aiming for wealth or is leisure time now more important? Another significant list is what leisure activities you enjoy the most, then rating them by cost, whether they are solitary or social activities, and whether you've been able to fit them in to your life lately. Are you a risk taker or do you proceed with caution? Even the most cautious of us can be successful entrepreneurs, but your own slant between these two types is important for you to know.

Getting the Whole Culinary Picture

An easy and enjoyable way to learn about the spectrum of food-related jobs is by joining one or more professional organizations. Among the largest are the American Culinary Federation (ACF), the International Association of Culinary Professionals (IACP), and Roundtable for Women in Foodservice (RWF). For contact

Stephanie Hersh

Stephanie Hersh defined her career goal of being a pastry chef at the age of 6 with the gift of a Betty Crocker Easy-Bake Oven. It was simplistic cause and effect: she produced the sweet offerings from batters, and everyone fluttered around telling her she was "terrific." She figured this could last her whole lifetime if she just kept on making cake. Her granny lived nearby and regularly let the diminutive yet determined youngster bake alongside her, making family desserts and good, sweet stuff.

Her parents were a harder nut to crack, insisting on scholastic accomplishment, first in her private high school, then in a four-year college. Looking back, she thinks it was the best for her because "I needed to grow up before going to culinary school." She worked part-time in restaurants, both front of the house and back, making some headway in the cooking hierarchy as she became more experienced. The work was everything she dreamed: it gave her pleasure, satisfaction, self-esteem, self confidence. Her personality drove her to "always be the best," and she knew that to be best she had to have professional training. The restaurant business was changing at that time, and she knew it would no longer be possible to work up in kitchens from dishwasher to executive chef. Restaurant owners were hiring the applicant with the best culinary education. In 1985, Stephanie graduated from the Culinary Institute of America in Hyde Park, New York.

Her first professional job as a hotel pastry chef in Boston was a rude awakening. For the first time, pastries became work to do and get done, and it wasn't fun. She still had her pastry shop dream, and to feed her savings faster she devised a dual plan, based on her new goals—"I just wanted to cook and enjoy it and make money." Stephanie took a job as a private chef, live-in, for a small family with two professional working parents and two children. With her daytimes freed up, she enrolled at Katharine Gibbs School, figuring she could still private chef and work days as a secretary, with almost no living expenses. Then something happened.

Julia Child phoned the school asking if they had a graduate to recommend, commenting that it would be nice if the person knew something about cooking. When they described Stephanie's culinary background, she turned her down saying that she really wanted a secretary not a chef. Stephanie was in the school office when it happened and asked permission to call Julia back so she could press her case for herself. There was serious persuasion involved on Stephanie's part and the repeated statement that "I just want a secretary; I don't need anyone to work in my kitchen." Stephanie agreed that she just wanted to be a secretary. The next morning, minutes after arriving at the Cambridge house, Stephanie was in the kitchen prepping three recipes for demonstration stages and cooking the aromatic fish stew to finish for serving. Julia had forgotten she had agreed to a television taping/interview and greeted Stephanie with a fistful of recipe copies and almost no instruction except "just wiggle a finger at me when you've got it all ready," while she went back to the camera crew.

Suddenly Stephanie Hersh was an administrator, a facilitator, an essential sidekick, and accepted—smack, dab, in the center of the high-profile food industry. She loves her job still, years later. Her schooling continued; with Julia's encouragement, Stephanie was the first graduate in Boston University's master's program in gastronomy. Her interest turned to teaching children, and, if her current career should diminish, she knows exactly what she wants to do next.

Her ultimate goal is to teach cooking to children, integrated into the existing school curriculum. She would teach culture with cuisine and teach kids to appreciate flavors, tastes, and colors. "Cooking and sharing good food creates harmony," she says. It's not surprising that two unrelated television production teams who are considering after-school cooking shows for children have both contacted Stephanie to serve as a consultant: she knows children and she knows television. The rest she can and will learn.

information, see page 23. There are regional culinary groups—guilds, societies, alliances—in many large urban areas, and if the school you choose doesn't have the information, someone at IACP headquarters will be able to give you a current name and address near you to contact. Even if you are not yet a bona fide culinarian, as soon as you are enrolled in a professional program, you can usually join in the student-member category—at a lower annual dues rate. Most organizations allow guests to come to their meetings and programs—a good way to get connected and see if you feel comfortable in the group before joining.

Among the rewards of joining a local culinary group are friendships; mentors; learning from varied guest speakers; job leads; customer referrals when another member is too booked to take the work; learning unrelated skills when you volunteer to work on program, membership, and communication committees; contributing to the community when you volunteer to work on a food-related benefit; and the lifelong asset of connections.

Take an inventory of "whom do I know that I can call about this?" Culinary groups provide a wealth of leads and good food and wine to enjoy. Get the name and phone number, call to find out when and where the next meeting will be, and ask if they welcome guests. When you get there tell the greeter "I'm new here; who can I talk to about (baking, catering kitchens, ethnic ingredient stores, this organization, volunteering)—pick a possible topic and start listening. Bring your business cards (not having them is unforgivable), give them out, and be sure to take cards from members you meet. Write the date of the meeting and what you talked about on the back before you go to sleep that night. Thus begins the building of a personal network, the invaluable channel to your peers.

The first time I met a friend of mine she was working as a waitress at a sort-of-Italian cafe where our Roundtable chapter was having a program titled "Networking." I was moderating the panel, and the waitress was mesmerized by the dynamic group of professional women who were the audience. I noticed her enthusiasm, talked to her a bit, and encouraged her to join our chapter. With a university degree in soils science, she was working as a server "because that is what I like to do more than anything else"—and her people skills were what I continued to notice as she joined the chapter, came to meetings, and changed jobs a few times. Within a year, she was hired by the oldest established winery in our region as Tasting Room Manager and now is their local Sales Manager. Some of her very limited free time is filled serving on the Roundtable Board as Vice President and Communications Coordinator. Her education in soils and geology gave her a head start in understanding wine production for her job. "Who you know" only opens the door, but "what you know" gets you the job.

Travel Stages for a Career

If you have already identified some role models in the food industry whom you admire and have learned a little about their lives and careers, you know that a long stretch of steady, hard work is the story of their success. We can break that stretch in sections, though, and understand ways that your own success can be a realistic goal.

Beginner

Focus on a career plan for yourself as early as you can—you will make changes, take detours, and acquire unrelated skills that you want to

Courtesy The French Culinary Institute

use—but having a predetermined route tells you whether you're lost or just on a scenic loop. Use the professional network you are gathering right away. At first that may be primarily your fellow students and your teachers, but they are an important network for you to maintain. How do you use them? As questions arise in your mind, ask "who do I know who might answer this?" Make contact with the person, ask your questions, strengthen your bond. Ask your teachers or your mentors about industry conferences and trade shows you can go to, and make an effort to do it. The more you know what is going on in the food industry, the better you can steer yourself to success. Donate your time and skills to publicized events—does your school or your restaurant put on fund-raisers for community projects? Volunteer to assist, cook, serve, or do whatever is needed, and talk to your peers at the event. Remember, always take your business cards and give them out as you are collecting new ones. Write on the backs! As soon as you become a "head chef"—whether it is your own restaurant or you're an employee—create some public appearance opportunities for yourself. Participate in community benefits that feature a group of local chefs providing the food. If you are developing a product through the restaurant or on your own, find opportunities to have guests taste it at local events. If you author a cookbook, offer to do book signings at local bookstores. Work closely with your culinary peers, participate in public events as much as possible, and barter your services for product. Keep your name out there, and it will become your billboard.

Intermediate

This is the stage to position yourself for publicity. The first step is to run your business (whether self-owned or profit-sharing status) so well you can be absent on tour. Go to Beard House dinners in New York City, and talk to them about scheduling you to cook one. Contact the nearest chapter of the American Institute of Wine & Food and ask if they will set up a program using you as a guest chef. Develop your public speaking skills; if you need help with public speaking, contact your local adult education program for workshops and local coaches. The better you can hold your audience's attention while you speak (and this includes table side in your restaurant), the more you will promote your success. As soon as you are confident speaking to medium-size groups (50 to 200 people) and have something to talk about, offer to be a speaker at professional conferences: American Culinary Federation, International Association of Culinary Professionals, and regional culinary organizations. After a few more experiences, and when word of your entertainment value gets around, you will be paid travel and lodging expenses to be a speaker (and in time you'll be paid an honorarium, as well). At this mid-career stage, you can search out ways to market your name and offers will come to you

Michael Roberts

Michael Roberts' original motivation to become a chef grew out of his frustration at not being able to express himself as a musician. He had a bachelor's degree in music theory and composition from New York University and was preparing to enter graduate school somewhat reluctantly. He realized that his "joyous hobby" of cooking could be an alternative career. He read "Larousse Gastronomique" as avidly as if it were a steamy novel. When he was cooking, he says "it was just me and the food and chopping—it was like being in a trance, a pure connection with the food, and feeling good." His parents told him he was crazy.

Michael chose a slightly unusual direction for culinary training: instead of selecting from the U.S. schools and academies, he wanted to be French-trained and found that the government school in Paris would take foreign residents. An uncle living in Paris agreed to let him come, and Michael intensified his study of the French language.

The French program is a vocational high school equivalence course, established for students who do not pass the examinations to be prepared for university. He had to wait for a place in the Paris school and during the 15 months it took he got a restaurant job, found an apartment, made friends, and achieved fluency. He calls it all ideal preparation, because by the time he started school he had worked in French kitchens and knew how they were run. He never wavered in his steady desire to be a restaurant chef. Then something happened.

While Michael was still pursuing his music education, he was diagnosed with Spinal Muscular Atrophy. Doctors didn't know the speed with which it would develop; there was no diet regime or physical therapy course and it affected his arms and legs. He knew it would eventually affect his ability to cook in the kitchen, but that was in the future; he would not be deterred from his dream.

Michael says there are really only thirty basic recipes, and that's what the Paris school taught. His own focus was that professional cooking was about personal creativity. With his French vocational school credential in hand, he moved back to New York City. His physical condition had not yet slowed him down: he was able to stand in the kitchen, travel on the rocking New York subway, and walk the city. He raced through lifetime experiences: cooked at "a happening place" named One Fifth Avenue, made the wild New York club scene several nights a week, and got married. By the age of 27, he was bored, had moved to Los Angeles, got divorced, and started cooking at a small restaurant on Sunset Boulevard. The celebrated Los Angeles restaurant critic, Lois Dwan, wrote about Michael's cooking, and it paid for his independence. He has owned every restaurant he has cooked in since then; investors find him and bankroll his creative ideas. Twin Palms in Pasadena (with partner Kevin Costner) was his last restaurant, sold in early 1997.

Cookbook-writing was Michael's first detour to compensate for a future loss of physical capacity and brought him moderate success, with "Secret Ingredients" (1988), "Fresh from the Freezer" (1990), and others. The mid-90s have brought the first restaurant consulting opportunities. A recent one is the total remodeling of a resort hotel and dining rooms in Cabo San Lucas, Baja Sur, Mexico. He is immersed in designing the culinary aspects of the Cabo hotel. He has built a herd of dairy goats for hand-crafted cheese and planted an organic garden for fresh produce; the creation of these on-site products frees two kitchen walk-in refrigerators that eventually can serve the cheese business or any other new enterprises that he may yet conceive.

unsolicited: consider allowing your name to be used on aprons or chef's clothing labels (this can be either your own merchandise line or being paid for the use of your name). Newcomers to the restaurant business looking for help and advice may turn to you, and you can decide whether to give it freely or charge as a consultant (probably a little of both depending on the circumstances). By now, you recognize the need for a support network to help you manage some of these outside activities: a lawyer, an accountant, a marketing assistant, and possibly a booking agent. Don't sit back thinking that when you need them they will be there. As with everything, you have to look ahead and look out for yourself.

Advanced

If you're doing it right, now is the stage to get paid for having fun. If you still want to cook, you'll be doing it, probably with one or two trained cooks behind you so you can take care of the peripheral business you've created. Here are some ways you'll find to stimulate your creative juices and make money at the same time: You'll be paid an honorarium and expenses as a speaker. You'll be recruited to head business development teams for other culinary start-ups. You'll be paid for product and service endorsements. You may spin off your name or your label on merchandise for royalties. You'll attract potential investors and/or buyers for expansion or retirement from your own restaurant or company. If this is fun for you, you'll find the time to do it.

Graduate

This is the time to be a mentor and a philanthropist within your culinary community. When you were a beginner in professional training, your school probably brought in the best local chefs to inspire you. You may have received a culinary scholarship from one of professional organizations. Now it's your turn to be on the giving side. You'll still get requests to speak and be paid well for most of the gigs, but consider giving some time to smaller groups of the next generation of chefs. The appearances you'll get paid for will be keynoter, industry spokesperson, and expert; consider being on a panel or a roundtable to answer questions one-on-one. You will be offered an investment position in food companies solely as an adviser. The fee you get for endorsements will be higher than ever. To truly be a graduate in this career field, you will consciously find, promote, and mentor promising individuals who can advance the industry in the future. Well done!

Keeping Your Options Open

The future of any career, say, ten or twenty years ahead, is excruciatingly difficult to find in focus. Whether you look through a camera's viewer or through eyeglasses customized to your needs, you make physical adjustments to bring a far-away object into focus.

Once you have chosen a culinary school for your training and started instruction, it's already time to start asking about future opportunities. Bombard your chefs at school with questions about what you need to know for jobs that sound enticing to you. You may not act on that information for several years, but you've started to adjust your focus whenever you gather more knowledge about future opportunities.

The speck on the horizon that is your future career is barely visible now, but as you move toward it or look for it through a magnifier, you will develop your own vision, and it will become excitingly clear to you. Good luck to every one of you.

SOME VALUABLE RESOURCES

The American Culinary Federation (ACF)
P.O. Box 3466
St. Augustine, FL 32084
Phone: 904-824-4468

The American Institute of Wine & Food (AIWF)
1550 Bryant St., Suite 700
San Francisco, CA 94103
Phone: 415-255-3000

James Beard Foundation
Beard House
167 West 12th St.
New York, NY 10011
Phone: 212-675-4984

International Association of Culinary Professionals (IACP)
304 West Liberty St., Suite 201
Louisville, KY 40202
Phone: 502-581-9786

National Career Development Association (NCDA)
317 Kertess Ave.
Worthington, OH 43085
Phone: 614-436-6116

Roundtable for Women in Foodservice (RWF)
1372 La Colina Dr. #B
Tustin, CA 92780
Phone: 714-838-2749

How to Use This Book

Peterson's Culinary Schools is a comprehensive guide to culinary schools in the United States and abroad. The guide provides detailed descriptions of 342 professional degree and apprenticeship programs.

Information about specific culinary programs listed in this book can be located many ways.

The book is organized into two main sections, each arranged by state within the United States and by country. The first section includes professional programs and the second, apprenticeship programs.

Professional programs are educational programs that offer formalized instruction in a class setting. A diploma, degree, or certificate is awarded to the student at the end of successful completion of a predetermined curriculum of courses and a minimum number of credit hours. Workplace training in the form of an externship or work-study program may be an option offered to students but is not usually required. An apprenticeship essentially is an on-the-job training program. Typical apprenticeship programs entail completion of a specific term (typically, three years or 6,000 hours) of full-time employment for wages in a food-service kitchen under a qualified chef. Classroom culinary instruction is usually required in addition to the scheduled work, and a certificate may be awarded.

Beginning on page 27, a "Quick Reference Chart" lists programs by state and country, indicates what degrees or awards are offered, and notes if the program offers degree specializations in the areas of culinary arts, baking and pastry, or management. Please be aware that there are other degree specializations, and you will have to refer to individual profiles to discover what an individual program may offer beyond these popular ones.

Two indexes are available at the end of the book. One index lists all programs alphabetically by name of the program or institutuion. The second is organized by state or country and shows the type of degree or award offered by each program.

Professional Program Profiles

Program Information indicates the year the program started offering classes, if the program is accredited by the American Culinary Federation Educational Institute, organizations that the program belongs to, the program calendar (semester, quarter, etc.), the type of degrees and awards offered, degree and award specializations, and the length of time needed to complete the degree or award.

Areas of Study include the courses available in the program.

Facilities list the number and types of facilities available to students.

Culinary Student Profile provides information on the total number of students enrolled in the program, the number of full- and part-time students enrolled, and the age range of students.

Faculty lists the number of full- and part-time faculty, including the number of faculty members who are culinary-accredited, industry professionals, or master chefs, and the names of prominent faculty members.

Special Programs note special educational opportunities offered by the program.

Expenses include information on full-time, part-time, in-state, and out-of-state tuition costs; special program-related fees; and application fees. Dollar signs without further notation refer to U.S. currency.

Financial Aid provides information on the number and amount of program-specific loans and scholarships awarded during the 1996–97 academic year and unique financial aid opportunities available to culinary students. This section covers only culinary-related financial aid and does not include types of financial aid that are open to all students, such as Pell Grants and Stafford Loans.

Housing indicates the type of on-campus housing available, its cost, and the number of culinary students housed on-campus, as well as the typical cost of off-campus housing in the area.

Application Information provides information on application deadlines, application materials a prospective student is required to submit to be considered for the program, the number of students who applied for admission to the program, and the number of students accepted to the program for the 1996–97 academic year.

Contact includes the name, address, telephone number, fax number, and e-mail address of the contact person for the program and the World Wide Web address of the program or institution.

Apprenticeship Profiles

Program Information indicates if the apprenticeship program is directly sponsored by a college, university, or culinary institute; if the apprenticeship program is approved by the American Culinary Federation; if an apprentice is eligible to receive a degree from a college or university upon successful completion of the program; and if any special apprenticeships are available.

Placement Information provides the number and types of locations where apprentices may be placed and lists the most popular placement locations of participants.

Apprentice Profile indicates the number of apprenticeships offered, the age range of participants, and the application materials a prospective apprentice must submit to be considered for the program.

Expenses provides information on the basic costs of participating in the program as well as the application fee and special program-related fees.

Entry-level compensation indicates the typical salary for an apprentice at the beginning of the apprenticeship program.

Contact includes the name, address, telephone number, and, where available, fax number and e-mail and World Wide Web addresses of the contact person for the apprenticeship program.

ABOUT THE DATA

The information in these listings was collected during the summer of 1997 using questionnaire mailings and telephone interviews with representatives of each of the programs. Although every entry has been thoroughly documented, you should be aware that changes in the programs may occur after publication of the book. Contact the institutions directly for the most current information on their culinary programs.

QUICK REFERENCE CHART FOR DEGREE TYPES AND PROGRAMS

School	Degree Types Offered	Culinary Arts	Baking and Pastry	Management
Alabama				
Bishop State Community College	C, A	■		■
James H. Faulkner State Community College	A			■
Jefferson State Community College	A	■		■
Lawson State Community College	C	■		
Wallace State Community College	D			
Alaska				
University of Alaska Fairbanks	C, A	■	■	
Arizona				
Arizona Western College	C	■		
The Art Institute of Phoenix	A	■		
Maricopa Skill Centers	C		■	
Scottsdale Community College	C, A	■		■
Scottsdale Culinary Institute	A	■		■
Arkansas				
Ozarka Technical College	C	■		
California				
Cabrillo College	C, A	■		
California Culinary Academy	C, A	■	■	
Chaffey College	C, A	■		■
City College of San Francisco	A			■
College of the Desert	C, A	■		■
College of the Sequoias	C			■
Contra Costa College	C	■	■	

Degree Types: ***C*** *= Certificate;* ***D*** *= Diploma;* ***A*** *= Associate;* ***B*** *= Bachelor's;* ***M*** *= Master's;* ***Ph.D.*** *= Doctorate.*

School	Degree Types Offered	Culinary Arts	Baking and Pastry	Management
California ***continued***				
The Culinary Institute of America	C		■	
Cypress College	C, A	■		■
Diablo Valley College	C	■	■	■
Epicurean School of Culinary Arts	C	■		
Grossmont College	C	■		
Laney College	A	■	■	
Los Angeles Trade-Technical College	C, A	■	■	
Mission College	C, A			■
Napa Valley College	C	■		
Opportunities Industrialization Center-West	C		■	■
Orange Coast College	C, A	■		■
Oxnard College	C, A	■		■
Richardson Researches, Inc.	D			
San Joaquin Delta College	C	■		
Santa Barbara City College	C, A	■		
Santa Rosa Junior College	C	■		
Shasta College	C	■		
Southern California School of Culinary Arts	D	■	■	
Tante Marie's Cooking School	C	■	■	
Westlake Culinary Institute	C	■	■	
Colorado				
Colorado Institute of Art	A	■		
Colorado Mountain College	A	■		
Cooking School of the Rockies	D	■		
Johnson & Wales University	A	■		
Pikes Peak Community College	A			■
Pueblo Community College	C, A	■		
School of Natural Cookery	C	■	■	
Warren Tech	C, A			■

Degree Types: ***C*** *= Certificate;* ***D*** *= Diploma;* ***A*** *= Associate;* ***B*** *= Bachelor's;* ***M*** *= Master's;* ***Ph.D.*** *= Doctorate.*

School	Degree Types Offered	Culinary Arts	Baking and Pastry	Management
Connecticut				
Connecticut Culinary Institute	D	■	■	
Gateway Community-Technical College	C	■		
Manchester Community-Technical College	C, A	■		■
Naugatuck Valley Community-Technical College	C, A	■		■
Norwalk Community-Technical College	C, A	■		■
Delaware				
Delaware Technical and Community College	D, A	■		■
Florida				
The Art Institute of Fort Lauderdale	D, A	■		
Charlotte Vocational Technical Center	C	■		
Daytona Beach Community College	A	■		■
Florida Community College at Jacksonville	A	■		■
Florida Culinary Institute	A	■	■	■
Gulf Coast Community College	A	■		■
Johnson & Wales University	A, B	■	■	
Lindsey Hopkins Technical Education Center	C	■	■	
Manatee Technical Institute	C	■		
McFatter Vocational Technical Center	C	■	■	
Miami Lakes Technical Education Center	C			
Pensacola Junior College	C, A	■		■
Pinellas Technical Education Center-Clearwater Campus	D	■		
Robert Morgan Vocational-Technical Center	C	■	■	
South Florida Community College	C, A			■

Degree Types: ***C*** *= Certificate;* ***D*** *= Diploma;* ***A*** *= Associate;* ***B*** *= Bachelor's;* ***M*** *= Master's;* ***Ph.D.*** *= Doctorate.*

School	Degree Types Offered	Culinary Arts	Baking and Pastry	Management
Florida *continued*				
South Technical Education Center	C	■		
The Southeast Institute of the Culinary Arts	D		■	
Southeastern Academy	D	■		
Georgia				
The Art Institute of Atlanta	A	■		
Atlanta Technical Institute	D	■		
Augusta Technical Institute	C, D	■		■
Savannah Technical Institute	D	■		
Hawaii				
University of Hawaii-Kapiolani Community College	C, A	■	■	
University of Hawaii-Kauai Community College	C, A	■		
University of Hawaii-Maui Community College	C, A	■	■	■
Idaho				
Boise State University	C, A	■		
Idaho State University	C	■		
Illinois				
College of DuPage	C, A	■	■	■
College of Lake County	C, A	■		■
Cooking Academy of Chicago	C	■	■	■
The Cooking and Hospitality Institute of Chicago	C, A	■	■	■
Elgin Community College	C, A	■	■	■
Kendall College	C, A	■	■	
Lexington College	A			■
Lincoln Land Community College	C, A			■
Moraine Valley Community College	C, A	■	■	■
Triton College	C, A	■		
Washburne Trade School	C	■		
William Rainey Harper College	C, A	■	■	■

Degree Types: ***C*** *= Certificate;* ***D*** *= Diploma;* ***A*** *= Associate;* ***B*** *= Bachelor's;* ***M*** *= Master's;* ***Ph.D.*** *= Doctorate.*

School	Degree Types Offered	Culinary Arts	Baking and Pastry	Management
Wilton School of Cake Decorating	C			
Indiana				
Ball State University	A, B			■
Indiana University-Purdue University Fort Wayne	A			■
Ivy Tech State College-Central Indiana	A	■	■	■
Ivy Tech State College-Northwest	A	■		
Vincennes University	A			
Iowa				
Des Moines Area Community College	A	■		
Iowa Lakes Community College	A			■
Kirkwood Community College	C, D, A	■	■	■
Kansas				
American Institute of Baking	C		■	
Kansas City Kansas Area Vocational Technical School	C			■
Wichita Area Technical College	D, A			■
Kentucky				
Kentucky Tech Elizabethtown	D			■
Kentucky Tech-Jefferson Campus	D			■
Sullivan College	D, A	■	■	■
University of Kentucky, Jefferson Community College	A	■		
West Kentucky State Vocational Technical School	C, D			
Louisiana				
Culinary Arts Institute of Louisiana	A	■		■
Delgado Community College	C, A	■		
Louisiana Technical College, Lafayette Campus	D, A	■		■
Louisiana Technical College-Baton Rouge Campus	C, D	■		

Degree Types: ***C*** *= Certificate;* ***D*** *= Diploma;* ***A*** *= Associate;* ***B*** *= Bachelor's;* ***M*** *= Master's;* ***Ph.D.*** *= Doctorate.*

School	Degree Types Offered	Culinary Arts	Baking and Pastry	Management
Louisiana *continued*				
Louisiana Technical College-Sidney N. Collier Campus	C	■		
Louisiana Technical College-Sowela Campus	D	■		
Nicholls State University	A, B	■		
Sclafani's Cooking School, Inc.	C	■	■	
Maine				
Southern Maine Technical College	D, A	■		■
Maryland				
Baltimore International College	C, D, A	■	■	■
L'Academie de Cuisine	C	■	■	
Massachusetts				
Berkshire Community College	C, A	■		■
Boston University	C	■		
Bristol Community College	C	■		
Bunker Hill Community College	C, A	■		
The Cambridge School of Culinary Arts	D			
Endicott College	B			■
Essex Agricultural and Technical Institute	C, A	■	■	
Holyoke Community College	C	■		
Newbury College	C, A	■	■	
Michigan				
Charles Stewart Mott Community College	C, A	■	■	■
Grand Rapids Community College	C, A	■	■	■
Henry Ford Community College	C, A	■		■
Macomb Community College	C, A	■		■
Monroe County Community College	C, A	■		■
Northern Michigan University	C, A, B	■		■
Northwestern Michigan College	A	■		■
Oakland Community College	C, A	■		■

Degree Types: ***C*** *= Certificate;* ***D*** *= Diploma;* ***A*** *= Associate;* ***B*** *= Bachelor's;* ***M*** *= Master's;* ***Ph.D.*** *= Doctorate.*

School	Degree Types Offered	Culinary Arts	Baking and Pastry	Management
Schoolcraft College	C, A	■		■
Washtenaw Community College	C, A	■		■
Minnesota				
Alexandria Technical College	C, D			
Hennepin Technical College	C	■		
National Baking Center	C		■	
Northwest Technical College	D			
South Central Technical College	D, A	■		
Mississippi				
Mississippi University for Women	C, B	■		
Missouri				
Missouri Culinary Institute	C	■		
St. Louis Community College at Forest Park	A			■
Montana				
The University of Montana-Missoula	C, A	■		■
Nebraska				
Central Community College-Hastings Campus	A	■		■
Metropolitan Community College	A	■		■
Southeast Community College, Lincoln Campus	A	■		■
Nevada				
Community College of Southern Nevada	C, A	■		
Truckee Meadows Community College	C, A			■
University of Nevada, Las Vegas	B	■		■
New Hampshire				
New Hampshire College	A, B	■		■
New Hampshire Community Technical College	C, A	■		
University of New Hampshire	A			■

Degree Types: ***C*** *= Certificate;* ***D*** *= Diploma;* ***A*** *= Associate;* ***B*** *= Bachelor's;* ***M*** *= Master's;* ***Ph.D.*** *= Doctorate.*

School	Degree Types Offered	Culinary Arts	Baking and Pastry	Management
New Jersey				
Atlantic Community College	C, A	■	■	■
Bergen Community College	C, A	■		■
Burlington County College	A			■
Hudson County Community College	C, A	■	■	
Middlesex County College	C, A	■		■
Morris County School of Technology	C			■
New Mexico				
Santa Fe Community College	C, A	■		
Southwestern Indian Polytechnic Institute	C, A	■		
Technical Vocational Institute Community College	C, A	■	■	■
New York				
Adirondack Community College	C, A	■		■
Culinary Institute of America	C, A, B	■	■	■
Erie Community College, City Campus	A	■		
The French Culinary Institute	C	■	■	
Jefferson Community College	A			■
Julie Sahni's School of Indian Cooking	D	■		
Mohawk Valley Community College	C, A			■
Monroe Community College	A			■
New School for Social Research	C	■	■	
New York Institute of Technology	C, A	■	■	
New York Restaurant School	C, A	■	■	■
New York University	B, M, Ph.D.			■
Onondaga Community College	C, A	■		■
Paul Smith's College of Arts and Sciences	C, A	■	■	■
Peter Kump's New York Cooking School	D	■	■	

Degree Types: ***C*** *= Certificate;* ***D*** *= Diploma;* ***A*** *= Associate;* ***B*** *= Bachelor's;* ***M*** *= Master's;* ***Ph.D.*** *= Doctorate.*

School	Degree Types Offered	Culinary Arts	Baking and Pastry	Management
Schenectady County Community College	C, A	■		
State University of New York College of Agriculture and Technology at Cobleskill	C, A	■		
State University of New York College of Technology at Alfred	A		■	■
State University of New York College of Technology at Delhi	A	■		
Sullivan County Community College	C, A			■
Syracuse University	B			■
Westchester Community College	A			■
North Carolina				
Asheville-Buncombe Technical Community College	A	■		■
Central Piedmont Community College	C, A	■		■
Guilford Technical Community College	C, D, A	■		
Sandhills Community College	C, D	■		
Wake Technical Community College	A	■		
North Dakota				
North Dakota State College of Science	D, A			■
Ohio				
Ashland County-West Holmes Career Center	D	■		
Cincinnati State Technical and Community College	C, A	■		
Columbus State Community College	A			■
Hocking College	A	■		
The Loretta Paganini School of Cooking	D	■	■	

Degree Types: ***C*** *= Certificate;* ***D*** *= Diploma;* ***A*** *= Associate;* ***B*** *= Bachelor's;* ***M*** *= Master's;* ***Ph.D.*** *= Doctorate.*

School	Degree Types Offered	Culinary Arts	Baking and Pastry	Management
Ohio *continued*				
Owens Community College	C, A			■
Sinclair Community College	C, A	■		■
University of Akron	C, A	■		■
Oklahoma				
Great Plains Area Vocational Technical Center	C			
Meridian Technology Center	C			
Metro Area Vocational Technical School District 22	C	■		
Oklahoma State University, Okmulgee	A	■	■	
Oregon				
Central Oregon Community College	C	■		
International School of Baking	C		■	
Linn-Benton Community College	A			■
Western Culinary Institute	D	■		
Pennsylvania				
The Art Institute of Philadelphia	A	■		
Bucks County Community College	C, A		■	■
Community College of Allegheny County Allegheny Campus	A	■		
Community College of Philadelphia	A			■
Drexel University	B	■		■
Harrisburg Area Community College	C, D, A	■		
Hiram G. Andrews Center	D, A	■		
Indiana University of Pennsylvania	C, B	■		■
JNA Institute of Culinary Arts	D, A	■		■
Keystone College	A	■		■
Mercyhurst College	A			■

Degree Types: ***C*** *= Certificate;* ***D*** *= Diploma;* ***A*** *= Associate;* ***B*** *= Bachelor's;* ***M*** *= Master's;* ***Ph.D.*** *= Doctorate.*

School	Degree Types Offered	Culinary Arts	Baking and Pastry	Management
Northampton County Area Community College	D, A	■		■
Pennsylvania College of Technology	C, A	■	■	■
Pennsylvania Institute of Culinary Arts	A	■		■
The Restaurant School	A			
Westmoreland County Community College	C, A	■	■	■
Rhode Island				
Johnson & Wales University	A, B, M	■	■	■
South Carolina				
Greenville Technical College	C, A	■		■
Horry-Georgetown Technical College	A	■		
Johnson & Wales University	A, B	■	■	■
Trident Technical College	D, A	■		■
South Dakota				
Mitchell Technical Institute	D	■		
Tennessee				
Opryland Hotel Culinary Institute	A	■		
Texas				
The Art Institute of Houston	A	■		
Del Mar College	C, A	■		■
El Centro College	C, A		■	■
El Paso Community College	C, A	■		■
Galveston College	C, A	■		■
Houston Community College System	C	■	■	
Le Chef College of Hospitality Careers	D, A	■		■
Odessa College	C, A	■		
St. Philip's College	A	■		■
Texas State Technical College-Waco/Marshall Campus	C	■		■

Degree Types: ***C*** *= Certificate;* ***D*** *= Diploma;* ***A*** *= Associate;* ***B*** *= Bachelor's;* ***M*** *= Master's;* ***Ph.D.*** *= Doctorate.*

School	Degree Types Offered	Culinary Arts	Baking and Pastry	Management
Utah				
Salt Lake Community College	C, A			■
Utah Valley State College	C, A, B	■		■
Vermont				
New England Culinary Institute	C, A, B	■		■
Virginia				
ATI-Career Institute-School of Culinary Arts	C	■		■
J. Sargeant Reynolds Community College	A	■		■
Johnson & Wales University	C, A	■		
Washington				
The Art Institute of Seattle	A		■	
Bates Technical College	A	■		
Bellingham Technical College	C	■	■	
Clark College	C, A	■	■	■
Edmonds Community College	A	■		
North Seattle Community College	C, A	■		■
Olympic College	C, A	■		
Renton Technical College	A	■		
Seattle Central Community College	C, A	■		■
Skagit Valley College	C, A	■		■
South Puget Sound Community College	C, A		■	■
South Seattle Community College	C, A		■	■
Spokane Community College	C, A	■	■	■
West Virginia				
Shepherd College	A	■		
West Virginia Northern Community College	C, A	■		
Wisconsin				
Chippewa Valley Technical College	A	■		
Fox Valley Technical College	C, A	■		

*Degree Types: **C** = Certificate; **D** = Diploma; **A** = Associate; **B** = Bachelor's; **M** = Master's; **Ph.D.** = Doctorate.*

School	Degree Types Offered	Culinary Arts	Baking and Pastry	Management
Madison Area Technical College	D, A	■	■	■
Moraine Park Technical College	C, A	■	■	■
Nicolet Area Technical College	C, A	■	■	■
The Postilion School of Culinary Art	D	■		
Waukesha County Technical College	C, D, A	■		■
Australia				
Canberra Institute of Technology	D	■		■
Crow's Nest College of Tafe	C	■		■
Le Cordon Bleu	C	■	■	
William Angliss Institute	C, D	■	■	■
Canada				
Canadore College of Applied Arts & Technology	C, D	■		■
Culinary Institute of Canada	C, D	■	■	
George Brown College of Applied Arts & Technology	C, D	■	■	■
Le Cordon Bleu Paris Cooking School	C	■	■	
Niagara College of Applied Arts and Technology	D	■		
Pacific Institute of Culinary Arts	D	■	■	
Southern Alberta Institute of Technology	D	■		
St. Clair College	D	■		
China				
Chopsticks Cooking Centre	C	■		
France				
Ecole des Arts Culinaires et de l'Hôtellerie	C, D	■		■
Ecole Lenotre	D	■	■	
Ecole Superieure de Cuisine Francaise Groupe Ferrandi	D	■		
La Varenne	D	■		
Le Cordon Bleu	C, D	■	■	

Degree Types: ***C*** *= Certificate;* ***D*** *= Diploma;* ***A*** *= Associate;* ***B*** *= Bachelor's;* ***M*** *= Master's;* ***Ph.D.*** *= Doctorate.*

School	Degree Types Offered	Culinary Arts	Baking and Pastry	Management
France *continued*				
Ritz-Escoffier Ecole de Gastronomie Francaise	D	■	■	
Ireland				
Ballymaloe Cookery School	C	■		
Italy				
The International Cooking School of Italian Food and Wine	C	■		
Scoula di Arte Culinaria "Cordon Bleu"	C	■	■	
Japan				
Le Cordon Bleu	C	■	■	
New Zealand				
The New Zealand School of Food and Wine	C	■		
South Africa				
Silwood Kitchen Cordons Bleus Cookery School	C, D	■		
United Kingdom				
Butlers Wharf Chef School	C, D	■		
Cookery at the Grange	C	■		
Le Cordon Bleu	C, D	■	■	
Leith's School of Food and Wine	C, D	■		
The Manor School of Fine Cuisine	C			
Tante Marie School of Cookery	C, D	■		
Thames Valley University	D	■		

Degree Types: ***C*** *= Certificate;* ***D*** *= Diploma;* ***A*** *= Associate;* ***B*** *= Bachelor's;* ***M*** *= Master's;* ***Ph.D.*** *= Doctorate.*

Professional PROGRAM Profiles

U.S. PROGRAMS

BISHOP STATE COMMUNITY COLLEGE

Culinary Arts Department

Mobile, Alabama

GENERAL INFORMATION
Public, coeducational, two-year college. Suburban campus. Founded in 1965. Accredited by Southern Association of Colleges and Schools.

PROGRAM INFORMATION
Accredited by American Culinary Federation Education Institute. Program calendar is divided into quarters. 18-month Certificate in Commercial Food Service. 2-year Associate degree in Culinary Arts.

AREAS OF STUDY
Baking; beverage management; buffet catering; computer applications; controlling costs in food service; convenience cookery; culinary French; culinary skill development; food preparation; food purchasing; food service communication; food service math; garde-manger; international cuisine; introduction to food service; kitchen management; management and human resources; meal planning; meat cutting; meat fabrication; menu and facilities design; nutrition; patisserie; restaurant opportunities; sanitation; saucier; seafood processing; soup, stock, sauce, and starch production; wines and spirits.

CULINARY STUDENT PROFILE
29 total: 24 full-time; 5 part-time.

FACULTY
2 full-time.

EXPENSES
In-state tuition: $39 per credit hour. Out-of-state tuition: $78 per credit hour.

APPLICATION INFORMATION
Students are accepted for enrollment in April, June, September, and December. Applicants must have a high school diploma or GED.

CONTACT
Levi Ezell, Program Coordinator, Culinary Arts Department, 414 Stanton Street, Mobile, AL 36617; 334-473-8692; Fax: 334-471-5961.

JAMES H. FAULKNER STATE COMMUNITY COLLEGE

School of Culinary Arts

Gulf Shores, Alabama

GENERAL INFORMATION
Public, coeducational, two-year college. Urban campus. Founded in 1965. Accredited by Southern Association of Colleges and Schools.

PROGRAM INFORMATION
Offered since 1994. Member of American Culinary Federation; American Culinary Federation Educational Institute; American Institute of Baking; Confrerie de la Chaine des Rotisseurs; Council on Hotel, Restaurant, and Institutional Education; National Restaurant Association. Program calendar is divided into quarters. 2-year Associate degree in Food Service Management.

AREAS OF STUDY
Baking; controlling costs in food service; culinary French; food preparation; food purchasing; food service math; garde-manger; international cuisine; introduction to food service; meal planning; meat cutting; menu and facilities design; nutrition; patisserie; sanitation; saucier; soup, stock, sauce, and starch production; wines and spirits; spices and aromatics.

FACILITIES
Bakery; 4 classrooms; computer laboratory; 2 demonstration laboratories; food production kitchen; gourmet dining room; 2 laboratories; 2 lecture rooms; library; 2 student lounges.

CULINARY STUDENT PROFILE
82 total: 70 full-time; 12 part-time.

James H. Faulkner State Community College *(continued)*

Faculty
3 full-time; 6 part-time. 6 are industry professionals; 1 is a master chef; 2 are culinary-accredited teachers. Prominent faculty: Gerhard Brill and Andy Carmadella.

Expenses
Tuition: $37.50 per quarter hour.

Financial Aid
Program-specific awards include American Culinary Federation scholarship ($400). Employment placement assistance is available. Employment opportunities within the program are available.

Housing
Single-sex housing available. Average on-campus housing cost per month: $425. Average off-campus housing cost per month: $300.

Application Information
Students are accepted for enrollment in June, September, and December. In 1996, 80 applied; 70 were accepted. Applicants must submit a formal application, letters of reference, an essay, and an application to Society of Hosteurs.

Contact
Ron Koetter, Coordinator, School of Culinary Arts, 3301 Gulf Shores Parkway, Gulf Shores, AL 36542; 334-968-3104; Fax: 334-968-3120; E-mail: rkoetter@faulkner.cc.al.us

Jefferson State Community College

Hospitality Management Division

Birmingham, Alabama

General Information
Public, coeducational, two-year college. Suburban campus. Founded in 1965. Accredited by Southern Association of Colleges and Schools.

Program Information
Offered since 1988. Accredited by American Culinary Federation Education Institute. Member of American Culinary Federation; Confrerie de la Chaine des Rotisseurs. Program calendar is divided into quarters. 2-year Associate degrees in Hotel/Motel Management; Food Service Management; Culinary Arts.

Areas of Study
Baking; beverage management; buffet catering; computer applications; confectionery show pieces; controlling costs in food service; convenience cookery; culinary French; culinary skill development; food preparation; food purchasing; food service communication; food service math; garde-manger; introduction to food service; kitchen management; management and human resources; meal planning; meat cutting; meat fabrication; nutrition; patisserie; sanitation; saucier; seafood processing; soup, stock, sauce, and starch production; wines and spirits.

Facilities
Bake shop; classroom; computer laboratory; demonstration laboratory; food production kitchen; gourmet dining room; laboratory; learning resource center; library; snack shop; student lounge.

Culinary Student Profile
35 total: 28 full-time; 7 part-time.

Faculty
2 full-time; 5 part-time. 3 are industry professionals; 1 is a culinary-accredited teacher; 2 are registered dietitians.

Special Programs
3-year culinary apprenticeship leading to certification with American Culinary Federation.

Expenses
Tuition: $34 per credit hour.

Financial Aid
Employment placement assistance is available.

Housing
Average off-campus housing cost per month: $350.

Application Information
Students are accepted for enrollment in September. In 1996, 20 applied; 20 were accepted. Applicants must submit a formal application, letters of reference, and an essay.

Contact
George White, Director, Hospitality Management Division, 2601 Carson Road, Birmingham, AL 35215-3098; 205-856-7898; Fax: 205-853-0340.

Lawson State Community College

Birmingham, Alabama

General Information
Public, coeducational, two-year college. Suburban campus. Founded in 1965. Accredited by Southern Association of Colleges and Schools.

Program Information
Program calendar is divided into quarters. 15-month Certificate in Commercial Foods/ Culinary Arts.

Areas of Study
Baking; buffet catering; controlling costs in food service; culinary skill development; food preparation; food purchasing; introduction to food service; kitchen management; meal planning; nutrition; nutrition and food service; sanitation; seafood processing; soup, stock, sauce, and starch production.

Facilities
Classroom; computer laboratory; demonstration laboratory; food production kitchen; learning resource center; lecture room; library; public restaurant; snack shop; student lounge; teaching kitchen.

Culinary Student Profile
17 total: 10 full-time; 7 part-time. 15 are under 25 years old; 1 is between 25 and 44 years old; 1 is over 44 years old.

Faculty
1 part-time. 1 is an industry professional. Prominent faculty: Deborah Ann Harris.

Expenses
Tuition: $441 per quarter full-time, $151 per quarter part-time.

Financial Aid
In 1996, 5 scholarships were awarded (average award was $1440). Employment placement assistance is available. Employment opportunities within the program are available.

Application Information
Students are accepted for enrollment in March, June, September, and December. Application deadline for fall is September 19. Application deadline for spring is March 21. Application deadline for summer is June 20. In 1996, 19 applied; 17 were accepted. Applicants must submit a formal application.

Contact
Deborah A. Harris, Instructor, Commercial Food Preparation, 3060 Wilson Road, SW, Birmingham, AL 35221-1798; 205-929-6378; Fax: 205-929-6316.

Wallace State Community College

Commercial Foods and Nutrition

Hanceville, Alabama

General Information
Public, coeducational, two-year college. Rural campus. Founded in 1966. Accredited by Southern Association of Colleges and Schools.

Program Information
Offered since 1973. Program calendar is divided into quarters. 18-month Diploma in Commercial Foods and Nutrition.

Areas of Study
Baking; buffet catering; controlling costs in food service; convenience cookery; food preparation; food purchasing; food service communication; food service math; introduction to food service;

Wallace State Community College ***(continued)***

kitchen management; meal planning; nutrition and food service; sanitation.

Facilities

Cafeteria; catering service; 2 classrooms; 3 computer laboratories; demonstration laboratory; food production kitchen; laboratory; learning resource center; lecture room; library; public restaurant; snack shop; 3 student lounges; teaching kitchen.

Culinary Student Profile

27 total: 18 full-time; 9 part-time.

Faculty

1 full-time; 1 part-time. Prominent faculty: Donna Jackson and Mary Lamar.

Expenses

In-state tuition: $438 per quarter full-time, $20 per credit hour part-time. Out-of-state tuition: $876 per quarter full-time, $40 per credit hour part-time.

Financial Aid

In 1996, 6 scholarships were awarded (average award was $1752). Employment placement assistance is available. Employment opportunities within the program are available.

Housing

Apartment-style and single-sex housing available.

Application Information

Students are accepted for enrollment in January, April, June, and September. In 1996, 35 applied; 20 were accepted. Applicants must submit a formal application.

Contact

Donna Jackson, Food Service Director, Commercial Foods and Nutrition, PO Box 2000, Hanceville, AL 35077-2000; 201-352-8227; Fax: 201-352-8228.

University of Alaska Fairbanks

Culinary Arts/Hospitality

Fairbanks, Alaska

General Information

Public, coeducational, university. Suburban campus. Founded in 1917. Accredited by Northwest Association of Schools and Colleges.

Program Information

Member of American Culinary Federation Educational Institute. Program calendar is divided into semesters. 1-year Certificates in Culinary Arts; Cooking; Baking. 2-year Associate degree in Culinary Arts.

Areas of Study

Baking; buffet catering; confectionery show pieces; controlling costs in food service; convenience cookery; culinary French; culinary skill development; food preparation; food purchasing; food service math; garde-manger; international cuisine; introduction to food service; kitchen management; meal planning; meat cutting; meat fabrication; nutrition; patisserie; sanitation; saucier; seafood processing; soup, stock, sauce, and starch production; wines and spirits; vegetarian cooking.

Culinary Student Profile

185 total: 45 full-time; 140 part-time.

Faculty

3 full-time; 8 part-time. 3 are certified executive chefs.

Expenses

In-state tuition: $71 per credit hour. Out-of-state tuition: $221 per credit hour.

Housing

Single-sex housing available.

Application Information

Students are accepted for enrollment in January and September. Applicants must submit a formal application.

Contact
Frank U. Davis, Associate Professor, Culinary Arts/Hospitality, 510 Second Avenue, Fairbanks, AK 99701; 907-474-5196; Fax: 907-474-7335.

Arizona Western College

Culinary Arts Program

Yuma, Arizona

General Information
Public, coeducational, two-year college. Rural campus. Founded in 1962. Accredited by North Central Association of Colleges and Schools.

Program Information
Offered since 1996. Program calendar is divided into semesters. 8-month Certificate in Culinary Arts.

Areas of Study
Baking; food preparation; food purchasing; garde-manger; international cuisine; management and human resources; meal planning; nutrition; sanitation; soup, stock, sauce, and starch production.

Facilities
Classroom; computer laboratory; food production kitchen; gourmet dining room; learning resource center.

Culinary Student Profile
20 total: 15 full-time; 5 part-time. 16 are under 25 years old; 3 are between 25 and 44 years old; 1 is over 44 years old.

Faculty
1 full-time; 2 part-time. 2 are industry professionals; 1 is a registered dietitian. Prominent faculty: Nancy L. Meister.

Special Programs
Placement in local restaurants for field experience.

Expenses
Tuition: $336 per semester full-time, $28 per credit hour part-time. Program-related fees include: $220 for lab food fees; $200 for books; $30 for uniforms.

Financial Aid
In 1996, 4 scholarships were awarded (average award was $300). Employment placement assistance is available.

Housing
Coed housing available. Average on-campus housing cost per month: $250. Average off-campus housing cost per month: $300.

Application Information
Students are accepted for enrollment in January and August. In 1996, 15 applied; 15 were accepted. Applicants must submit a formal application.

Contact
Nancy L. Meister, Coordinator, Culinary Arts Program, PO Box 929, Yuma, AZ 85366-0929; 520-344-7779; Fax: 520-344-7730; E-mail: aw_meistern@awc.cc.az.us

The Art Institute of Phoenix

The School of Culinary Arts

Phoenix, Arizona

General Information
Private, coeducational, two-year college. Urban campus. Founded in 1995.

Program Information
Offered since 1996. Member of American Culinary Federation; International Association of Culinary Professionals. Program calendar is divided into quarters. 18-month Associate degree in Culinary Arts.

Areas of Study
Baking; beverage management; buffet catering; computer applications; confectionery show pieces; controlling costs in food service; culinary French; culinary skill development; food preparation; food purchasing; food service math; garde-manger; international cuisine; introduction to food service; kitchen management; management and human resources; meal planning; meat cutting; meat fabrication; menu and facilities design; nutrition; nutrition and food service; patisserie; restaurant opportunities; sanitation; saucier; seafood

The Art Institute of Phoenix *(continued)*

processing; soup, stock, sauce, and starch production; wines and spirits.

Facilities
Bakery; 4 classrooms; 4 computer laboratories; 2 food production kitchens; laboratory; learning resource center; 4 lecture rooms; public restaurant; student lounge; 3 teaching kitchens.

Culinary Student Profile
120 full-time. 35 are under 25 years old; 65 are between 25 and 44 years old; 20 are over 44 years old.

Faculty
3 full-time; 6 part-time. 7 are industry professionals; 1 is a master chef; 1 is a culinary-accredited teacher. Prominent faculty: Walter Leible and Anthony Rea.

Expenses
Application fee: $150. Tuition: $3488 per quarter. Program-related fees include: $595 for knives, uniforms, and first-quarter books.

Financial Aid
Employment placement assistance is available.

Housing
Average off-campus housing cost per month: $395.

Application Information
Students are accepted for enrollment in January, April, July, and October. In 1996, 110 applied; 106 were accepted. Applicants must submit a formal application and an essay.

Contact
Timothy Dengler, Assistant Director of Admissions, The School of Culinary Arts, 2233 West Dunlap Avenue, Phoenix, AZ 85021-2859; 800-474-2479; Fax: 602-216-0439; E-mail: denglert@aii.edu

See affiliated programs: Colorado Institute of Art; New York Restaurant School; The Art Institute of Atlanta; The Art Institute of Fort Lauderdale; The Art Institute of Houston; The Art Institute of Philadelphia; The Art Institute of Seattle.

See display on page 48.

Maricopa Skill Centers

Phoenix, Arizona

General Information
Public, coeducational, adult vocational school. Urban campus. Founded in 1962.

Program Information
Offered since 1977. Program calendar is year-round. 14-week Certificate in Pastry Goods Maker. 18-week Certificates in Baker's Helper; Kitchen Helper. 27-week Certificate in Cook's Apprentice. 6-month Certificate in Restaurant Food Preparation.

Areas of Study
Baking; food preparation; introduction to food service; meat cutting; restaurant opportunities; soup, stock, sauce, and starch production.

Facilities
Cafeteria; catering service; classroom; computer laboratory; demonstration laboratory; food production kitchen; learning resource center; lecture room; teaching kitchen.

Culinary Student Profile
25 full-time.

Faculty
3 full-time. 1 is a certified working chef.

Expenses
Application fee: $5. Tuition: $90 per week full-time, $75 per week part-time. Program-related fees include: $200 for lab fees.

Financial Aid
In 1996, 1 scholarship was awarded (award was $500). Employment opportunities within the program are available.

Application Information
Students are accepted for enrollment year-round. In 1996, 60 applied; 60 were accepted. Applicants must submit a formal application.

Contact
Dave Bochicchio, Instructor, Food Preparation Program, 1245 East Buckeye Road, Phoenix, AZ

Maricopa Skill Centers *(continued)*

85034-4101; 602-238-4300; Fax: 602-238-4307; World Wide Web: http://www.gwc.maricopa.edu/msc/

See affiliated apprenticeship program.

Scottsdale Community College

Culinary Arts Program

Scottsdale, Arizona

General Information
Public, coeducational, two-year college. Suburban campus. Founded in 1969. Accredited by North Central Association of Colleges and Schools.

Program Information
Offered since 1984. Program calendar is divided into semesters. 2-year Associate degrees in Hospitality Management; Culinary Arts. 9-month Certificate in Culinary Arts.

Areas of Study
Baking; controlling costs in food service; culinary skill development; food preparation; food service math; garde-manger; international cuisine; introduction to food service; kitchen management; management and human resources; meat cutting; meat fabrication; menu and facilities design; nutrition; patisserie; sanitation; saucier; soup, stock, sauce, and starch production; dining room service.

Facilities
Bake shop; cafeteria; 3 classrooms; 3 computer laboratories; food production kitchen; gourmet dining room; learning resource center; library; public restaurant.

Culinary Student Profile
36 total: 33 full-time; 3 part-time.

Faculty
2 full-time; 4 part-time. 4 are industry professionals; 2 are culinary-accredited teachers. Prominent faculty: Sarah R. Labensky.

Expenses
Tuition: $1800 per 9 months full-time, $37 per credit hour part-time.

Financial Aid
In 1996, 4 scholarships were awarded (average award was $250). Employment placement assistance is available.

Application Information
Students are accepted for enrollment in August. Application deadline for fall is continuous with a recommended date of April 15. In 1996, 45 applied; 35 were accepted. Applicants must submit a formal application, an essay, and ASSET scores.

Contact
Sarah R. Labensky, Culinary Arts Program, Culinary Arts Program, 9000 East Chaparral Road, Scottsdale, AZ 85250-2699; 602-423-6244; Fax: 602-423-6200.

Scottsdale Culinary Institute

Culinary Arts and Sciences and Restaurant Management

Scottsdale, Arizona

General Information
Private, coeducational, culinary institute. Urban campus. Founded in 1986.

Program Information
Offered since 1986. Accredited by American Culinary Federation Education Institute. Member of American Culinary Federation; American Culinary Federation Educational Institute; American Institute of Wine & Food; Council on Hotel, Restaurant, and Institutional Education; International Association of Culinary Professionals; National Restaurant Association; Women Chefs and Restaurateurs; Phoenix Restaurant Association. Program calendar is divided into 6-week cycles. 15-month Associate degree in Culinary Arts/Science/Restaurant Management.

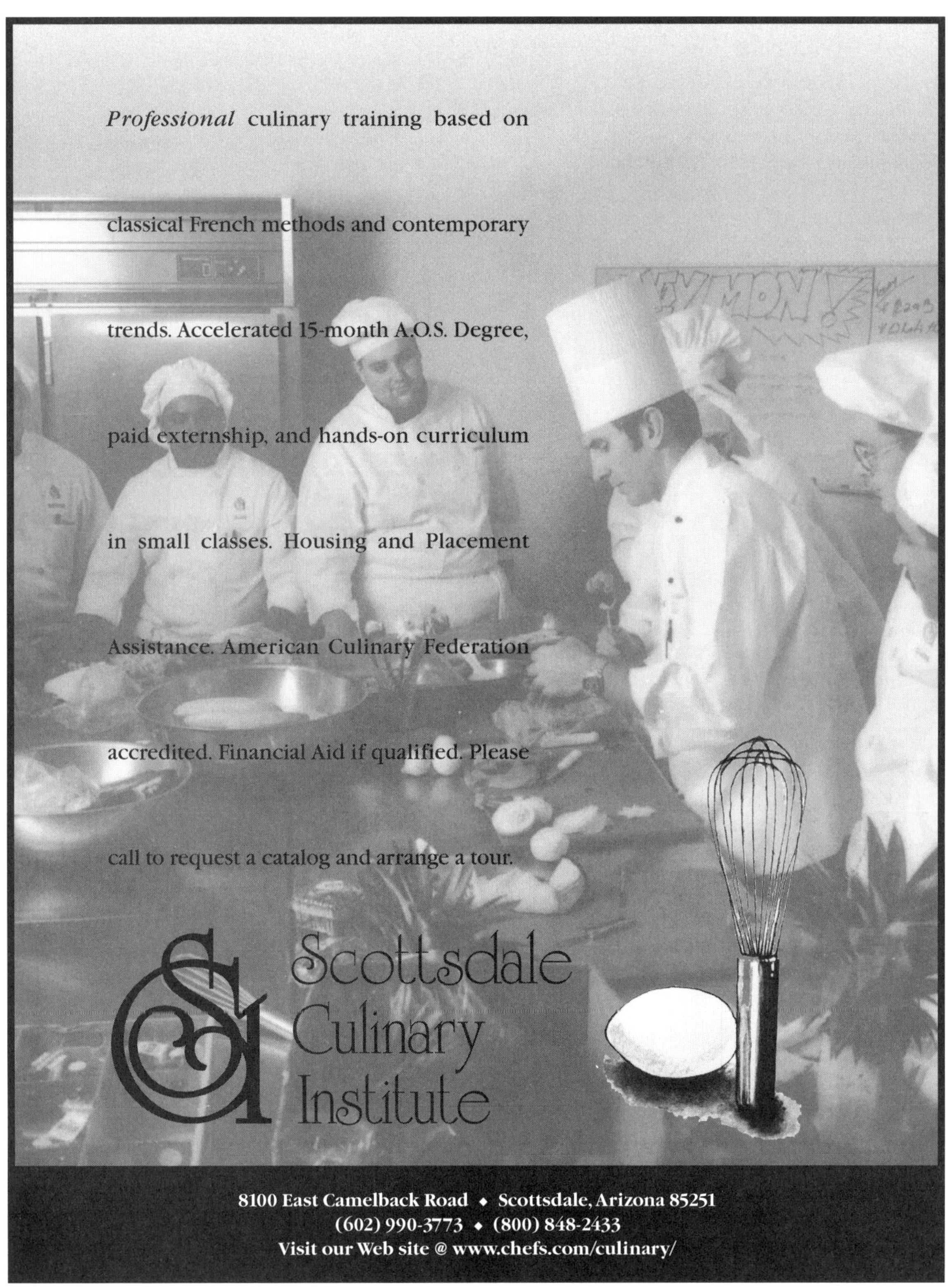
Professional culinary training based on classical French methods and contemporary trends. Accelerated 15-month A.O.S. Degree, paid externship, and hands-on curriculum in small classes. Housing and Placement Assistance. American Culinary Federation accredited. Financial Aid if qualified. Please call to request a catalog and arrange a tour.
Scottsdale Culinary Institute
8100 East Camelback Road ♦ Scottsdale, Arizona 85251
(602) 990-3773 ♦ (800) 848-2433
Visit our Web site @ www.chefs.com/culinary/

Scottsdale Culinary Institute *(continued)*

Areas of Study
Baking; beverage management; buffet catering; confectionery show pieces; controlling costs in food service; culinary French; culinary skill development; food preparation; food purchasing; food service communication; food service math; garde-manger; international cuisine; introduction to food service; kitchen management; management and human resources; meal planning; meat cutting; meat fabrication; menu and facilities design; nutrition; nutrition and food service; patisserie; restaurant opportunities; sanitation; saucier; seafood processing; soup, stock, sauce, and starch production; wines and spirits.

Facilities
Bake shop; demonstration laboratory; food production kitchen; gourmet dining room; 4 lecture rooms; library; public restaurant; student lounge; 4 teaching kitchens.

Culinary Student Profile
320 full-time.

Faculty
17 full-time; 3 part-time. 20 are culinary-accredited teachers.

Expenses
Application fee: $25. Tuition: $15,385 per degree. Program-related fees include: $785 for textbooks, knives, uniforms, and utensil set.

Financial Aid
In 1996, 4 scholarships were awarded (average award was $2000); 250 loans were granted (average loan was $2611). Employment placement assistance is available. Employment opportunities within the program are available.

Housing
Average off-campus housing cost per month: $350.

Application Information
Students are accepted for enrollment in January, February, April, May, July, August, October, and November. In 1996, 351 applied; 320 were accepted. Applicants must submit a formal application, letters of reference, and academic transcripts and have a high school diploma or GED.

Contact
Darren S. Leite, Director of Admissions, Culinary Arts and Sciences and Restaurant Management, 8100 East Camelback Road, Scottsdale, AZ 85251; 602-990-3773; Fax: 602-990-0351; World Wide Web: http://chefs.com/culinary/

See display on page 51.

Ozarka Technical College

Culinary Arts

Melbourne, Arkansas

General Information
Public, coeducational, two-year college. Rural campus. Founded in 1973. Accredited by North Central Association of Colleges and Schools.

Program Information
Offered since 1973. Program calendar is divided into semesters. 9-month Certificate in Culinary Arts.

Areas of Study
Baking; computer applications; food preparation; food service math; introduction to food service; nutrition; sanitation.

Facilities
Cafeteria; catering service; classroom; computer laboratory; demonstration laboratory; food production kitchen; learning resource center; lecture room; library; public restaurant.

Faculty
1 full-time.

Expenses
Tuition: $444 per semester full-time, $37 per credit hour part-time. Program-related fees include: $25 for consumables used in program.

Financial Aid
Employment placement assistance is available.

Application Information
Students are accepted for enrollment in August. Applicants must submit a formal application.

Contact
Jean Hall, Counselor, Culinary Arts, PO Box 10, Melbourne, AR 72556; 870-368-7371; Fax: 870-368-4733; World Wide Web: http://www.ozarka.tec.ar.us

Cabrillo College

Culinary Arts and Hospitality Management

Aptos, California

General Information
Public, coeducational, two-year college. Suburban campus. Founded in 1959. Accredited by Western Association of Schools and Colleges.

Program Information
Accredited by American Culinary Federation Education Institute. Member of American Culinary Federation; National Restaurant Association; Santa Cruz Area Restaurant Association. Program calendar is divided into semesters. 1-year Certificate in Culinary Arts. 2-year Associate degree in Culinary Arts.

Areas of Study
Baking; beverage management; buffet catering; computer applications; controlling costs in food service; culinary skill development; food preparation; food purchasing; food service math; garde-manger; introduction to food service; kitchen management; management and human resources; meat cutting; meat fabrication; nutrition; patisserie; sanitation; soup, stock, sauce, and starch production; wines and spirits; food service management; ethnic cuisine.

Facilities
Bakery; cafeteria; 2 catering services; 3 classrooms; 3 computer laboratories; demonstration laboratory; 2 food production kitchens; garden; gourmet dining room; 3 laboratories; learning resource center; lecture room; library; public restaurant; snack shop; student lounge; teaching kitchen.

Culinary Student Profile
230 total: 30 full-time; 200 part-time.

Faculty
3 full-time; 3 part-time. 6 are industry professionals.

Expenses
Tuition: $210 per semester full-time, $77 per unit part-time.

Financial Aid
In 1996, 8 scholarships were awarded (average award was $500). Employment placement assistance is available. Employment opportunities within the program are available.

Application Information
Students are accepted for enrollment in January, June, and August. Applicants must submit a formal application and academic transcripts.

Contact
Katherine Niven, Director of Culinary Arts, Culinary Arts and Hospitality Management, 6500 Soquel Drive, Aptos, CA 95003-3194; 408-479-5749; Fax: 408-479-5769; E-mail: kaniven@cabrillo.cc.ca.us

California Culinary Academy

San Francisco, California

General Information
Private, coeducational, culinary institute. Urban campus. Founded in 1977.

Program Information
Offered since 1977. Accredited by American Culinary Federation Education Institute. Member of American Culinary Federation; American Institute of Wine & Food; Educational Foundation of the NRA; International Association of Culinary Professionals; James Beard Foundation, Inc.; National Restaurant Association; The Bread Bakers Guild of America; Women Chefs and Restaurateurs. Program calendar is divided into semesters. 18-month Associate degree in Culinary Arts. 30-week Certificate in Baking and Pastry.

California Culinary Academy *(continued)*

Areas of Study

Baking; beverage management; computer applications; confectionery show pieces; controlling costs in food service; culinary skill development; food preparation; food purchasing; food service math; garde-manger; international cuisine; introduction to food service; kitchen management; management and human resources; meat cutting; meat fabrication; menu and facilities design; nutrition; patisserie; restaurant opportunities; sanitation; saucier; seafood processing; soup, stock, sauce, and starch production; wines and spirits.

Facilities

Bake shop; 4 bakeries; cafeteria; coffee shop; computer laboratory; demonstration laboratory; 5 food production kitchens; 5 laboratories; 4 lecture rooms; library; 2 public restaurants.

Culinary Student Profile

575 full-time.

Faculty

36 full-time; 4 part-time. 37 are industry professionals; 1 is a master chef; 2 are culinary-accredited teachers. Prominent faculty: Jurgen Weise and Peter Reinhart.

Expenses

Application fee: $35. Tuition: $27,250 per degree.

Application Information

Students are accepted for enrollment year-round. Applicants must submit a formal application and an essay, a resume, and academic transcripts and take a Wonderlic aptitude test.

Contact

Admissions Office, 625 Polk Street, San Francisco, CA 94102; 800-BAY-CHEF; Fax: 415-771-2194; World Wide Web: http://www.baychef.com

---

The Academy has provided quality professional training in culinary arts for 20 years. Currently offered are an 18-month Associate of Occupational Studies (AOS) degree program in culinary arts, a 30-week Baking and Pastry Arts Certificate program, and Weekend Professional Skills programs for part-time students in culinary arts, baking and pastry arts, and continuing education. Year-round enrollment is available in all programs. The Academy has two satellite campuses—the College of Food/Salinas and the College of Food/San Diego. Program offerings include a certificate course of study in basic culinary professional skills. To obtain a catalog, students should call 800-BAY-CHEF or visit the Academy's Web site at http://www.baychef.com.

Chaffey College

Culinary Arts Program

Rancho Cucamonga, California

General Information

Public, coeducational, two-year college. Suburban campus. Founded in 1883. Accredited by Western Association of Schools and Colleges.

Program Information

Member of American Culinary Federation; American Dietetic Association; Council on Hotel, Restaurant, and Institutional Education; National Restaurant Association; Women Chefs and Restaurateurs. Program calendar is divided into semesters. 2-semester Certificates in Dietetic Supervisor; Restaurant Management; Hotel Management; Culinary Arts. 2-year Associate degrees in Dietetic Technician; Food Service Management.

Areas of Study

Baking; buffet catering; computer applications; controlling costs in food service; culinary skill development; food preparation; food purchasing; introduction to food service; management and human resources; meal planning; menu and facilities design; nutrition; nutrition and food service; restaurant opportunities; sanitation; soup, stock, sauce, and starch production.

Facilities

Cafeteria; catering service; classroom; computer laboratory; demonstration laboratory; food production kitchen; gourmet dining room; laboratory; learning resource center; lecture room; library; public restaurant; teaching kitchen.

Culinary Student Profile

454 total: 154 full-time; 300 part-time.

Faculty

1 full-time; 17 part-time. 2 are culinary-accredited teachers.

Expenses

Tuition: $156 per semester full-time, $13 per unit part-time.

Financial Aid

In 1996, 2 scholarships were awarded (average award was $250). Employment placement assistance is available. Employment opportunities within the program are available.

Application Information

Students are accepted for enrollment in January, May, and August. Application deadline for fall is August 15. Application deadline for spring is January 20. Application deadline for summer is May 31.

Contact

Suzanne Johnson, Department Chairperson, Culinary Arts Program, 5885 Haven Avenue, Rancho Cucamonga, CA 91737-3002; 909-941-2711; Fax: 909-466-2831.

City College of San Francisco

Hotel and Restaurant Operation

San Francisco, California

General Information

Public, coeducational, two-year college. Urban campus. Founded in 1935. Accredited by Western Association of Schools and Colleges.

Program Information

Offered since 1936. Accredited by American Culinary Federation Education Institute. Member of American Culinary Federation; Council on Hotel, Restaurant, and Institutional Education; Women Chefs and Restaurateurs; California Restaurant Association. Program calendar is divided into semesters. 4-semester Associate degree in Hotel and Restaurant Operation.

Areas of Study

Baking; beverage management; buffet catering; computer applications; confectionery show pieces; controlling costs in food service; culinary French; culinary skill development; food preparation; food purchasing; food service communication; food service math; garde-manger; international cuisine; kitchen management; management and human resources; meat cutting; meat fabrication; menu and facilities design; nutrition and food service; patisserie; restaurant opportunities; sanitation; saucier; seafood processing; soup, stock, sauce, and starch production; wines and spirits; orientation to hospitality; hospitality accounting.

Facilities

Bake shop; cafeteria; catering service; 2 classrooms; computer laboratory; demonstration laboratory; gourmet dining room; learning resource center; library; snack shop; 3 teaching kitchens.

Culinary Student Profile

254 total: 168 full-time; 86 part-time. 101 are under 25 years old; 142 are between 25 and 44 years old; 11 are over 44 years old.

Faculty

11 full-time; 6 part-time. 17 are industry professionals.

Special Programs

240-hour internship at one of 100 hotels/ restaurants in the Bay Area.

Expenses

In-state tuition: $13 per unit. Out-of-state tuition: $130 per unit. Program-related fees include: $150 for uniforms; $130 for kitchen tools; $150 for books per semester.

Financial Aid

Program-specific awards include 20 "Hotel and Restaurant Foundation" scholarships. Employment placement assistance is available. Employment opportunities within the program are available.

City College of San Francisco ***(continued)***

Application Information
Students are accepted for enrollment in January and August. Application deadline for fall is continuous with a recommended date of April 14. Application deadline for spring is continuous with a recommended date of November 10. In 1996, 140 applied; 130 were accepted. Applicants must submit a formal application and interview, and international students must submit TOEFL scores.

Contact
Hotel and Restaurant Department, Hotel and Restaurant Operation, 50 Phelan Avenue, San Francisco, CA 94112-1821; 415-239-3908; Fax: 415-239-3913; E-mail: kmanning@ccsf.cc.ca.us

---

The Hotel and Restaurant Department at City College of San Francisco is a program that enables students to gain skills, contacts in the industry, and an education that prepares them for a long-term career in the hospitality field. By learning about cooking as well as the sales, management, and business aspects of the food world, students have a broad and flexible foundation on which to build their careers. Graduates have become well-known chefs and restaurateurs, general managers, and successful business owners. This is an affordable program in a city with unlimited opportunities.

College of the Desert

Culinary Arts Program

Palm Desert, California

General Information
Public, coeducational, two-year college. Suburban campus. Founded in 1959. Accredited by Western Association of Schools and Colleges.

Program Information
Offered since 1985. Program calendar is divided into semesters. 2-semester Certificate in Culinary Arts. 2-year Associate degree in Culinary Management.

Areas of Study
Baking; beverage management; computer applications; controlling costs in food service; culinary skill development; food preparation; food purchasing; garde-manger; introduction to food service; kitchen management; management and human resources; meal planning; nutrition; patisserie; sanitation; seafood processing; soup, stock, sauce, and starch production.

Culinary Student Profile
40 total: 20 full-time; 20 part-time.

Faculty
1 full-time; 3 part-time.

Expenses
In-state tuition: $13 per unit. Out-of-state tuition: $135.25 per unit.

Application Information
Students are accepted for enrollment in January and August. Applicants must submit a formal application.

Contact
Admissions Office, Culinary Arts Program, 43-500 Monterey Avenue, Palm Desert, CA 92260-9305; 760-346-8041 Ext. 517; Fax: 760-341-8678.

College of the Sequoias

Consumer Family Studies

Visalia, California

General Information
Public, coeducational, two-year college. Rural campus. Founded in 1925. Accredited by Western Association of Schools and Colleges.

Program Information
Offered since 1992. Program calendar is divided into semesters. 13-unit Certificate in Basic Food Service. 18-unit Certificate in Dietetic Service Supervisor. 28-month Certificate in Food Service Management.

Areas of Study
Baking; buffet catering; confectionery show pieces; controlling costs in food service; culinary skill

development; food preparation; food purchasing; food service communication; food service math; introduction to food service; kitchen management; management and human resources; meal planning; menu and facilities design; nutrition; nutrition and food service; sanitation; saucier; soup, stock, sauce, and starch production.

Facilities
Classroom; computer laboratory; demonstration laboratory; food production kitchen; laboratory; learning resource center; lecture room; library; student lounge; teaching kitchen.

Culinary Student Profile
42 total: 12 full-time; 30 part-time. 10 are under 25 years old; 25 are between 25 and 44 years old; 7 are over 44 years old.

Faculty
2 full-time; 4 part-time. 1 is a master chef; 3 are registered dietitians. Prominent faculty: Millie Owens and Laura Baker.

Expenses
Tuition: $13 per unit.

Application Information
Students are accepted for enrollment in January, June, and August. Applicants must submit a formal application.

Contact
Admissions, Consumer Family Studies, 915 South Mooney Boulevard, Visalia, CA 93277-2234; 209-730-3727; Fax: 209-730-3894.

Contra Costa College

Culinary Arts

San Pablo, California

General Information
Public, coeducational, two-year college. Urban campus. Founded in 1948. Accredited by Western Association of Schools and Colleges.

Program Information
Program calendar is divided into semesters. 2-year Certificates in Cooking; Baking.

Areas of Study
Baking; buffet catering; confectionery show pieces; controlling costs in food service; culinary French; culinary skill development; food preparation; food purchasing; food service communication; food service math; garde-manger; international cuisine; introduction to food service; kitchen management; management and human resources; meal planning; meat cutting; meat fabrication; menu and facilities design; nutrition; patisserie; sanitation; saucier; seafood processing; soup, stock, sauce, and starch production; wines and spirits.

Culinary Student Profile
60 full-time.

Faculty
2 full-time; 3 part-time.

Expenses
In-state tuition: $13 per unit. Out-of-state tuition: $135 per unit.

Application Information
Students are accepted for enrollment in January and August. Applicants must submit a formal application.

Contact
Culinary Arts Department, Culinary Arts, 2600 Mission Bell Drive, San Pablo, CA 94806-3195; 510-235-7800 Ext. 3111; Fax: 510-236-6768.

The Culinary Institute of America

The Culinary Institute of America at Greystone

St. Helena, California

General Information
Private, coeducational, culinary institute. Rural campus. Founded in 1946.

Program Information
Offered since 1995. Accredited by American Culinary Federation Education Institute. Member of American Culinary Federation; American Institute of Wine & Food; Council on Hotel,

The Culinary Institute of America *(continued)*

Restaurant, and Institutional Education; James Beard Foundation, Inc.; Napa Valley Wine Library Association; National Restaurant Association; The Bread Bakers Guild of America; Women Chefs and Restaurateurs. Program calendar is year-round. Continuing Education Units. 30-week Certificate in Baking and Pastry Arts.

Areas of Study

Baking; buffet catering; confectionery show pieces; controlling costs in food service; culinary skill development; food preparation; food purchasing; food service communication; garde-manger; international cuisine; introduction to food service; kitchen management; nutrition; nutrition and food service; patisserie; sanitation; saucier; soup, stock, sauce, and starch production; wines and spirits.

Facilities

2 bake shops; 2 bakeries; 3 catering services; 10 classrooms; computer laboratory; demonstration laboratory; food production kitchen; 2 gardens; learning resource center; library; public restaurant; 4 teaching kitchens; vineyard.

Culinary Student Profile

3,060 total: 60 full-time; 3,000 part-time.

Faculty

9 full-time; 40 part-time. 23 are industry professionals; 1 is a master chef; 25 are culinary-accredited teachers. Prominent faculty: Catherine Brandell and Bo Friberg.

Special Programs

21-week Culinary Comprehensive Program.

Expenses

Application fee: $30. Tuition: $13,990 per certificate full-time, $695 per week part-time. Program-related fees include: $695 for equipment; $95 for second-term practical exam.

Financial Aid

In 1996, 10 scholarships were awarded (average award was $1500). Program-specific awards include work-study program. Employment placement assistance is available. Employment opportunities within the program are available.

Housing

Apartment-style housing available. Average off-campus housing cost per month: $500.

Application Information

Students are accepted for enrollment year-round. Applicants must submit a formal application, letters of reference, and academic transcripts.

Contact

Marsha Chism, Education Assistant, The Culinary Institute of America at Greystone, 2555 Main Street, St. Helena, CA 94574; 800-333-9242; Fax: 707-967-2410.

See affiliated program in Hyde Park, New York.

See display on page 168.

Cypress College

Hospitality Management/ Culinary Arts

Cypress, California

General Information

Public, coeducational, two-year college. Suburban campus. Founded in 1966. Accredited by Western Association of Schools and Colleges.

Program Information

Offered since 1971. Member of Council on Hotel, Restaurant, and Institutional Education; Interntional Food Service Executives Association; National Restaurant Association. Program calendar is divided into semesters. 2-year Certificates in Hotel Operations; Food Service Management; Culinary Arts. 3-year Associate degrees in Culinary Arts; Hotel Operations; Food Service Management.

Areas of Study

Baking; buffet catering; computer applications; controlling costs in food service; food purchasing; garde-manger; international cuisine; management and human resources; menu and facilities design; nutrition; sanitation.

Facilities
Bakery; coffee shop; 2 computer laboratories; food production kitchen; laboratory; 4 lecture rooms; library; public restaurant; 2 snack shops; student lounge.

Culinary Student Profile
160 total: 115 full-time; 45 part-time.

Faculty
3 full-time; 7 part-time. 10 are industry professionals.

Expenses
Tuition: $13 per unit. Program-related fees include: $60 for uniforms; $75 for knives.

Financial Aid
Employment placement assistance is available. Employment opportunities within the program are available.

Application Information
Students are accepted for enrollment in January, June, and August. In 1996, 160 applied; 160 were accepted. Applicants must submit a formal application.

Contact
David Schweiger, Department Coordinator, Hospitality Management/Culinary Arts, 9200 Valley View Street, Cypress, CA 90630-5897; 714-826-2220 Ext. 208; Fax: 714-527-1077.

Diablo Valley College

Culinary Arts Department

Pleasant Hill, California

General Information
Public, coeducational, two-year college. Suburban campus. Founded in 1949. Accredited by Western Association of Schools and Colleges.

Program Information
Offered since 1971. Accredited by American Culinary Federation Education Institute. Member of American Culinary Federation; American Culinary Federation Educational Institute; Interntional Food Service Executives Association; National Restaurant Association. Program calendar is divided into semesters. 18-month Certificate in Hotel Administration. 2-year Certificates in Baking and Pastry; Restaurant Management; Culinary Arts.

Areas of Study
Baking; beverage management; computer applications; food preparation; food purchasing; garde-manger; international cuisine; management and human resources; meat cutting; meat fabrication; menu and facilities design; patisserie; restaurant opportunities; sanitation; soup, stock, sauce, and starch production; wines and spirits.

Facilities
Bake shop; cafeteria; catering service; 6 classrooms; coffee shop; computer laboratory; demonstration laboratory; 2 food production kitchens; gourmet dining room; learning resource center; 6 lecture rooms; library; public restaurant; snack shop; 2 teaching kitchens.

Culinary Student Profile
250 total: 150 full-time; 100 part-time. 75 are under 25 years old; 100 are between 25 and 44 years old; 75 are over 44 years old.

Faculty
5 full-time; 13 part-time. Prominent faculty: Jack Hendrickson and Nader Sharkes.

Expenses
Tuition: $13 per unit. Program-related fees include: $260 for uniforms and books (first semester).

Financial Aid
In 1996, 30 scholarships were awarded (average award was $500). Employment placement assistance is available.

Application Information
Students are accepted for enrollment in January, June, and August. In 1996, 100 applied; 100 were accepted. Applicants must submit a formal application.

Contact
Jack Hendrickson, Department Chair, Culinary Arts Department, 321 Golf Club Road, Pleasant Hill, CA 94523-1544; 510-685-1230 Ext. 555; Fax: 510-825-8412.

Epicurean School of Culinary Arts

Los Angeles, California

General Information
Private, coeducational, culinary institute. Urban campus. Founded in 1985.

Program Information
Offered since 1985. Program calendar is divided into semesters. 20-week Certificate in Culinary Arts.

Areas of Study
Baking; culinary skill development; international cuisine; meal planning.

Facilities
Classroom.

Culinary Student Profile
15 full-time.

Faculty
3 full-time; 3 part-time. 3 are industry professionals; 3 are culinary-accredited teachers.

Expenses
Tuition: $1600 per certificate.

Application Information
Students are accepted for enrollment year-round.

Contact
Shelley Janson, Director, 8759 Melrose Avenue, Los Angeles, CA 90069; 310-659-5990; Fax: 310-659-0302.

Grossmont College

Culinary Arts

El Cajon, California

General Information
Public, coeducational, two-year college. Suburban campus. Founded in 1961. Accredited by Western Association of Schools and Colleges.

Program Information
Member of American Culinary Federation; Council on Hotel, Restaurant, and Institutional Education. Program calendar is divided into semesters. 1-year Certificate in Culinary Arts.

Areas of Study
Baking; buffet catering; controlling costs in food service; food purchasing; nutrition; soup, stock, sauce, and starch production; healthy professional cooking; hot foods.

Facilities
Cafeteria; catering service; classroom; 6 gardens; learning resource center; 3 lecture rooms; library; snack shop; student lounge; teaching kitchen.

Culinary Student Profile
75 total: 25 full-time; 50 part-time. 15 are under 25 years old; 45 are between 25 and 44 years old; 15 are over 44 years old.

Faculty
3 full-time; 6 part-time. 1 is an industry professional; 1 is a master chef; 1 is a culinary-accredited teacher; 1 is a registered dietitian. Prominent faculty: Catherine Robertson and Joseph Orate.

Expenses
In-state tuition: $156 per semester full-time, $13 per unit part-time. Out-of-state tuition: $1212 per semester full-time, $101 per unit part-time. Program-related fees include: $15 for lab fees.

Financial Aid
Employment placement assistance is available. Employment opportunities within the program are available.

Application Information
Students are accepted for enrollment in January and August. Application deadline for fall is June 10. Application deadline for spring is November 10. In 1996, 200 applied; 75 were accepted. Applicants must submit a formal application.

Contact
Catherine Robertson, Coordinator, Culinary Arts, 8800 Grossmont College Drive, El Cajon, CA 92020-1799; 619-644-7327; Fax: 619-461-3396; E-mail: croberts@mail.gcced.cc.ca.us

Grossmont College is located in suburban San Diego. The Culinary Arts Program is the redefinition of a former program that better fits industry needs for well-prepared chefs and kitchen personnel. The 24-unit certificate includes course work in buffet/catering, baking and pastry making, hot foods, soups, stocks and sauces, nutrition, and food purchasing, with options in several areas to complete the certificate. A 120-hour work experience/apprentice component includes placement in fine restaurants that enables students a professional experience to apply the knowledge they acquire. Students may participate in a yearly competition sponsored by ACIF.

Laney College

Culinary Arts

Oakland, California

General Information
Public, coeducational, two-year college. Urban campus. Founded in 1953. Accredited by Western Association of Schools and Colleges.

Program Information
Member of American Institute of Baking; American Vegan Society; Council on Hotel, Restaurant, and Institutional Education; Educational Foundation of the NRA; National Restaurant Association; The Bread Bakers Guild of America; Retail Bakers of America. Program calendar is divided into semesters. 2-year Associate degrees in Baking; Cooking.

Areas of Study
Baking; confectionery show pieces; controlling costs in food service; culinary skill development; food preparation; food purchasing; food service math; garde-manger; international cuisine; introduction to food service; kitchen management; meal planning; meat cutting; nutrition; patisserie; sanitation; saucier; wines and spirits.

Facilities
Bake shop; bakery; cafeteria; catering service; 4 classrooms; coffee shop; computer laboratory; demonstration laboratory; 2 food production kitchens; gourmet dining room; 4 laboratories; learning resource center; lecture room; library; 3 public restaurants; snack shop; student lounge; 4 teaching kitchens.

Culinary Student Profile
190 total: 90 full-time; 100 part-time.

Faculty
5 full-time; 4 part-time. 4 are industry professionals; 3 are master chefs; 2 are master bakers. Prominent faculty: Wayne Stoker and Cleo Ross.

Expenses
Tuition: $120 per semester full-time, $10 per unit part-time. Program-related fees include: $120 for knives and books.

Financial Aid
In 1996, 2 scholarships were awarded (average award was $200); 80 loans were granted. Employment placement assistance is available. Employment opportunities within the program are available.

Application Information
Students are accepted for enrollment in January and August. Application deadline for spring is December 1. Application deadline for fall is June 1. In 1996, 120 applied; 90 were accepted. Applicants must submit a formal application.

Contact
Wayne Stoker, Co-Chair, Culinary Arts, 900 Fallon Street, Oakland, CA 94607-4893; 510-464-3407; Fax: 510-464-3240.

Los Angeles Trade-Technical College

Culinary Arts

Los Angeles, California

General Information
Public, coeducational, two-year college. Urban campus. Founded in 1925. Accredited by Western Association of Schools and Colleges.

Los Angeles Trade-Technical College *(continued)*

Program Information
Offered since 1925. Accredited by American Culinary Federation Education Institute. Member of American Culinary Federation; American Culinary Federation Educational Institute; American Institute of Baking; Interntional Food Service Executives Association; National Restaurant Association; The Bread Bakers Guild of America; Women Chefs and Restaurateurs. Program calendar is divided into semesters. 2-year Associate degrees in Culinary Arts; Professional Baking. 2-year Certificates in Culinary Arts; Professional Baking.

Areas of Study
Baking; culinary French; culinary skill development; food preparation; food purchasing; food service math; garde-manger; international cuisine; introduction to food service; kitchen management; meal planning; meat cutting; meat fabrication; nutrition and food service; patisserie; sanitation; saucier; soup, stock, sauce, and starch production.

Facilities
Bake shop; bakery; cafeteria; catering service; classroom; coffee shop; computer laboratory; demonstration laboratory; food production kitchen; garden; learning resource center; lecture room; library; public restaurant.

Culinary Student Profile
270 full-time.

Faculty
8 full-time; 7 part-time. 12 are industry professionals; 3 are culinary-accredited teachers. Prominent faculty: Nancy Berkoff and Robert Wemischner.

Expenses
Tuition: $156 per semester. Program-related fees include: $350 for uniforms and books for 4 semesters.

Financial Aid
In 1996, 10 scholarships were awarded (average award was $500). Employment placement assistance is available. Employment opportunities within the program are available.

Housing
Average off-campus housing cost per month: $600.

Application Information
Students are accepted for enrollment in January and August. In 1996, 90 applied; 90 were accepted.

Contact
Culinary Arts Office, Culinary Arts, 400 West Washington Boulevard, Los Angeles, CA 90015-4108; 213-744-9480; Fax: 213-748-7334; E-mail: lungck@laccd.cc.ca.us

Mission College

Santa Clara, California

General Information
Public, coeducational, two-year college. Urban campus. Founded in 1967. Accredited by Western Association of Schools and Colleges.

Program Information
Offered since 1967. Accredited by American Culinary Federation Education Institute. Member of Council on Hotel, Restaurant, and Institutional Education; Educational Foundation of the NRA. Program calendar is divided into semesters. 4-semester Associate degree in Food Service Management. 4-semester Certificate in Food Service.

Areas of Study
Baking; beverage management; food preparation; food purchasing; meal planning; menu and facilities design; nutrition; sanitation; wines and spirits; restaurant operation.

Facilities
4 classrooms; demonstration laboratory; food production kitchen; gourmet dining room.

Culinary Student Profile
160 total: 100 full-time; 60 part-time.

Faculty
3 full-time; 6 part-time. 8 are industry professionals; 1 is a master chef.

Expenses
Tuition: $13 per unit.

Financial Aid
In 1996, 5 scholarships were awarded (average award was $250). Employment placement assistance is available. Employment opportunities within the program are available.

Application Information
Students are accepted for enrollment in January, May, and August. In 1996, 80 applied; 80 were accepted.

Contact
Dietrich Amarell, Department Chair, 3000 Mission College Boulevard, Santa Clara, CA 95054-1897; 408-748-2753; Fax: 408-496-0462; E-mail: dietrich_amarell@wvmccd.cc.ca.us; World Wide Web: http://www.wvmccd.cc.ca.us/mc

Napa Valley College

Napa Valley Cooking School

St. Helena, California

General Information
Public, coeducational, culinary institute. Suburban campus. Founded in 1996.

Program Information
Offered since 1996. Program calendar is divided into semesters. 13-month Certificate in Culinary Arts.

Areas of Study
Baking; buffet catering; computer applications; controlling costs in food service; culinary French; culinary skill development; food preparation; food purchasing; food service math; garde-manger; international cuisine; introduction to food service; meat cutting; meat fabrication; nutrition; nutrition and food service; patisserie; restaurant opportunities; sanitation; saucier; seafood processing; soup, stock, sauce, and starch production; wines and spirits; wine and food; vegetarian cookery.

Facilities
Classroom; computer laboratory; food production kitchen; garden; lecture room; library; student lounge; teaching kitchen; vineyard.

Culinary Student Profile
16 full-time. 3 are under 25 years old; 10 are between 25 and 44 years old; 3 are over 44 years old.

Faculty
1 full-time; 10 part-time. 10 are industry professionals. Prominent faculty: George Terrassa.

Expenses
Application fee: $25. Tuition: $10,550 per certificate.

Financial Aid
Employment placement assistance is available.

Housing
Average off-campus housing cost per month: $400.

Application Information
Students are accepted for enrollment in August. In 1996, 18 applied; 16 were accepted. Applicants must submit a formal application, letters of reference, academic transcripts, and a letter describing their career interest.

Contact
Lauren Deblois, Culinary Coordinator, Napa Valley Cooking School, 1088 College Avenue, St. Helena, CA 94574; 707-967-2930; Fax: 707-967-2909.

Opportunities Industrialization Center-West

Culinary Arts Program

Redwood City, California

General Information
Private, coeducational, adult vocational school. Urban campus. Founded in 1965. Accredited by Western Association of Schools and Colleges.

Opportunities Industrialization Center-West *(continued)*

Program Information

Offered since 1965. Program calendar is year-round. 4-month Certificates in Pastry; Sous Chef/Entry Management; Line Cook; Prep/Pantry Cook.

Areas of Study

Baking; buffet catering; controlling costs in food service; culinary French; culinary skill development; food preparation; food purchasing; food service communication; food service math; international cuisine; introduction to food service; kitchen management; meal planning; meat cutting; nutrition; nutrition and food service; patisserie; restaurant opportunities; sanitation; saucier; seafood processing; soup, stock, sauce, and starch production; wines and spirits.

Facilities

Cafeteria; catering service; classroom; food production kitchen; gourmet dining room; learning resource center; lecture room; library; public restaurant.

Culinary Student Profile

70 total: 45 full-time; 25 part-time.

Faculty

6 full-time; 1 part-time. 4 are industry professionals. Prominent faculty: Scott M. Brunson and Sean Choyer.

Expenses

Tuition: $5703 per certificate.

Financial Aid

Employment placement assistance is available. Employment opportunities within the program are available.

Application Information

Students are accepted for enrollment year-round. In 1996, 50 applied; 45 were accepted. Applicants must submit a formal application.

Contact

Derrick Sheppard, Counselor/Job Developer, Culinary Arts Program, 2050 Broadway, Redwood City, CA 94063; 650-568-2883; Fax: 650-568-2884.

Orange Coast College

Hospitality Department

Costa Mesa, California

General Information

Public, coeducational, two-year college. Suburban campus. Founded in 1947. Accredited by Western Association of Schools and Colleges.

Program Information

Offered since 1964. Accredited by American Culinary Federation Education Institute. Program calendar is divided into semesters. 1-year Certificates in Institutional Dietetic Service Supervisor; Child Nutrition Programs; Fast Food Service; Catering; Culinary Arts. 2-year Associate degrees in Hotel Management; Food Service Management; Culinary Arts. 2-year Certificates in Institutional Dietetic Technician; Restaurant Supervision. 3-year Certificate in Cook Apprentice. 30-month Certificate in Institutional Dietetic Service Manager.

Areas of Study

Baking; beverage management; buffet catering; computer applications; controlling costs in food service; convenience cookery; culinary skill development; food preparation; food purchasing; food service communication; food service math; garde-manger; international cuisine; introduction to food service; kitchen management; management and human resources; meal planning; meat cutting; meat fabrication; menu and facilities design; nutrition; patisserie; restaurant opportunities; sanitation; saucier; seafood processing; soup, stock, sauce, and starch production; dining room management; hotel and restaurant law.

Culinary Student Profile

400 total: 220 full-time; 180 part-time.

Faculty

6 full-time; 14 part-time.

Expenses

In-state tuition: $13 per unit. Out-of-state tuition: $140 per unit.

Application Information
Students are accepted for enrollment in January and August. Applicants must submit a formal application.

Contact
Dan Beard, Program Coordinator, Hospitality Department, 2701 Fairview Road, PO Box 5005, Costa Mesa, CA 92628-5005; 714-432-5835.

Oxnard College

Hotel and Restaurant Management

Oxnard, California

General Information
Public, coeducational, two-year college. Suburban campus. Founded in 1975. Accredited by Western Association of Schools and Colleges.

Program Information
Offered since 1985. Member of American Culinary Federation; American Culinary Federation Educational Institute; Council on Hotel, Restaurant, and Institutional Education; Educational Foundation of the NRA. Program calendar is divided into semesters. 2-year Associate degrees in Restaurant Management; Hotel Management; Culinary Arts. 2-year Certificates in Restaurant Management; Hotel Management; Culinary Arts.

Areas of Study
Baking; beverage management; buffet catering; computer applications; controlling costs in food service; convenience cookery; culinary French; culinary skill development; food preparation; food purchasing; food service communication; food service math; garde-manger; international cuisine; introduction to food service; kitchen management; management and human resources; meal planning; menu and facilities design; nutrition; nutrition and food service; sanitation; soup, stock, sauce, and starch production.

Facilities
Bake shop; cafeteria; catering service; 2 classrooms; computer laboratory; demonstration laboratory; 2 food production kitchens; gourmet dining room; laboratory; learning resource center; lecture room; library; public restaurant; teaching kitchen.

Culinary Student Profile
60 total: 20 full-time; 40 part-time. 15 are under 25 years old; 35 are between 25 and 44 years old; 10 are over 44 years old.

Faculty
2 full-time; 5 part-time. 3 are industry professionals; 1 is a culinary-accredited teacher. Prominent faculty: Frank Haywood and Abdallah Al-Sadek.

Special Programs
153-hour hotel/motel internship.

Expenses
Tuition: $156 per semester full-time, $13 per unit part-time. Program-related fees include: $350 for tools and uniforms; $65 for textbooks per class.

Financial Aid
In 1996, 8 scholarships were awarded (average award was $500). Employment placement assistance is available. Employment opportunities within the program are available.

Application Information
Students are accepted for enrollment in January, June, and August. Application deadline for fall is August 18. Application deadline for spring is January 12. In 1996, 70 applied; 70 were accepted. Applicants must submit a formal application.

Contact
Frank Haywood, Hotel and Restaurant Management, 4000 South Rose Avenue, Oxnard, CA 93033-6699; 805-986-5869; Fax: 805-986-5865.

Richardson Researches, Inc.

Hayward, California

General Information
Private, coeducational, confectionery food consultancy company. Urban campus. Founded in 1972.

Richardson Researches, Inc. *(continued)*

Program Information
Offered since 1977. Member of Institute of Food Technologists; National Confectioner's Association. 1-week Diplomas in Continental Chocolates; Chocolate Technology; Confectionery Technology.

Facilities
Computer laboratory; demonstration laboratory; laboratory; lecture room.

Culinary Student Profile
18 full-time.

Faculty
2 full-time; 1 part-time. 3 are industry professionals. Prominent faculty: Terry Richardson and Margaret Knight.

Expenses
Tuition: $1450 per diploma.

Application Information
Students are accepted for enrollment in March, July, and October. Applicants must submit a formal application.

Contact
Terry Richardson, President, 23449 Foley Street, Hayward, CA 94545; 510-785-1350; Fax: 510-785-6857; E-mail: info@richres.com; World Wide Web: http://www.richres.com

San Joaquin Delta College

Culinary Arts Department

Stockton, California

General Information
Public, coeducational, two-year college. Urban campus. Founded in 1935. Accredited by Western Association of Schools and Colleges.

Program Information
Offered since 1979. Program calendar is divided into semesters. 18-month Certificate in Advanced Culinary Arts. 5-month Certificates in Basic Culinary Arts; Dietetic Service Supervisor.

Areas of Study
Baking; buffet catering; computer applications; controlling costs in food service; convenience cookery; culinary French; culinary skill development; food preparation; food purchasing; garde-manger; introduction to food service; management and human resources; meal planning; menu and facilities design; nutrition; nutrition and food service; restaurant opportunities; sanitation; saucier; seafood processing; soup, stock, sauce, and starch production.

Facilities
Bakery; cafeteria; catering service; 3 classrooms; coffee shop; computer laboratory; demonstration laboratory; food production kitchen; gourmet dining room; laboratory; learning resource center; 3 lecture rooms; library; public restaurant; snack shop; 4 student lounges; teaching kitchen; 2 vineyards.

Culinary Student Profile
125 total: 45 full-time; 80 part-time. 35 are under 25 years old; 65 are between 25 and 44 years old; 25 are over 44 years old.

Faculty
2 full-time; 6 part-time. 1 is a culinary-accredited teacher; 1 is a registered dietitian. Prominent faculty: Char Britto and John Britto.

Special Programs
Culinary Arts Skills Workshop.

Expenses
Tuition: $195 per semester full-time, $13 per unit part-time.

Financial Aid
In 1996, 12 scholarships were awarded (average award was $400). Employment placement assistance is available.

Application Information
Students are accepted for enrollment in January, June, and August. Applications are accepted continuously for fall, spring, and summer. In 1996, 125 applied; 125 were accepted.

Contact
Hazel Hill, Division Chairperson, Culinary Arts Department, 5151 Pacific Avenue, Stockton, CA

95207-6370; 209-954-5516; Fax: 209-954-5600; E-mail: hhill@sjdccd.cc.ca.us; World Wide Web: http://www.sjccd.cc.ca.us/FCHS/sjdc.html

Santa Barbara City College

Hotel/Restaurant and Culinary Department

Santa Barbara, California

General Information

Public, coeducational, two-year college. Suburban campus. Founded in 1908. Accredited by Western Association of Schools and Colleges.

Program Information

Offered since 1970. Accredited by American Culinary Federation Education Institute. Member of American Culinary Federation; American Culinary Federation Educational Institute; American Institute of Wine & Food; Council on Hotel, Restaurant, and Institutional Education; Educational Foundation of the NRA; National Restaurant Association; California Restaurant Association. Program calendar is divided into semesters. 4-semester Certificate in Culinary Arts. 5-semester Associate degree in Culinary Arts.

Areas of Study

Baking; beverage management; buffet catering; computer applications; confectionery show pieces; controlling costs in food service; convenience cookery; culinary French; culinary skill development; food preparation; food purchasing; food service communication; food service math; garde-manger; international cuisine; introduction to food service; kitchen management; management and human resources; meal planning; meat cutting; meat fabrication; menu and facilities design; nutrition; nutrition and food service; patisserie; restaurant opportunities; sanitation; saucier; seafood processing; soup, stock, sauce, and starch production; wines and spirits; restaurant ownership; bartending.

Facilities

Bake shop; cafeteria; catering service; 3 classrooms; coffee shop; computer laboratory; demonstration laboratory; 5 food production kitchens; garden; gourmet dining room; laboratory; learning resource center; lecture room; library; 2 public restaurants; 2 snack shops; student lounge; 3 teaching kitchens; 2 food preparation laboratories.

Culinary Student Profile

130 full-time. 35 are under 25 years old; 85 are between 25 and 44 years old; 10 are over 44 years old.

Faculty

14 full-time; 5 part-time. 13 are industry professionals; 2 are master chefs; 4 are culinary-accredited teachers. Prominent faculty: John Dunn and Randy Bublitz.

Special Programs

Student-run food operation, opportunity to cook for students on field trip.

Expenses

Tuition: $250 per semester. Program-related fees include: $280 for uniforms; $85 for tools and case.

Financial Aid

In 1996, 60 scholarships were awarded (average award was $500). Program-specific awards include additional scholarships from private sources. Employment placement assistance is available. Employment opportunities within the program are available.

Housing

Average off-campus housing cost per month: $400.

Application Information

Students are accepted for enrollment in January and August. Application deadline for fall is continuous with a recommended date of May 1. Application deadline for spring is continuous with a recommended date of September 1. In 1996, 120 applied; 48 were accepted. Applicants must submit a formal application and interview.

Contact

John Dunn, Chairperson, Hotel/Restaurant and Culinary Department, 721 Cliff Drive, Santa Barbara, CA 93109-2394; 805-965-0851 Ext. 2457; Fax: 805-963-7222.

Santa Rosa Junior College

Consumer and Family Studies Department

Santa Rosa, California

General Information
Public, coeducational, two-year college. Urban campus. Founded in 1918. Accredited by Western Association of Schools and Colleges.

Program Information
Program calendar is divided into semesters. 1-year Certificate in Culinary Training.

Areas of Study
Baking; buffet catering; computer applications; controlling costs in food service; culinary French; culinary skill development; food preparation; food purchasing; food service communication; food service math; garde-manger; international cuisine; introduction to food service; kitchen management; meat fabrication; nutrition; nutrition and food service; patisserie; restaurant opportunities; sanitation; saucier; seafood processing; soup, stock, sauce, and starch production.

Facilities
Bake shop; bakery; 3 classrooms; computer laboratory; 2 demonstration laboratories; 7 food production kitchens; garden; gourmet dining room; laboratory; learning resource center; 3 lecture rooms; library; public restaurant; teaching kitchen.

Culinary Student Profile
425 total: 75 full-time; 350 part-time.

Faculty
1 full-time; 16 part-time. 12 are industry professionals. Prominent faculty: Michael Salinger and Cathy Burgett.

Expenses
Tuition: $13 per unit.

Financial Aid
Employment placement assistance is available. Employment opportunities within the program are available.

Housing
Coed housing available.

Application Information
Students are accepted for enrollment in January and August.

Contact
Michael Salinger, Program Coordinator, Consumer and Family Studies Department, 1501 Mendocino Avenue, Santa Rosa, CA 95401-4395; 707-527-4591; Fax: 707-527-4816; E-mail: msalinge@santarosa.edu

Shasta College

Culinary Arts

Redding, California

General Information
Public, coeducational, two-year college. Urban campus. Founded in 1948. Accredited by Western Association of Schools and Colleges.

Program Information
Program calendar is divided into semesters. 12-month Certificate in Culinary Arts.

Areas of Study
Baking; buffet catering; computer applications; controlling costs in food service; convenience cookery; culinary French; culinary skill development; food preparation; food purchasing; food service communication; food service math; garde-manger; introduction to food service; restaurant opportunities; sanitation; soup, stock, sauce, and starch production.

Facilities
Bakery; cafeteria; catering service; classroom; food production kitchen; gourmet dining room; laboratory; learning resource center; lecture room; library; public restaurant; teaching kitchen.

Faculty
1 full-time; 1 part-time. 1 is an industry professional; 1 is a culinary-accredited teacher. Prominent faculty: Michael A. Piccinino.

Expenses
In-state tuition: $13 per unit. Out-of-state tuition: $125 per unit.

Financial Aid
In 1996, 2 scholarships were awarded (average award was $100); 2 loans were granted (average loan was $100). Employment placement assistance is available. Employment opportunities within the program are available.

Housing
Coed housing available. Average on-campus housing cost per month: $125.

Application Information
Students are accepted for enrollment in January and August. Applicants must submit a formal application.

Contact
Michael Piccinino, Instructor, Culinary Arts, PO Box 496006, Redding, CA 96049-6006; 916-225-4829; Fax: 916-225-4706.

Southern California School of Culinary Arts

Professional Culinary Arts

South Pasadena, California

General Information
Private, coeducational, culinary institute. Urban campus. Founded in 1994.

Program Information
Offered since 1994. Accredited by American Culinary Federation Education Institute. Member of American Culinary Federation; American Institute of Baking; Educational Foundation of the NRA; International Association of Culinary Professionals. Program calendar is divided into semesters. 15-month Diploma in Professional Culinary Arts. 15-week Diplomas in Advanced Professional Baking; Advanced Professional Cookery.

Areas of Study
Baking; beverage management; confectionery show pieces; controlling costs in food service; convenience cookery; culinary French; culinary skill development; food preparation; food purchasing; food service communication; food service math; garde-manger; kitchen management; meat cutting; meat fabrication; nutrition; sanitation; seafood processing; soup, stock, sauce, and starch production; wines and spirits.

Facilities
Bake shop; bakery; 6 classrooms; coffee shop; computer laboratory; food production kitchen; garden; 6 laboratories; learning resource center; library; public restaurant; 6 teaching kitchens.

Culinary Student Profile
140 full-time. 18 are under 25 years old; 112 are between 25 and 44 years old; 10 are over 44 years old.

Faculty
10 full-time. 10 are consumer educations teachers. Prominent faculty: Robert Danhi and Leslie Bilderbach.

Special Programs
Culinary tours of Italy, field trips to local Los Angeles food and beverage venues.

Expenses
Application fee: $25. Tuition: $19,580 per 15 months. Program-related fees include: $700 for uniforms, books, and tool kit.

Financial Aid
In 1996, 6 scholarships were awarded (average award was $250). Employment placement assistance is available. Employment opportunities within the program are available.

Housing
Average off-campus housing cost per month: $700.

Application Information
Students are accepted for enrollment in February, June, and October. Application deadline for fall is October 1. Application deadline for spring is Febuary 2. Application deadline for summer is May 25. In 1996, 175 applied; 102 were accepted. Applicants must submit a formal application,

Southern California School of Culinary Arts *(continued)*

letters of reference, a resume, academic transcripts, and a personal statement.

Contact
Cristina Williams, Director, Marketing and Admissions, Professional Culinary Arts, 1420 El Centro, South Pasadena, CA 91030; 888-900-CHEF; Fax: 818-403-8494; E-mail: scsca@earthlink.net

Tante Marie's Cooking School

San Francisco, California

General Information
Private, coeducational, culinary institute. Urban campus. Founded in 1979.

Program Information
Offered since 1979. Member of American Institute of Wine & Food; International Association of Culinary Professionals; James Beard Foundation, Inc.; Women Chefs and Restaurateurs; San Francisco Professional Food Society. 6-month Certificates in Professional Pastry Program; Professional Culinary Program.

Areas of Study
Baking; culinary French; culinary skill development; food preparation; food purchasing; garde-manger; international cuisine; introduction to food service; meal planning; meat cutting; nutrition; patisserie; restaurant opportunities; sanitation; saucier; seafood processing; soup, stock, sauce, and starch production; wines and spirits.

Facilities
Demonstration laboratory; garden; 2 teaching kitchens.

Culinary Student Profile
24 total: 12 full-time; 12 part-time.

Faculty
2 full-time; 2 part-time. 1 is an industry professional; 1 is a master chef; 2 are culinary-accredited teachers. Prominent faculty: Catherine Pantsios and Cathy Burgett.

Expenses
Application fee: $75. Tuition: $13,200 per 6 months full-time, $5280 per 6 months part-time.

Financial Aid
In 1996, 2 scholarships were awarded (average award was $5280). Employment placement assistance is available. Employment opportunities within the program are available.

Housing
Average off-campus housing cost per month: $1000.

Application Information
Students are accepted for enrollment in April and October. In 1996, 26 applied; 24 were accepted. Applicants must submit a formal application.

Contact
Peggy Lynch, Administrator, 271 Francisco Street, San Francisco, CA 94133; 415-788-6699; Fax: 415-788-8924.

Westlake Culinary Institute

Let's Get Cookin'

Westlake Village, California

General Information
Private, coeducational, culinary institute. Suburban campus. Founded in 1988.

Program Information
Offered since 1978. Member of American Institute of Wine & Food; International Association of Culinary Professionals; James Beard Foundation, Inc.; National Association for the Specialty Food Trade, Inc.; Women Chefs and Restaurateurs; Southern California Culinary Guild. Program calendar is divided into trimesters. 20-week Certificate in Baking. 24-week Certificate in Professional Cooking. 3-week Certificate in Catering, Beginning Course.

Areas of Study
Baking; culinary French; culinary skill development; food preparation; international cuisine; meal planning; sanitation; saucier; seafood processing; soup, stock, sauce, and starch production; wines and spirits.

Facilities
Classroom; teaching kitchen.

Culinary Student Profile
36 part-time. 3 are under 25 years old; 28 are between 25 and 44 years old; 5 are over 44 years old.

Faculty
3 part-time. 2 are industry professionals; 1 is a culinary-accredited teacher. Prominent faculty: Cecilia DeCastro.

Special Programs
Tours of produce market, special restaurant dinners.

Expenses
Application fee: $100. Tuition: $2400 per 24 weeks part-time.

Financial Aid
Employment placement assistance is available.

Application Information
Students are accepted for enrollment in January and July. Applicants must submit a formal application.

Contact
Phyllis Vaccarelli, Owner/Director, Let's Get Cookin', 4643 Lakeview Canyon Road, Westlake Village, CA 91361; 818-991-3940; Fax: 805-495-2554.

Colorado Institute of Art

School of Culinary Arts

Denver, Colorado

General Information
Private, coeducational, four-year college. Urban campus. Founded in 1952.

Program Information
Offered since 1994. Accredited by American Culinary Federation Education Institute. Member of American Culinary Federation; American Culinary Federation Educational Institute; American Dietetic Association; American Institute of Wine & Food; Council on Hotel, Restaurant, and Institutional Education; Educational Foundation of the NRA; International Association of Culinary Professionals; Interntional Food Service Executives Association; National Restaurant Association; Sommelier Society of America. Program calendar is divided into quarters. 18-month Associate degree in Culinary Arts.

Areas of Study
Baking; beverage management; buffet catering; computer applications; confectionery show pieces; controlling costs in food service; culinary French; culinary skill development; food preparation; food purchasing; food service communication; food service math; garde-manger; international cuisine; introduction to food service; kitchen management; management and human resources; meat fabrication; menu and facilities design; nutrition; nutrition and food service; patisserie; restaurant opportunities; sanitation; saucier; seafood processing; soup, stock, sauce, and starch production; wines and spirits; hospitality law.

Facilities
Bakery; catering service; 3 classrooms; computer laboratory; 4 food production kitchens; garden; gourmet dining room; learning resource center; library; public restaurant; student lounge.

Culinary Student Profile
460 total: 400 full-time; 60 part-time. 110 are under 25 years old; 325 are between 25 and 44 years old; 25 are over 44 years old.

Faculty
13 full-time; 12 part-time. 11 are industry professionals; 12 are culinary-accredited teachers; 2 are registered dietitians. Prominent faculty: Gary J. Prell and Chris Dejohn.

Colorado Institute of Art *(continued)*

Expenses

Application fee: $50. Tuition: $3420 per quarter. Program-related fees include: $250 for lab fees per quarter; $605 for supply kit (knives, uniforms, textbooks).

Financial Aid

In 1996, 4 scholarships were awarded (average award was $9765). Employment placement assistance is available. Employment opportunities within the program are available.

Housing

168 culinary students housed on campus. Coed and apartment-style housing available. Average on-campus housing cost per month: $400. Average off-campus housing cost per month: $400.

Application Information

Students are accepted for enrollment in January, April, July, and October. In 1996, 465 applied; 460 were accepted. Applicants must submit a formal application and an essay and interview.

Contact

Barbara Browning, Vice President, Director of Admissions, School of Culinary Arts, 200 East Ninth Avenue, Denver, CO 80203; 800-275-2420; Fax: 303-860-8520; World Wide Web: http://www.aii.edu

See affiliated programs: New York Restaurant School; The Art Institute of Atlanta; The Art Institute of Fort Lauderdale; The Art Institute of Houston; The Art Institute of Philadelphia; The Art Institute of Phoenix; The Art Institute of Seattle.

See display on page 48.

Colorado Mountain College

Keystone, Colorado

General Information

Public, coeducational, two-year college. Rural campus. Founded in 1967.

Program Information

Offered since 1993. Member of American Culinary Federation; American Culinary Federation Educational Institute; Council on Hotel, Restaurant, and Institutional Education; Educational Foundation of the NRA; National Restaurant Association. Program calendar is divided into semesters. 3-year Associate degree in Culinary Arts.

Areas of Study

Baking; beverage management; buffet catering; computer applications; controlling costs in food service; convenience cookery; culinary French; culinary skill development; food preparation; food purchasing; food service communication; food service math; garde-manger; international cuisine; introduction to food service; kitchen management; management and human resources; meal planning; meat cutting; meat fabrication; menu and facilities design; nutrition; nutrition and food service; patisserie; restaurant opportunities; sanitation; saucier; seafood processing; soup, stock, sauce, and starch production.

Facilities

Bake shop; bakery; 6 cafeterias; 4 catering services; 10 classrooms; coffee shop; 2 computer laboratories; 4 demonstration laboratories; 12 food production kitchens; 5 gourmet dining rooms; learning resource center; 12 lecture rooms; library; 12 public restaurants; snack shop; teaching kitchen.

Culinary Student Profile

36 full-time. 12 are under 25 years old; 24 are between 25 and 44 years old.

Faculty

1 full-time; 8 part-time. 4 are industry professionals; 1 is a culinary-accredited teacher. Prominent faculty: Doug Schwartz and Chris Wing.

Special Programs

ACF apprenticeship.

Expenses

Tuition: $1500 per year. Program-related fees include: $750 for tools and texts.

Financial Aid
In 1996, 6 scholarships were awarded (average award was $800). Employment placement assistance is available. Employment opportunities within the program are available.

Housing
Coed and apartment-style housing available. Average on-campus housing cost per month: $275. Average off-campus housing cost per month: $350.

Application Information
Students are accepted for enrollment in June. Application deadline for summer is April 15. In 1996, 75 applied; 12 were accepted. Applicants must submit a formal application, letters of reference, and an essay.

Contact
Admissions, PO Box 10001, Glenwood Springs, CO 81602; 800-621-8559; Fax: 970-945-7279.

Cooking School of the Rockies

Professional Culinary Arts Program

Boulder, Colorado

General Information
Private, coeducational, culinary institute. Urban campus. Founded in 1991.

Program Information
Offered since 1996. Member of American Culinary Federation; American Institute of Wine & Food; International Association of Culinary Professionals. Program calendar is divided into semesters. 6-month Diploma in Culinary Arts.

Areas of Study
Baking; confectionery show pieces; controlling costs in food service; culinary French; culinary skill development; food preparation; food service math; international cuisine; meal planning; meat cutting; menu and facilities design; nutrition and food service; patisserie; restaurant opportunities; saucier; soup, stock, sauce, and starch production; wines and spirits.

Facilities
Catering service; demonstration laboratory; food production kitchen; garden; teaching kitchen; gourmet take-out and cafe.

Culinary Student Profile
21 full-time.

Faculty
2 full-time; 1 part-time. 3 are culinary-accredited teachers. Prominent faculty: Robert Reynolds and Michael Comstedt.

Special Programs
One-month study in Carpentras, France.

Expenses
Tuition: $17,500 per diploma. Program-related fees include: $350 for books, clothing, and knives.

Financial Aid
Employment placement assistance is available. Employment opportunities within the program are available.

Housing
Average off-campus housing cost per month: $550.

Application Information
Students are accepted for enrollment in January and July. In 1996, 21 applied; 21 were accepted. Applicants must submit a formal application and an essay and interview.

Contact
Joan Brett, Director, Professional Culinary Arts Program, 637 South Broadway, Suite H, Boulder, CO 80303; 303-494-7988; Fax: 303-494-7999; E-mail: jbrett3768@aol.com

Johnson & Wales University

College of Culinary Arts

Vail, Colorado

General Information
Private, coeducational, two-year college. Rural campus. Founded in 1993. Accredited by New England Association of Schools and Colleges.

Johnson & Wales University *(continued)*

Program Information
Offered since 1993. Member of American Culinary Federation; American Dietetic Association; American Institute of Baking; American Institute of Wine & Food; Confrerie de la Chaine des Rotisseurs; Council on Hotel, Restaurant, and Institutional Education; Educational Foundation of the NRA; Institute of Food Technologists; International Association of Culinary Professionals; International Foodservice Editorial Council; Interntional Food Service Executives Association; James Beard Foundation, Inc.; National Restaurant Association; Oldways Preservation and Exchange Trust; The Bread Bakers Guild of America. Program calendar is modular. 1-year Associate degree in Culinary Arts.

Areas of Study
Culinary skill development; food preparation; garde-manger; international cuisine; kitchen management; meal planning; meat cutting; menu and facilities design; nutrition; nutrition and food service; sanitation; saucier; soup, stock, sauce, and starch production; wines and spirits.

Facilities
Classroom; 2 food production kitchens; learning resource center.

Culinary Student Profile
40 full-time. 16 are under 25 years old; 23 are between 25 and 44 years old; 1 is over 44 years old.

Faculty
17 part-time. 17 are industry professionals.

Expenses
Tuition: $18,432 per year.

Financial Aid
Employment placement assistance is available. Employment opportunities within the program are available.

Application Information
Students are accepted for enrollment in June. In 1996, 131 applied; 123 were accepted. Applicants must have a bachelor's degree.

Contact
Licia Dwyer, Director of Culinary Admissions, College of Culinary Arts, 8 Abbott Park Place, Providence, RI 02903-3703; 800-DIAL-JWU; Fax: 401-598-4712; E-mail: admissions@jwu.edu; World Wide Web: http://www.jwu.edu

See affiliated programs at Charleston, South Carolina; North Miami, Florida; Providence, Rhode Island; Norfolk, Virginia.

See display on page 216.

Pikes Peak Community College

Culinary Institute of Colorado Springs

Colorado Springs, Colorado

General Information
Public, coeducational, two-year college. Urban campus. Founded in 1968. Accredited by North Central Association of Colleges and Schools.

Program Information
Offered since 1986. Member of American Culinary Federation; National Restaurant Association. Program calendar is divided into semesters. 2-year Associate degree in Food Management.

Areas of Study
Baking; beverage management; buffet catering; computer applications; controlling costs in food service; culinary skill development; food preparation; food purchasing; food service math; garde-manger; introduction to food service; kitchen management; management and human resources; meal planning; meat cutting; meat fabrication; menu and facilities design; nutrition; nutrition and food service; restaurant opportunities; sanitation; saucier; soup, stock, sauce, and starch production; wines and spirits.

Facilities
Bakery; cafeteria; catering service; classroom; coffee shop; computer laboratory; food production

kitchen; laboratory; learning resource center; lecture room; library; snack shop; student lounge; teaching kitchen.

Culinary Student Profile
40 full-time.

Faculty
3 full-time. 3 are culinary-accredited teachers. Prominent faculty: George J. Bissonnette and Robert Hudson.

Expenses
In-state tuition: $4000 per year. Out-of-state tuition: $4800 per year.

Financial Aid
In 1996, 15 scholarships were awarded (average award was $1500).

Housing
Average off-campus housing cost per month: $250.

Application Information
Students are accepted for enrollment in January, May, and August. Applications are accepted continuously for fall, spring, and summer. In 1996, 25 applied; 25 were accepted. Applicants must submit a formal application.

Contact
George Bissonnette, Chef Instructor, Culinary Institute of Colorado Springs, 5675 South Academy Boulevard, Colorado Springs, CO 80906; 719-540-7371; Fax: 719-540-7453.

Pueblo Community College

Culinary Arts Department

Pueblo, Colorado

General Information
Public, coeducational, two-year college. Urban campus. Founded in 1979. Accredited by North Central Association of Colleges and Schools.

Program Information
Offered since 1984. Member of American Culinary Federation; American Culinary Federation Educational Institute; Educational Foundation of the NRA. Program calendar is divided into semesters. 1-year Certificate in Culinary Arts. 2-year Associate degree in Culinary Arts.

Areas of Study
Baking; beverage management; computer applications; controlling costs in food service; culinary skill development; food preparation; food purchasing; food service communication; garde-manger; management and human resources; meal planning; nutrition; nutrition and food service; sanitation; soup, stock, sauce, and starch production; legal aspects of food service management.

Facilities
Bake shop; bakery; cafeteria; catering service; classroom; computer laboratory; food production kitchen; gourmet dining room; learning resource center; library.

Culinary Student Profile
60 total: 40 full-time; 20 part-time. 15 are under 25 years old; 40 are between 25 and 44 years old; 5 are over 44 years old.

Faculty
2 full-time; 5 part-time. 2 are industry professionals; 5 are culinary-accredited teachers. Prominent faculty: Carol S. Himes and Charles Becker.

Expenses
Application fee: $5. Tuition: $930 per semester full-time, $61.30 per credit hour part-time. Program-related fees include: $100 for knife set; $50 for garde manger kit; $50 for chef's coat and hat.

Financial Aid
In 1996, 7 scholarships were awarded (average award was $250). Employment placement assistance is available. Employment opportunities within the program are available.

Application Information
Students are accepted for enrollment in January, May, and September. In 1996, 60 applied; 60 were accepted. Applicants must submit a formal application.

Pueblo Community College *(continued)*

Contact

Carol S. Himes, Department Chair, Culinary Arts Department, 900 West Orman Avenue, Pueblo, CO 81004-1499; 719-549-3071; Fax: 719-549-3070; E-mail: himes@pcc.cccoes.edu

School of Natural Cookery

Boulder, Colorado

General Information

Private, coeducational, culinary institute. Urban campus. Founded in 1983.

Program Information

Offered since 1983. Program calendar is divided into trimesters. Certificates in Teacher Training; Personal Chef Training. 2-week Certificate in Baking/Pastry. 5-week Certificate in Fundamentals of Cooking.

Areas of Study

Baking; meal planning; energetic nutrition; personal chef repertoire.

Culinary Student Profile

8 full-time.

Faculty

2 full-time; 4 part-time. 4 are industry professionals; 1 is a master chef; 1 is a culinary-accredited teacher. Prominent faculty: Joanne Saltzman and Mary Bowman.

Expenses

Tuition: $3500 per 5 weeks.

Financial Aid

Employment placement assistance is available. Employment opportunities within the program are available.

Housing

Average off-campus housing cost per month: $450.

Application Information

Students are accepted for enrollment in January, June, and September. Applicants must submit a formal application and letters of reference and interview.

Contact

Joanne Saltzman, Director, PO Box 19466, Boulder, CO 80308; 303-444-8068; E-mail: snc@sprynet.com

Warren Tech

Vocational Culinary Arts

Lakewood, Colorado

General Information

Private, coeducational, two-year college. Urban campus. Founded in 1974.

Program Information

Offered since 1974. Member of American Culinary Federation; Food Service Instructors of Colorado. Program calendar is divided into semesters. 2-year Associate degree in Commercial Food Service Management. 9-month Certificate in Restaurant Arts.

Areas of Study

Baking; beverage management; buffet catering; computer applications; controlling costs in food service; culinary skill development; food preparation; food purchasing; food service communication; food service math; garde-manger; international cuisine; management and human resources; meal planning; meat cutting; nutrition; nutrition and food service; restaurant opportunities; sanitation; saucier; seafood processing; soup, stock, sauce, and starch production.

Facilities

Bake shop; bakery; catering service; classroom; coffee shop; demonstration laboratory; food production kitchen; gourmet dining room; learning resource center; lecture room; library; public restaurant; teaching kitchen; delicatessen.

Culinary Student Profile
80 total: 40 full-time; 40 part-time. 73 are under 25 years old; 5 are between 25 and 44 years old; 2 are over 44 years old.

Faculty
5 full-time; 1 part-time. 3 are industry professionals; 3 are culinary-accredited teachers. Prominent faculty: Sharron Pizzuto and Dennis Gomes.

Special Programs
Food service site visitations, Culinary Club.

Expenses
Tuition: $1400 per semester full-time, $700 per semester part-time. Program-related fees include: $50 for uniforms.

Financial Aid
In 1996, 3 scholarships were awarded (average award was $500); 3 loans were granted (average loan was $500). Program-specific awards include Alliant Food Service scholarship ($200). Employment placement assistance is available. Employment opportunities within the program are available.

Application Information
Students are accepted for enrollment in January and September. Application deadline for fall is August 18. Application deadline for winter is January 5. In 1996, 86 applied; 83 were accepted. Applicants must submit a formal application.

Contact
Sharron Pizzuto, Service Instructor, Vocational Culinary Arts, 13300 West 2nd Place, Lakewood, CO 80228-1256; 303-982-8555; Fax: 303-982-8547.

Connecticut Culinary Institute

Farmington, Connecticut

General Information
Private, coeducational, culinary institute. Suburban campus. Founded in 1987.

Program Information
Offered since 1987. Member of James Beard Foundation, Inc. Program calendar is year-round. 10-week Diploma in Pastry Arts. 3-month Diploma in Culinary Arts.

Areas of Study
Baking; culinary skill development; food preparation; garde-manger; international cuisine; meal planning; nutrition; patisserie; sanitation; saucier; seafood processing; soup, stock, sauce, and starch production.

Facilities
Bake shop; bakery; 4 classrooms; 4 food production kitchens; 2 learning resource centers; 2 libraries; public restaurant; 4 teaching kitchens.

Culinary Student Profile
206 total: 162 full-time; 44 part-time.

Faculty
7 full-time; 3 part-time. 10 are industry professionals.

Expenses
Application fee: $25. Tuition: $5300 per diploma. Program-related fees include: $366 for tools; $684 for tools (pastry and baking).

Financial Aid
Employment placement assistance is available.

Application Information
Students are accepted for enrollment year-round. Applicants must submit a formal application.

Contact
Admissions Department, 230 Farmington Avenue, Farmington, CT 06032; 860-677-7869; Fax: 860-676-0679.

Gateway Community-Technical College

Hospitality Management

New Haven, Connecticut

General Information
Public, coeducational, two-year college. Urban campus. Founded in 1968. Accredited by New England Association of Schools and Colleges.

Gateway Community-Technical College (continued)

Program Information
Offered since 1968. Member of American Wine Society; Council on Hotel, Restaurant, and Institutional Education; Society of Wine Educators; Amenti del Vino. Program calendar is divided into semesters. 1-year Certificate in Culinary Arts.

Areas of Study
Baking; beverage management; buffet catering; computer applications; confectionery show pieces; controlling costs in food service; convenience cookery; culinary skill development; food preparation; food purchasing; food service math; international cuisine; introduction to food service; kitchen management; management and human resources; meal planning; nutrition; nutrition and food service; restaurant opportunities; sanitation; seafood processing; soup, stock, sauce, and starch production; wines and spirits.

Facilities
Bake shop; 2 cafeterias; 2 catering services; 4 computer laboratories; demonstration laboratory; 2 food production kitchens; gourmet dining room; laboratory; 2 libraries; 2 student lounges; teaching kitchen.

Culinary Student Profile
160 total: 110 full-time; 50 part-time.

Faculty
2 full-time; 5 part-time. 1 is a master chef. Prominent faculty: Peter Cisek.

Special Programs
One-day visit to the International Hotel-Restaurant show in New York City.

Expenses
Application fee: $20. Tuition: $907 per semester full-time, $81 per credit hour part-time. Program-related fees include: $100 for uniforms; $150 for textbooks.

Financial Aid
Employment placement assistance is available. Employment opportunities within the program are available.

Application Information
Students are accepted for enrollment in January and September. Application deadline for fall is September 6. Application deadline for spring is January 16. In 1996, 140 applied; 94 were accepted. Applicants must submit a formal application.

Contact
Eugene J. Spaziani, Director, Hospitality Management, 60 Sargent Drive, New Haven, CT 06511-5918; 203-789-7067; Fax: 203-789-6510.

Manchester Community-Technical College

Culinary Arts Department

Manchester, Connecticut

General Information
Public, coeducational, two-year college. Suburban campus. Founded in 1963. Accredited by New England Association of Schools and Colleges.

Program Information
Offered since 1967. Accredited by American Culinary Federation Education Institute. Member of American Culinary Federation; American Culinary Federation Educational Institute; American Wine Society; Council on Hotel, Restaurant, and Institutional Education; Educational Foundation of the NRA; National Restaurant Association; Society of Wine Educators. Program calendar is divided into semesters. 1-year Certificate in Culinary Arts. 2-year Associate degrees in Hotel Tourism; Food Service Management.

Areas of Study
Baking; beverage management; buffet catering; computer applications; confectionery show pieces; controlling costs in food service; culinary skill development; food preparation; food purchasing; food service math; garde-manger; international cuisine; introduction to food service; management and human resources; meal planning; nutrition; nutrition and food service; sanitation; seafood

processing; soup, stock, sauce, and starch production; wines and spirits.

Facilities
20 classrooms; 2 food production kitchens.

Culinary Student Profile
200 total: 50 full-time; 150 part-time.

Faculty
7 full-time; 4 part-time. 7 are culinary-accredited teachers. Prominent faculty: Glenn S. Lemaire and Michael P. Hiza.

Expenses
Application fee: $20. Tuition: $804 per semester full-time, $67 per credit part-time. Program-related fees include: $50 for uniforms; $300 for books.

Financial Aid
In 1996, 4 scholarships were awarded (average award was $500). Employment placement assistance is available. Employment opportunities within the program are available.

Application Information
Students are accepted for enrollment in January and September. Applicants must submit a formal application.

Contact
Glenn S. Lemaire, Department Chair, Culinary Arts Department, PO Box 1046, Manchester, CT 06045-1046; 860-647-6136; Fax: 860-647-6238; E-mail: ma_lemaire@commnet.edu

Naugatuck Valley Community-Technical College

Hospitality Management Programs

Waterbury, Connecticut

General Information
Public, coeducational, two-year college. Urban campus. Founded in 1967. Accredited by New England Association of Schools and Colleges.

Program Information
Offered since 1982. Member of Council on Hotel, Restaurant, and Institutional Education; Educational Foundation of the NRA; National Restaurant Association. Program calendar is divided into semesters. Certificate in Culinary Arts Supervision. 2-year Associate degrees in Hotel Management; Food Services Management.

Areas of Study
Buffet catering; computer applications; culinary skill development; food preparation; food purchasing; food service communication; food service math; garde-manger; introduction to food service; kitchen management; management and human resources; meal planning; menu and facilities design; nutrition; nutrition and food service; restaurant opportunities; sanitation.

Facilities
Catering service; 2 classrooms; 6 computer laboratories; demonstration laboratory; food production kitchen; gourmet dining room; laboratory; learning resource center; library; student lounge; teaching kitchen.

Culinary Student Profile
125 total: 75 full-time; 50 part-time. 60 are under 25 years old; 45 are between 25 and 44 years old; 20 are over 44 years old.

Faculty
3 full-time; 8 part-time. 8 are industry professionals; 3 are culinary-accredited teachers. Prominent faculty: Todd Jones and Karen Russo.

Special Programs
Cooperative education/work experience.

Expenses
Application fee: $20. Tuition: $907 per semester full-time, $243 per 3-credit course part-time.

Financial Aid
Employment placement assistance is available. Employment opportunities within the program are available.

Application Information
Students are accepted for enrollment in January, June, and September. Applications are accepted continuously for fall, spring, and summer. Applicants must submit a formal application.

Naugatuck Valley Community-Technical College *(continued)*

Contact
Todd Jones, Program Coordinator, Hospitality Management Programs, 750 Chase Parkway, Waterbury, CT 06708-3000; 203-575-8175.

Norwalk Community-Technical College

Culinary Arts Program

Norwalk, Connecticut

General Information
Public, coeducational, two-year college. Suburban campus. Founded in 1961. Accredited by New England Association of Schools and Colleges.

Program Information
Offered since 1992. Member of Council on Hotel, Restaurant, and Institutional Education; Educational Foundation of the NRA; National Restaurant Association; Connecticut Restaurant Association. Program calendar is divided into semesters. 1-year Certificate in Culinary Arts. 2-year Associate degrees in Hotel/Motel Management; Restaurant/Food Service Management.

Areas of Study
Baking; buffet catering; computer applications; controlling costs in food service; culinary skill development; food preparation; food purchasing; garde-manger; international cuisine; introduction to food service; management and human resources; nutrition; sanitation; soup, stock, sauce, and starch production.

Facilities
Bakery; cafeteria; catering service; 2 classrooms; computer laboratory; demonstration laboratory; food production kitchen; garden; gourmet dining room; laboratory; 2 lecture rooms; library; snack shop; student lounge; teaching kitchen.

Culinary Student Profile
80 total: 20 full-time; 60 part-time. 35 are under 25 years old; 36 are between 25 and 44 years old; 9 are over 44 years old.

Faculty
1 full-time; 4 part-time. 3 are industry professionals. Prominent faculty: Thomas J. Connolly.

Special Programs
Educational exchange program with two French culinary schools.

Expenses
Application fee: $20. Tuition: $907 per semester full-time, $243 per 3-credit course part-time. Program-related fees include: $50 for uniforms.

Financial Aid
In 1996, 1 scholarship was awarded (award was $850). Employment placement assistance is available. Employment opportunities within the program are available.

Application Information
Students are accepted for enrollment in January and September. In 1996, 22 applied; 22 were accepted. Applicants must submit a formal application.

Contact
Tom Connolly, Coordinator, Culinary Arts Program, 188 Richards Avenue, Norwalk, CT 06854-1655; 203-857-7355; Fax: 203-857-3327; E-mail: nk_connolly@commnet.edu

Delaware Technical and Community College

Culinary Arts/Food Service Management

Newark, Delaware

General Information
Public, coeducational, two-year college. Suburban campus. Founded in 1967. Accredited by Middle States Association of Colleges and Schools.

Program Information
Offered since 1993. Member of American Culinary Federation; National Restaurant Association. Program calendar is divided into semesters. 2-year Associate degrees in Food Service Management; Culinary Arts. 2-year Diploma in Food Service Management.

Areas of Study
Baking; beverage management; computer applications; controlling costs in food service; culinary skill development; food preparation; food purchasing; food service math; garde-manger; international cuisine; introduction to food service; management and human resources; menu and facilities design; nutrition; restaurant opportunities; sanitation; saucier; seafood processing; soup, stock, sauce, and starch production.

Facilities
Cafeteria; classroom; 10 computer laboratories; learning resource center; lecture room; library; teaching kitchen.

Culinary Student Profile
88 total: 48 full-time; 40 part-time. 22 are under 25 years old; 44 are between 25 and 44 years old; 22 are over 44 years old.

Faculty
3 full-time; 6 part-time. 9 are industry professionals. Prominent faculty: David Nolker and Ed Hennessy.

Expenses
Application fee: $10. In-state tuition: $630 per semester full-time, $52.50 per credit part-time. Out-of-state tuition: $1575 per semester full-time, $131.25 per credit part-time. Program-related fees include: $45 for materials; $7.50 for lab fees per credit hour.

Financial Aid
Employment placement assistance is available. Employment opportunities within the program are available.

Application Information
Students are accepted for enrollment in September. Application deadline for fall is April 15. In 1996, 38 applied; 24 were accepted. Applicants must submit a formal application, letters of reference, and a portfolio.

Contact
Admission Department, Culinary Arts/Food Service Management, 400 Stanton-Christiana Road, Newark, DE 19713; 302-454-3954; Fax: 302-453-3029.

The Art Institute of Fort Lauderdale

School of Culinary Arts

Fort Lauderdale, Florida

General Information
Private, coeducational, two-year college. Urban campus. Founded in 1968.

Program Information
Offered since 1991. Accredited by American Culinary Federation Education Institute. Member of American Culinary Federation; Council on Hotel, Restaurant, and Institutional Education; National Restaurant Association; Florida Restaurant Association. Program calendar is divided into quarters. 1-year Diploma in Culinary Arts. 18-month Associate degree in Culinary Arts.

Areas of Study
Baking; beverage management; buffet catering; computer applications; confectionery show pieces; controlling costs in food service; culinary French; culinary skill development; food preparation; food purchasing; food service math; garde-manger; international cuisine; introduction to food service; kitchen management; management and human resources; meat cutting; meat fabrication; menu and facilities design; nutrition; patisserie; sanitation; saucier; seafood processing; soup, stock, sauce, and starch production; wines and spirits.

Facilities
Bake shop; 2 classrooms; 6 computer laboratories; demonstration laboratory; food production kitchen; gourmet dining room; laboratory; learning resource center; public restaurant; snack shop; student lounge; teaching kitchen; vineyard.

The Art Institute of Fort Lauderdale *(continued)*

Culinary Student Profile
260 total: 240 full-time; 20 part-time.

Faculty
4 full-time; 6 part-time.

Expenses
Application fee: $50. Tuition: $226 per credit. Program-related fees include: $250 for lab fees per quarter; $550 for culinary kit (includes uniforms).

Financial Aid
In 1996, 6 scholarships were awarded (average award was $3000). Employment placement assistance is available. Employment opportunities within the program are available.

Housing
Coed housing available. Average off-campus housing cost per month: $500.

Application Information
Students are accepted for enrollment in January, April, July, and October. Applicants must submit a formal application and have a high school diploma or GED.

Contact
Janet Scott, Director of Culinary Admissions, School of Culinary Arts, 1799 Southeast 17th Street Causeway, Fort Lauderdale, FL 33316-3000; 800-275-7603 Ext. 431; Fax: 954-523-7676.

See affiliated programs: Colorado Institute of Art; New York Restaurant School; The Art Institute of Atlanta; The Art Institute of Houston; The Art Institute of Philadelphia; The Art Institute of Phoenix; The Art Institute of Seattle.

See display on page 48.

Charlotte Vocational Technical Center

Culinary Arts Program

Port Charlotte, Florida

General Information
Public, coeducational, two-year college. Urban campus. Founded in 1980.

Program Information
Program calendar is divided into quarters. 1,080-hour Certificate in Commercial Foods and Culinary Arts.

Areas of Study
Baking; computer applications; controlling costs in food service; convenience cookery; culinary French; culinary skill development; food preparation; food purchasing; food service communication; food service math; garde-manger; international cuisine; introduction to food service; kitchen management; management and human resources; meal planning; menu and facilities design; nutrition; patisserie; restaurant opportunities; sanitation; saucier; seafood processing; soup, stock, sauce, and starch production; wines and spirits; safety and first aid; ice sculpture.

Culinary Student Profile
30 total: 24 full-time; 6 part-time.

Faculty
3 full-time.

Expenses
Tuition: $799 per certificate.

Application Information
Students are accepted for enrollment in January, February, March, April, May, June, August, September, October, November, and December. Applicants must be at least 16 years of age.

Contact
Dick Santello, Student Counselor, Culinary Arts Program, 18300 Toledo Blade Boulevard, Port Charlotte, FL 33948-3399; 941-629-6819 Ext. 110; Fax: 941-629-2058.

Daytona Beach Community College

Culinary Arts Department

Daytona Beach, Florida

General Information
Public, coeducational, two-year college. Urban campus. Founded in 1958. Accredited by Southern Association of Colleges and Schools.

PROGRAM INFORMATION
Offered since 1997. Program calendar is divided into semesters. 2-year Associate degree in Culinary Management.

AREAS OF STUDY
Baking; beverage management; buffet catering; computer applications; controlling costs in food service; culinary skill development; food preparation; food purchasing; food service communication; food service math; garde-manger; international cuisine; introduction to food service; kitchen management; management and human resources; meat cutting; meat fabrication; menu and facilities design; nutrition; restaurant opportunities; sanitation; saucier; seafood processing; soup, stock, sauce, and starch production; wines and spirits; hospitality law.

CULINARY STUDENT PROFILE
50 total: 40 full-time; 10 part-time.

FACULTY
4 full-time; 1 part-time.

EXPENSES
In-state tuition: $41.50 per credit hour. Out-of-state tuition: $155.47 per credit hour.

APPLICATION INFORMATION
Students are accepted for enrollment in January, May, and August. Applicants must submit a formal application.

CONTACT
Jeff Conklin, Program Manager, Culinary Arts Department, PO Box 2811, Daytona Beach, FL 32120; 904-255-8131 Ext. 3735; Fax: 904-254-3063.

FLORIDA COMMUNITY COLLEGE AT JACKSONVILLE

Culinary Management

Jacksonville, Florida

GENERAL INFORMATION
Public, coeducational, two-year college. Urban campus. Founded in 1963. Accredited by Southern Association of Colleges and Schools.

PROGRAM INFORMATION
Offered since 1989. Accredited by American Culinary Federation Education Institute. Member of American Culinary Federation; American Culinary Federation Educational Institute; American Dietetic Association; Council on Hotel, Restaurant, and Institutional Education; National Restaurant Association. Program calendar is divided into semesters. 2-year Associate degree in Culinary Management.

AREAS OF STUDY
Baking; buffet catering; computer applications; controlling costs in food service; culinary skill development; food preparation; food purchasing; garde-manger; international cuisine; introduction to food service; management and human resources; meal planning; menu and facilities design; nutrition; nutrition and food service; restaurant opportunities; sanitation; saucier; seafood processing; soup, stock, sauce, and starch production; wines and spirits; American regional.

FACILITIES
Bake shop; cafeteria; 5 classrooms; computer laboratory; 2 food production kitchens; gourmet dining room; 2 learning resource centers; library; student lounge; teaching kitchen.

CULINARY STUDENT PROFILE
130 total: 120 full-time; 10 part-time. 13 are under 25 years old; 97 are between 25 and 44 years old; 20 are over 44 years old.

FACULTY
6 full-time; 12 part-time. 12 are industry professionals; 4 are culinary-accredited teachers; 1 is a dietetics professor. Prominent faculty: Joseph Harrold and Afred Fricke.

EXPENSES
Application fee: $25. In-state tuition: $498 per semester full-time, $41.50 per credit hour part-time. Out-of-state tuition: $1888.80 per semester full-time, $157.40 per credit hour part-time. Program-related fees include: $180 for knife set.

FINANCIAL AID
In 1996, 10 scholarships were awarded (average award was $500). Employment opportunities within the program are available.

Florida Community College at Jacksonville *(continued)*

Application Information
Students are accepted for enrollment in January, May, and August. Applications are accepted continuously for fall, winter, and spring. In 1996, 50 applied; 45 were accepted. Applicants must submit a formal application.

Contact
Alfred Fricke, Program Manager, Culinary Management, 4501 Capper Road, Jacksonville, FL 32218; 904-766-6652; Fax: 904-766-6654; E-mail: africke@fccj.cc.fl.us

Florida Culinary Institute

West Palm Beach, Florida

General Information
Private, coeducational, culinary institute. Suburban campus. Founded in 1984.

Program Information

Accredited by American Culinary Federation Education Institute. Member of American Culinary Federation; American Culinary Federation Educational Institute; Council on Hotel, Restaurant, and Institutional Education; Educational Foundation of the NRA; James Beard Foundation, Inc.; National Restaurant Association. Program calendar is divided into quarters. 18-month Specialized Associate degrees in Food and Beverage Management; International Baking and Pastry; Culinary Arts.

Areas of Study

Baking; beverage management; confectionery show pieces; controlling costs in food service; food preparation; food purchasing; garde-manger; international cuisine; meat fabrication; sanitation; soup, stock, sauce, and starch production; wines and spirits.

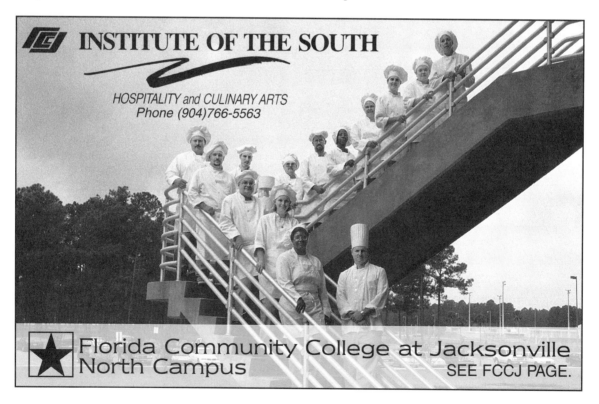

Facilities
2 bake shops; catering service; 9 classrooms; computer laboratory; garden; learning resource center; 2 lecture rooms; library; public restaurant; student lounge; 8 teaching kitchens.

Culinary Student Profile
600 full-time. 300 are under 25 years old; 250 are between 25 and 44 years old; 50 are over 44 years old.

Faculty
21 full-time. Prominent faculty: Manfred Schmidtke and Jack Marshall.

Expenses
Application fee: $25. Tuition: $3100 per quarter.

Financial Aid
Employment placement assistance is available.

Housing
Average off-campus housing cost per month: $400.

Application Information
Students are accepted for enrollment in January, April, July, and October. Applicants must submit a formal application.

Contact
Scott Spitolnick, Admissions Director, 1126 53rd Court, West Palm Beach, FL 33407; 561-842-8324; Fax: 561-842-9503; E-mail: neit@newenglandtech.com

---

Florida Culinary Institute's "Students First" philosophy encourages a one-to-one learning experience. Through close attention to detail, the chef instructors identify the areas that each student needs to develop. Open communication with each student allows the creation of a personal path to success. Together, the teacher and student identify the student's level of skill, the desired and attainable level of achievement, and the path to get there. The Institute believes in teaching students only the courses they need to get started with their careers. Training and learning are provided in an environment that not only exposes the student to the theory, but also stresses hands-on experience so that graduates have the maximum opportunity for employment upon graduation.

Gulf Coast Community College

Culinary Management

Panama City, Florida

General Information
Public, coeducational, two-year college. Founded in 1957. Accredited by Southern Association of Colleges and Schools.

Program Information
Accredited by American Culinary Federation Education Institute. Member of American Culinary Federation; American Culinary Federation Educational Institute; Confrerie de la Chaine des Rotisseurs. Program calendar is divided into semesters. 2-year Associate degree in Culinary Management.

Areas of Study
Baking; beverage management; buffet catering; computer applications; confectionery show pieces; controlling costs in food service; convenience cookery; culinary French; culinary skill development; food preparation; food purchasing; food service communication; food service math; garde-manger; international cuisine; introduction to food service; kitchen management; management and human resources; meal planning; meat cutting; meat fabrication; menu and facilities design; nutrition; nutrition and food service; patisserie; restaurant opportunities; sanitation; saucier; seafood processing; soup, stock, sauce, and starch production; wines and spirits.

Facilities
2 bake shops; cafeteria; catering service; classroom; 3 computer laboratories; demonstration laboratory; 2 food production kitchens; garden; gourmet dining room; 2 laboratories; learning

Gulf Coast Community College *(continued)*

resource center; lecture room; library; public restaurant; snack shop; student lounge; teaching kitchen.

Culinary Student Profile
32 full-time.

Faculty
3 full-time; 1 part-time. 3 are culinary-accredited teachers.

Special Programs
French exchange, Vocational Institute Clubs of America competitions, American Culinary Federation competitions.

Expenses
In-state tuition: $38.44 per credit hour. Out-of-state tuition: $144.16 per credit hour.

Financial Aid
In 1996, 8 scholarships were awarded (average award was $800). Employment placement assistance is available. Employment opportunities within the program are available.

Application Information
Students are accepted for enrollment in January and August. In 1996, 32 applied; 32 were accepted. Applicants must submit a formal application and an essay and interview.

Contact
Travis A. Herr, Coordinator, Culinary Management, 5230 West Highway 98, Panama City, FL 32401-1058; 850-872-3850; Fax: 850-872-3836.

Johnson & Wales University

College of Culinary Arts

North Miami, Florida

General Information
Private, coeducational, four-year college. Urban campus. Founded in 1992. Accredited by New England Association of Schools and Colleges.

Program Information
Offered since 1992. Member of American Culinary Federation; American Dietetic Association; American Institute of Baking; American Institute of Wine & Food; Confrerie de la Chaine des Rotisseurs; Council on Hotel, Restaurant, and Institutional Education; Educational Foundation of the NRA; Institute of Food Technologists; International Association of Culinary Professionals; International Foodservice Editorial Council; Interntional Food Service Executives Association; James Beard Foundation, Inc.; National Restaurant Association; Oldways Preservation and Exchange Trust; Tasters Guild International; The Bread Bakers Guild of America. Program calendar is divided into quarters. 2-year Associate degrees in Culinary Arts; Baking and Pastry Arts. 4-year Bachelor's degree in Culinary Arts.

Areas of Study
Baking; garde-manger; international cuisine; meat cutting; nutrition and food service; sanitation.

Facilities
Bake shop; cafeteria; 12 classrooms; computer laboratory; 8 food production kitchens; learning resource center; 6 lecture rooms; library; public restaurant; 2 student lounges; meatroom/butcher shop.

Culinary Student Profile
782 total: 676 full-time; 106 part-time.

Faculty
13 full-time; 2 part-time.

Expenses
Tuition: $14,376 per year.

Financial Aid
Employment placement assistance is available. Employment opportunities within the program are available.

Housing
319 culinary students housed on campus. Coed housing available. Average on-campus housing cost per month: $379.

Application Information
Students are accepted for enrollment in March, June, July, September, and December. In 1996,

1,237 applied; 934 were accepted. Applicants must submit a formal application.

Contact

Jeffery Greenip, Director of Admissions, College of Culinary Arts, 1701 Northeast 127th Street, North Miami, FL 33181; 800-BEA-CHEF; Fax: 305-892-7020; E-mail: admissions@jwu.edu; World Wide Web: http://www.jwu.edu

See affiliated programs at Charleston, South Carolina; Providence, Rhode Island; Vail, Colorado; Norfolk, Virginia.

See display on page 216.

Lindsey Hopkins Technical Education Center

Culinary Arts Program

Miami, Florida

General Information

Public, coeducational, technical institute. Urban campus. Founded in 1937.

Program Information

Program calendar is divided into trimesters. 1,440-hour Certificate of Completion in Commercial Baking and Cooking. 720-hour Certificate of Completions in Commercial Cooking; Commercial Baking.

Areas of Study

Baking; buffet catering; confectionery show pieces; convenience cookery; culinary French; culinary skill development; food preparation; food purchasing; food service communication; food service math; garde-manger; international cuisine; kitchen management; meal planning; meat cutting; patisserie; restaurant opportunities; sanitation; saucier; seafood processing; soup, stock, sauce, and starch production.

Culinary Student Profile

20 full-time.

Faculty

2 full-time.

Expenses

Tuition: $222 per trimester.

Application Information

Students are accepted for enrollment in January, May, and September.

Contact

Esteban Sardon, Assistant Principal, Culinary Arts Program, 750 Northwest 20th Street, Miami, FL 33127; 305-324-6070; Fax: 305-545-6397.

Manatee Technical Institute

Commercial Foods and Culinary Arts

Bradenton, Florida

General Information

Public, coeducational, technical institute. Suburban campus. Founded in 1961.

Program Information

Offered since 1981. Program calendar is divided into quarters. 10-month Certificate in Commercial Foods and Culinary Arts.

Areas of Study

Baking; culinary skill development; food preparation; food purchasing; food service math; introduction to food service; meal planning; meat cutting; menu and facilities design; sanitation; saucier; seafood processing; soup, stock, sauce, and starch production.

Culinary Student Profile

20 full-time. 8 are under 25 years old; 10 are between 25 and 44 years old; 2 are over 44 years old.

Faculty

1 full-time; 1 part-time. 2 are culinary-accredited teachers.

Expenses

Application fee: $5. Tuition: $784.40 per certificate. Program-related fees include: $60 for lab fees; $120 for knife set; $120 for books.

Manatee Technical Institute *(continued)*

Financial Aid
In 1996, 3 scholarships were awarded (average award was $300). Employment placement assistance is available.

Housing
Average off-campus housing cost per month: $500.

Application Information
Students are accepted for enrollment in January, February, March, April, August, September, October, November, and December. In 1996, 25 applied; 25 were accepted. Applicants must submit a formal application.

Contact
Marilyn Whatley, Counselor, Commercial Foods and Culinary Arts, 5603 34th Street, W, Bradenton, FL 34210; 941-751-7917; Fax: 941-751-7927.

McFatter Vocational Technical Center

McFatter School of Culinary Arts

Davie, Florida

General Information
Public, coeducational, occupational education center. Suburban campus. Founded in 1985.

Program Information
Offered since 1996. Member of American Culinary Federation. Program calendar is divided into quarters. 1-year Certificate in Culinary Arts. 9-week Certificates in Soap, Stock, and Sauce Production; Meat Fabrication; Garde Manger; Baking.

Areas of Study
Baking; food preparation; garde-manger; management and human resources; meat cutting; meat fabrication; nutrition; sanitation; soup, stock, sauce, and starch production.

Facilities
Bake shop; bakery; cafeteria; catering service; 2 classrooms; 2 coffee shops; computer laboratory; 3 demonstration laboratories; 3 food production kitchens; gourmet dining room; laboratory; learning resource center; lecture room; library; 2 public restaurants; snack shop; 2 teaching kitchens.

Culinary Student Profile
105 total: 85 full-time; 20 part-time. 30 are under 25 years old; 69 are between 25 and 44 years old; 6 are over 44 years old.

Faculty
2 full-time; 5 part-time. Prominent faculty: V. Paul Citrullo, Jr.

Expenses
Tuition: $420 per semester full-time, $218 per semester part-time.

Financial Aid
In 1996, 5 scholarships were awarded (average award was $2000). Program-specific awards include aid from private industry and service agencies. Employment placement assistance is available.

Application Information
Students are accepted for enrollment in January, April, June, August, and November. In 1996, 105 applied; 105 were accepted. Applicants must complete an informal application during orientation.

Contact
Evelyn Smith, Counselor, McFatter School of Culinary Arts, 6500 Nova Drive, Davie, FL 33317; 954-370-8324; Fax: 954-370-1647; World Wide Web: http://www.gate.net/~mcfatter

Miami Lakes Technical Education Center

Commercial Foods and Culinary Arts Department

Miami Lakes, Florida

General Information
Public, coeducational, technical institute. Urban campus. Founded in 1979.

Program Information
Program calendar is divided into trimesters. 1,440-hour Certificate in Commercial Foods.

Areas of Study
Baking; controlling costs in food service; food preparation; garde-manger; introduction to food service; kitchen management; patisserie; sanitation; soup, stock, sauce, and starch production.

Culinary Student Profile
48 full-time.

Faculty
3 full-time; 1 part-time.

Expenses
Tuition: $222 per trimester.

Application Information
Students are accepted for enrollment year-round. Applicants must take the Test of Adult Basic Education.

Contact
Dale Brubaker, Coordinator, Commercial Foods and Culinary Arts Department, 5780 Northwest 158th Street, Miami Lakes, FL 33014; 305-557-1100; Fax: 305-557-7391.

Pensacola Junior College

Culinary Management Program

Pensacola, Florida

General Information
Public, coeducational, two-year college. Suburban campus. Founded in 1948. Accredited by Southern Association of Colleges and Schools.

Program Information
Offered since 1995. Member of American Culinary Federation; American Culinary Federation Educational Institute; Council on Hotel, Restaurant, and Institutional Education. Program calendar is divided into semesters. 2-year Associate degree in Culinary Management. 420-hour Certificate in Food Production and Services.

Areas of Study
Baking; beverage management; buffet catering; computer applications; confectionery show pieces; controlling costs in food service; culinary skill development; food preparation; food purchasing; food service math; garde-manger; international cuisine; introduction to food service; kitchen management; management and human resources; meal planning; menu and facilities design; nutrition; patisserie; sanitation; saucier; soup, stock, sauce, and starch production; dining room management.

Facilities
Bake shop; classroom; computer laboratory; food production kitchen; gourmet dining room; learning resource center; public restaurant; student lounge.

Culinary Student Profile
100 full-time.

Faculty
1 full-time; 4 part-time. 4 are industry professionals; 1 is a culinary-accredited teacher.

Special Programs
Mystery box competitions.

Expenses
In-state tuition: $42 per credit hour. Out-of-state tuition: $126 per credit hour. Program-related fees include: $50 for lab fees per class; $60 for uniforms.

Financial Aid
Employment placement assistance is available. Employment opportunities within the program are available.

Application Information
Students are accepted for enrollment in January, May, and August. Applicants must submit a formal application.

Contact
Howard Aller, Director, Culinary Management Program, 1000 College Boulevard, Pensacola, FL 32504-8998; 850-484-1422; Fax: 850-484-1543; E-mail: haller@pjc.cc.fl.us

Pinellas Technical Education Center-Clearwater Campus

Culinary Arts/Commercial Food

Clearwater, Florida

General Information
Public, coeducational, two-year college. Urban campus. Founded in 1969.

Program Information
Offered since 1969. Accredited by American Culinary Federation Education Institute. Member of American Culinary Federation; American Culinary Federation Educational Institute; American Dietetic Association; Educational Foundation of the NRA; National Restaurant Association. Program calendar is divided into trimesters. 1,800-hour Diploma in Culinary Arts/Commercial Foods.

Areas of Study
Baking; buffet catering; computer applications; convenience cookery; culinary skill development; food preparation; food purchasing; food service communication; garde-manger; international cuisine; introduction to food service; kitchen management; meal planning; meat fabrication; nutrition; nutrition and food service; sanitation; saucier; soup, stock, sauce, and starch production.

Facilities
Bake shop; bakery; cafeteria; catering service; 2 classrooms; computer laboratory; demonstration laboratory; 2 food production kitchens; gourmet dining room; 3 laboratories; learning resource center; lecture room; library; teaching kitchen.

Culinary Student Profile
60 full-time. 30 are under 25 years old; 20 are between 25 and 44 years old; 10 are over 44 years old.

Faculty
4 full-time. 2 are industry professionals; 2 are culinary-accredited teachers.

Expenses
Tuition: $1440 per diploma.

Financial Aid
In 1996, 3 scholarships were awarded (average award was $100). Employment placement assistance is available.

Application Information
Students are accepted for enrollment year-round. In 1996, 22 applied; 22 were accepted. Applicants must submit a formal application and interview.

Contact
Vincent Calandra, Department Chair, Culinary Arts/Commercial Food, 6100 154th Avenue, N, Clearwater, FL 33760; 813-538-7167; Fax: 813-538-7203.

Robert Morgan Vocational-Technical Center

Culinary Arts Programs

Miami, Florida

General Information
Public, coeducational, technical institute. Suburban campus. Founded in 1979.

Program Information
Offered since 1979. Member of Interntional Food Service Executives Association. Program calendar is divided into trimesters. 212-hour Certificate in Cake Decoration. 720-hour Certificates in Baking; Cooking.

Areas of Study
Baking; culinary skill development.

Facilities
Bake shop; bakery; cafeteria; catering service; classroom; computer laboratory; food production kitchen; learning resource center; library; snack shop.

Culinary Student Profile
120 total: 50 full-time; 70 part-time.

Faculty
3 full-time; 5 part-time. 8 are culinary-accredited teachers. Prominent faculty: Luis Rivera and Malcolm Morgan.

Expenses
In-state tuition: $.60 per hour. Out-of-state tuition: $4.60 per hour. Program-related fees include: $40 for textbooks; $50 for uniforms.

Financial Aid
In 1996, 15 scholarships were awarded (average award was $500); 15 loans were granted (average loan was $700). Program-specific awards include scholarships from industry and organizations. Employment placement assistance is available. Employment opportunities within the program are available.

Application Information
Students are accepted for enrollment in January, April, and September. Application deadline for spring is continuous with a recommended date of January 10. Application deadline for summer is continuous with a recommended date of April 10. Application deadline for winter is continuous with a recommended date of September 10. In 1996, 230 applied; 200 were accepted. Applicants must submit a formal application.

Contact
Giorgio Moro, Food Services Coordinator, Culinary Arts Programs, 18180 Southwest 122nd Avenue, Miami, FL 33177; 305-253-9920 Ext. 197; Fax: 305-253-3023.

South Florida Community College

Hospitality Management

Avon Park, Florida

General Information
Public, coeducational, two-year college. Rural campus. Founded in 1965. Accredited by Southern Association of Colleges and Schools.

Program Information
Member of Educational Foundation of the NRA. Program calendar is divided into semesters. 10-month Certificate in Food Management and Production. 2-year Associate degree in Hospitality Management.

Areas of Study
Baking; beverage management; computer applications; controlling costs in food service; culinary skill development; food preparation; food purchasing; food service communication; food service math; garde-manger; introduction to food service; kitchen management; management and human resources; meal planning; sanitation.

Facilities
Catering service; classroom; computer laboratory; demonstration laboratory; food production kitchen; gourmet dining room; laboratory; learning resource center; lecture room; library; public restaurant; student lounge; teaching kitchen; vineyard.

Culinary Student Profile
36 total: 15 full-time; 21 part-time. 8 are under 25 years old; 25 are between 25 and 44 years old; 3 are over 44 years old.

Faculty
1 full-time; 3 part-time. 1 is an industry professional.

Expenses
Tuition: $126 per course. Program-related fees include: $125 for knives and case; $100 for uniforms; $150 for lab fees.

Financial Aid
Employment placement assistance is available. Employment opportunities within the program are available.

Housing
Coed housing available. Average on-campus housing cost per month: $200. Average off-campus housing cost per month: $350.

Application Information
Students are accepted for enrollment in January, May, August, and December. In 1996, 49 applied; 49 were accepted. Applicants must submit a formal application.

South Florida Community College *(continued)*

Contact
Admissions Office, Hospitality Management, 600 West College Drive, Avon Park, FL 33825; 941-382-6900; Fax: 941-453-2365.

South Technical Education Center

Culinary Arts

Boynton Beach, Florida

General Information
Public, coeducational, technical institute. Urban campus. Founded in 1966.

Program Information
Offered since 1966. Program calendar is divided into quarters. 2-year Certificate of Completion in Culinary Arts.

Areas of Study
Baking; buffet catering; computer applications; confectionery show pieces; controlling costs in food service; convenience cookery; culinary French; culinary skill development; food preparation; food purchasing; food service communication; food service math; garde-manger; international cuisine; introduction to food service; kitchen management; meal planning; meat cutting; meat fabrication; menu and facilities design; nutrition; patisserie; restaurant opportunities; sanitation; saucier; seafood processing; soup, stock, sauce, and starch production.

Culinary Student Profile
14 total: 8 full-time; 6 part-time.

Faculty
1 full-time. 1 is a certified executive chef.

Expenses
Tuition: $205.80 per quarter full-time, $102.90 per quarter part-time.

Application Information
Students are accepted for enrollment in January, April, August, and November.

Contact
Ron Zabkiewicz, Chef Instructor, Culinary Arts, 1300 Southwest 30th Avenue, Boynton Beach, FL 33426; 561-369-7000; Fax: 561-369-7024.

The Southeast Institute of the Culinary Arts

Commercial Foods/Culinary Arts

St. Augustine, Florida

General Information
Public, coeducational, culinary institute. Urban campus. Founded in 1970.

Program Information
Offered since 1970. Accredited by American Culinary Federation Education Institute. Member of American Culinary Federation; American Culinary Federation Educational Institute; Educational Foundation of the NRA. Program calendar is divided into semesters. 18-week Diploma in Baking/Pastry.

Areas of Study
Baking; buffet catering; confectionery show pieces; controlling costs in food service; culinary French; culinary skill development; food preparation; food purchasing; food service communication; food service math; garde-manger; international cuisine; introduction to food service; kitchen management; management and human resources; meal planning; meat cutting; nutrition; nutrition and food service; patisserie; restaurant opportunities; sanitation; saucier; seafood processing; soup, stock, sauce, and starch production; ice carving.

Facilities
Bake shop; bakery; cafeteria; catering service; classroom; computer laboratory; demonstration laboratory; food production kitchen; garden; gourmet dining room; laboratory; learning resource center; lecture room; library; public restaurant; teaching kitchen.

Culinary Student Profile
265 total: 100 full-time; 165 part-time.

Faculty
17 full-time. 17 are culinary-accredited teachers. Prominent faculty: Walter Achatz and David Bearl.

Expenses
Application fee: $15. Tuition: $179 per 9 weeks.

Financial Aid
In 1996, 10 scholarships were awarded (average award was $150). Employment placement assistance is available. Employment opportunities within the program are available.

Application Information
Students are accepted for enrollment in January, March, June, August, and October. In 1996, 150 applied. Applicants must submit a formal application.

Contact
David Bearl, Coordinator, Commercial Foods/Culinary Arts, 2980 Collins Avenue, St. Augustine, FL 32095; 904-829-1061; Fax: 904-824-6750.

Southeastern Academy

Culinary Training Center

Kissimmee, Florida

General Information
Private, coeducational, career school. Urban campus. Founded in 1974.

Program Information
Offered since 1990. Member of American Culinary Federation; Florida Restaurant Association. Program calendar is divided into 5-week cycles. 30-week Diploma in Culinary Arts.

Areas of Study
Baking; beverage management; buffet catering; controlling costs in food service; culinary skill development; food preparation; food purchasing; garde-manger; international cuisine; introduction to food service; meat cutting; nutrition; sanitation; soup, stock, sauce, and starch production; wines and spirits.

Facilities
Bakery; cafeteria; 4 classrooms; 2 food production kitchens; gourmet dining room; laboratory; learning resource center; lecture room; library; public restaurant; student lounge; 2 teaching kitchens.

Culinary Student Profile
15 full-time.

Faculty
10 full-time; 2 part-time. 4 are industry professionals; 2 are culinary-accredited teachers. Prominent faculty: Richard Woodring.

Special Programs
Local field trips.

Expenses
Application fee: $150. Tuition: $8695 per diploma.

Financial Aid
Employment placement assistance is available. Employment opportunities within the program are available.

Housing
Coed housing available. Average on-campus housing cost per month: $600.

Application Information
Students are accepted for enrollment year-round. Applicants must submit a formal application and interview.

Contact
Gary Gaetano, Vice President, Admissions, Culinary Training Center, PO Box 421768, Kissimmee, FL 34742-1768; 407-847-4444; Fax: 407-847-8793.

The Art Institute of Atlanta

School of Culinary Arts

Atlanta, Georgia

General Information
Private, coeducational, two-year college. Urban campus. Founded in 1949. Accredited by Southern Association of Colleges and Schools.

The Art Institute of Atlanta *(continued)*

Program Information
Offered since 1992. Accredited by American Culinary Federation Education Institute. Member of American Culinary Federation; American Culinary Federation Educational Institute. Program calendar is divided into quarters. 2-year Associate degree in Culinary Arts.

Areas of Study
Baking; buffet catering; computer applications; controlling costs in food service; culinary skill development; food preparation; food purchasing; food service communication; garde-manger; international cuisine; introduction to food service; kitchen management; management and human resources; nutrition; restaurant opportunities; sanitation; food styling.

Facilities
Bake shop; 3 classrooms; 5 computer laboratories; food production kitchen; learning resource center; 2 lecture rooms; library; public restaurant; student lounge; 3 teaching kitchens.

Culinary Student Profile
479 full-time. 263 are under 25 years old; 195 are between 25 and 44 years old; 21 are over 44 years old.

Faculty
9 full-time; 5 part-time. 11 are industry professionals; 2 are culinary-accredited teachers; 1 is a master craftsman.

Special Programs
1-day workshops in specialized culinary subjects.

Expenses
Application fee: $50. Tuition: $3536 per quarter full-time, $221 per credit hour part-time. Program-related fees include: $675 for culinary kit; $250 for lab fees per quarter.

Financial Aid
Employment placement assistance is available. Employment opportunities within the program are available.

Housing
Average off-campus housing cost per month: $450.

Application Information
Students are accepted for enrollment in January, April, July, and October. In 1996, 346 applied; 321 were accepted. Applicants must submit a formal application and academic transcripts.

Contact
Todd M. Knutson, Director of Admissions, School of Culinary Arts, 3376 Peachtree Road, NE, Atlanta, GA 30326-1018; 404-266-1383; Fax: 404-898-9551; E-mail: knutsont@aii.edu; World Wide Web: http://www.aii.edu

See affiliated programs: Colorado Institute of Art; New York Restaurant School; The Art Institute of Fort Lauderdale; The Art Institute of Houston; The Art Institute of Philadelphia; The Art Institute of Phoenix; The Art Institute of Seattle.

The School of Culinary Arts in Atlanta offers a 2-year, 6-quarter Associate in Arts degree in culinary arts, which is accredited by the American Culinary Federation Educational Institute (ACFEI). The program is offered both days and evenings. Students acquire a fundamental understanding of the process of cooking and finish with advanced food preparation techniques. The program includes a foundation of health, safety, and nutritional requirements. Business studies provide a background in cost control, supervision of food service personnel, and the operation of a commercial kitchen. Top professionals in the culinary field make up the faculty. More than 95% of graduates are generally employed in their field within 6 months.

See display on page 48.

Atlanta Technical Institute

Culinary Arts Program

Atlanta, Georgia

General Information
Public, coeducational, two-year college. Urban campus. Founded in 1968.

Program Information
Program calendar is divided into quarters. 18-month Diploma in Culinary Arts.

Areas of Study
Baking; beverage management; buffet catering; computer applications; controlling costs in food service; convenience cookery; food preparation; food purchasing; food service communication; food service math; garde-manger; international cuisine; introduction to food service; kitchen management; management and human resources; meal planning; meat cutting; menu and facilities design; nutrition; restaurant opportunities; sanitation; saucier; seafood processing; soup, stock, sauce, and starch production; wines and spirits.

Culinary Student Profile
85 full-time.

Faculty
5 full-time; 1 part-time.

Expenses
Application fee: $15. Tuition: $296 per quarter full-time, $21 per credit part-time.

Application Information
Students are accepted for enrollment in January, April, June, and September. Applicants must have a high school diploma or GED.

Contact
Barbara Boyd, Chair, Culinary Arts Program, 1560 Metropolitan Parkway, Atlanta, GA 30310; 404-756-3727; Fax: 404-756-0932.

Augusta Technical Institute

Culinary Arts

Augusta, Georgia

General Information
Public, coeducational, two-year college. Suburban campus. Founded in 1961. Accredited by Southern Association of Colleges and Schools.

Program Information
Offered since 1984. Member of American Culinary Federation; Educational Foundation of the NRA; National Restaurant Association. Program calendar is divided into quarters. 3-quarter Certificate in Food Service. 6-quarter Diploma in Culinary Arts.

Areas of Study
Baking; buffet catering; food preparation; food purchasing; garde-manger; menu and facilities design; nutrition and food service; sanitation; soup, stock, sauce, and starch production.

Facilities
Bake shop; cafeteria; catering service; food production kitchen; gourmet dining room; learning resource center; library; teaching kitchen.

Culinary Student Profile
30 total: 20 full-time; 10 part-time.

Faculty
2 full-time. Prominent faculty: Willie Mae Crittendem and Kathleen Fervan.

Expenses
Application fee: $15. In-state tuition: $274 per quarter full-time, $21 per credit hour part-time. Out-of-state tuition: $504 per quarter full-time, $42 per credit hour part-time. Program-related fees include: $100 for uniforms; $300 for books; $200 for equipment.

Financial Aid
Employment placement assistance is available.

Housing
Average off-campus housing cost per month: $500.

Application Information
Students are accepted for enrollment in March and September. Application deadline for fall is September 1. Application deadline for spring is March 1. In 1996, 50 applied; 40 were accepted. Applicants must submit a formal application and a portfolio.

Contact
Willie Mae Crittenden, Department Head, Culinary Arts, 3116 Deans Bridge Road, Augusta, GA 30906; 706-771-4083; Fax: 706-771-4016; E-mail: wcritten@augusta.tech.us

Savannah Technical Institute

Culinary Arts Program

Savannah, Georgia

General Information
Public, coeducational, two-year college. Urban campus. Founded in 1929. Accredited by Southern Association of Colleges and Schools.

Program Information
Offered since 1981. Accredited by American Culinary Federation Education Institute. Member of American Culinary Federation Educational Institute; Georgia Hospitality Association. Program calendar is divided into quarters. 12-month Diploma in Culinary Arts Option II. 18-month Diploma in Culinary Arts Option I.

Areas of Study
Baking; buffet catering; computer applications; controlling costs in food service; culinary skill development; food preparation; food purchasing; food service math; garde-manger; international cuisine; introduction to food service; kitchen management; meal planning; nutrition; sanitation; saucier; soup, stock, sauce, and starch production; dining room/guest services.

Facilities
Catering service; classroom; computer laboratory; gourmet dining room; library; student lounge; teaching kitchen.

Culinary Student Profile
24 total: 22 full-time; 2 part-time.

Faculty
1 full-time; 1 part-time. 1 is an industry professional; 1 is a culinary-accredited teacher.

Special Programs
Field trips to restaurants, special food production presentations, culinary competitions.

Expenses
Application fee: $15. Tuition: $276 per quarter full-time, $21 per credit hour part-time. Program-related fees include: $150 for knives; $60 for uniforms; $100 for books per quarter.

Financial Aid
Employment placement assistance is available.

Housing
Average off-campus housing cost per month: $350.

Application Information
Students are accepted for enrollment in January, April, July, and October. In 1996, 25 applied; 25 were accepted. Applicants must submit a formal application and have a high school diploma or GED.

Contact
Marvis Hinson, Program Director, Culinary Arts Program, 5717 White Bluff Road, Savannah, GA 31499; 912-351-4553; Fax: 912-351-4526.

University of Hawaii-Kapiolani Community College

Food Service and Hospitality Education Department

Honolulu, Hawaii

General Information
Public, coeducational, two-year college. Urban campus. Founded in 1957. Accredited by Western Association of Schools and Colleges.

Program Information
Offered since 1965. Accredited by American Culinary Federation Education Institute. Member of American Culinary Federation; American Culinary Federation Educational Institute; Council on Hotel, Restaurant, and Institutional Education; Educational Foundation of the NRA; National Restaurant Association. Program calendar is divided into semesters. 18-month Certificate in Culinary Arts. 2-year Associate degrees in Patisserie; Culinary Arts.

Areas of Study
Baking; beverage management; computer applications; confectionery show pieces; controlling costs in food service; food preparation; food service math; garde-manger; international cuisine; introduction to food service; meal planning; menu and facilities design; nutrition and food service; patisserie; sanitation; soup, stock, sauce, and starch production; foundations of guest services; Asian/Pacific cookery.

Facilities
2 bake shops; bakery; cafeteria; 12 classrooms; coffee shop; computer laboratory; demonstration laboratory; 4 food production kitchens; 2 gourmet dining rooms; learning resource center; library; 3 public restaurants; snack shop; student lounge.

Culinary Student Profile
750 total: 375 full-time; 375 part-time. 200 are under 25 years old; 400 are between 25 and 44 years old; 150 are over 44 years old.

Faculty
14 full-time; 8 part-time. 15 are industry professionals; 7 are culinary-accredited teachers.

Special Programs
2-week Christmas break paid internships, apprenticeship at the Governor's mansion.

Expenses
Application fee: $25. In-state tuition: $468 per semester full-time, $39 per credit hour part-time. Out-of-state tuition: $2856 per semester full-time, $238 per credit hour part-time.

Financial Aid
In 1996, 45 scholarships were awarded (average award was $600); 250 loans were granted (average loan was $1200). Employment placement assistance is available. Employment opportunities within the program are available.

Housing
Average off-campus housing cost per month: $700.

Application Information
Students are accepted for enrollment in January, May, and August. Application deadline for fall is July 1. Application deadline for spring is November 15. Application deadline for summer is April 15. In 1996, 432 applied; 432 were accepted. Applicants must submit a formal application.

Contact
Lori Maehara, Food Service Counselor, Food Service and Hospitality Education Department, 4303 Diamond Head Road, Honolulu, HI 96816-4421; 808-734-9466; Fax: 808-734-9212; E-mail: yonemori@kccada.kcc.hawaii.edu

University of Hawaii-Kauai Community College

Culinary Arts Department

Lihue, Hawaii

General Information
Public, coeducational, two-year college. Urban campus. Founded in 1965. Accredited by Western Association of Schools and Colleges.

Program Information
Accredited by American Culinary Federation Education Institute. Member of American Culinary Federation. Program calendar is divided into semesters. 1-semester Certificate of Completion in Culinary Arts. 1-year Certificate of Achievement in Culinary Arts. 2-year Associate degree in Culinary Arts.

Areas of Study
Baking; buffet catering; computer applications; controlling costs in food service; culinary skill development; food preparation; food purchasing; food service math; garde-manger; international cuisine; introduction to food service; management and human resources; nutrition and food service; sanitation; saucier; soup, stock, sauce, and starch production.

Facilities
Bake shop; cafeteria; classroom; computer laboratory; food production kitchen; laboratory; learning resource center; lecture room; library; public restaurant; student lounge.

University of Hawaii-Kauai Community College *(continued)*

Culinary Student Profile
70 total: 40 full-time; 30 part-time. 57 are under 25 years old; 12 are between 25 and 44 years old; 1 is over 44 years old.

Faculty
4 full-time. 3 are industry professionals. Prominent faculty: Clarence Nishi and Mark Oyama.

Expenses
In-state tuition: $473 per semester full-time, $39.50 per credit part-time. Out-of-state tuition: $2861 per semester full-time, $238.50 per credit part-time. Program-related fees include: $64 for uniforms; $94 for knives.

Financial Aid
In 1996, 11 scholarships were awarded (average award was $400). Employment placement assistance is available.

Application Information
Students are accepted for enrollment in January and August. Application deadline for fall is August 1. Application deadline for spring is December 1. In 1996, 93 applied; 93 were accepted. Applicants must submit a formal application.

Contact
Admissions and Records Office, Culinary Arts Department, 3-1901 Kaumualii Highway, Lihue, HI 96766-9591; 808-245-8225; Fax: 808-245-8297.

University of Hawaii-Maui Community College

Food Service Program

Kahului, Hawaii

General Information
Public, coeducational, two-year college. Founded in 1967.

Program Information
Offered since 1977. Accredited by American Culinary Federation Education Institute. Program calendar is divided into semesters. 1-year Certificate in Culinary Arts. 2-year Associate degrees in Food Service-Baking Specialty; Food Service-Culinary Arts Specialty.

Areas of Study
Baking; beverage management; buffet catering; computer applications; controlling costs in food service; culinary skill development; food preparation; food purchasing; food service communication; food service math; garde-manger; international cuisine; introduction to food service; management and human resources; nutrition; sanitation; soup, stock, sauce, and starch production.

Culinary Student Profile
140 total: 112 full-time; 28 part-time.

Faculty
5 full-time; 3 part-time.

Expenses
Tuition: $477 per semester full-time, $43.50 per credit part-time. Program-related fees include: $50 for knives; $125 for uniforms and shoes; $125 for books.

Application Information
Students are accepted for enrollment in January and August. Applicants must submit a formal application.

Contact
Karen K. Tanaka, Program Coordinator, Food Service Program, 310 Kaahumanu Avenue, Kahului, HI 96732; 808-984-3225; Fax: 808-984-3314.

Boise State University

Culinary Arts Program

Boise, Idaho

General Information
Public, coeducational, comprehensive institution. Urban campus. Founded in 1932. Accredited by Northwest Association of Schools and Colleges.

Program Information
Offered since 1979. Accredited by American Culinary Federation Education Institute. Member of American Culinary Federation; American Culinary Federation Educational Institute. Program calendar is divided into semesters. 18-month Certificate in Culinary Arts. 2-year Associate degree in Culinary Arts.

Areas of Study
Baking; buffet catering; controlling costs in food service; culinary skill development; food preparation; food purchasing; food service communication; food service math; garde-manger; international cuisine; introduction to food service; kitchen management; menu and facilities design; nutrition; patisserie; sanitation; seafood processing; soup, stock, sauce, and starch production; wines and spirits.

Facilities
Bake shop; bakery; catering service; 3 classrooms; 2 demonstration laboratories; food production kitchen; 3 laboratories; learning resource center; 3 lecture rooms; library; public restaurant; snack shop.

Culinary Student Profile
45 full-time.

Faculty
3 full-time; 2 part-time. 2 are industry professionals; 3 are culinary-accredited teachers. Prominent faculty: Julie Hosman-Kulm and Vern Hickman.

Expenses
Tuition: $1020 per semester.

Financial Aid
In 1996, 4 scholarships were awarded (average award was $300).

Housing
Coed, apartment-style, and single-sex housing available.

Application Information
Students are accepted for enrollment in January and August. Application deadline for fall is August 25. Application deadline for spring is January 15. Applicants must submit a formal application and complete an entrance test.

Contact
Student Services Department, Culinary Arts Program, 1910 University Drive, Boise, ID 83725-0399; 208-385-1431.

Idaho State University

Culinary Arts Program

Pocatello, Idaho

General Information
Public, coeducational, university. Rural campus. Founded in 1901. Accredited by Northwest Association of Schools and Colleges.

Program Information
Offered since 1967. Program calendar is divided into semesters. 2-semester Certificate in Culinary Arts Technology.

Areas of Study
Baking; buffet catering; controlling costs in food service; convenience cookery; culinary skill development; food preparation; food purchasing; international cuisine; kitchen management; management and human resources; meal planning; meat cutting; menu and facilities design; nutrition; nutrition and food service; patisserie; sanitation; saucier; seafood processing; soup, stock, sauce, and starch production.

Facilities
Bake shop; catering service; classroom; coffee shop; computer laboratory; food production kitchen; learning resource center; lecture room; library.

Culinary Student Profile
15 full-time.

Faculty
2 full-time. 1 is an industry professional; 1 is a master chef. Prominent faculty: David K. Hanson and Henri Nippert.

Expenses
Application fee: $20. Tuition: $1212 per semester full-time, $99 per credit part-time.

Idaho State University ***(continued)***

Financial Aid
Employment placement assistance is available.

Housing
Coed housing available. Average on-campus housing cost per month: $340.

Application Information
Students are accepted for enrollment in January and August. Applications are accepted continuously for fall and spring. In 1996, 6 applied; 6 were accepted. Applicants must submit a formal application.

Contact
David K. Hanson, Program Coordinator, Culinary Arts Program, Box 8380, Pocatello, ID 83209; 208-236-3327; Fax: 208-236-4641; E-mail: hansdavi@isu.edu

College of DuPage

Culinary Arts/Pastry Arts

Glen Ellyn, Illinois

General Information
Public, coeducational, two-year college. Suburban campus. Founded in 1967. Accredited by North Central Association of Colleges and Schools.

Program Information
Offered since 1967. Accredited by American Culinary Federation Education Institute. Member of American Culinary Federation; American Culinary Federation Educational Institute; American Institute of Baking; Council on Hotel, Restaurant, and Institutional Education; Educational Foundation of the NRA; Interntional Food Service Executives Association; National Restaurant Association. Program calendar is divided into quarters. 1-year Certificates in Pastry Chef; Hotel Operations; Beverage Management; Food Service Administration; Hotel Management; Culinary Arts. 2-year Associate degrees in Food Service Administration; Hotel Management; Culinary Arts.

Areas of Study
Baking; beverage management; buffet catering; controlling costs in food service; culinary skill development; food preparation; food purchasing; food service math; garde-manger; international cuisine; introduction to food service; kitchen management; management and human resources; menu and facilities design; nutrition; patisserie; sanitation.

Facilities
Bake shop; cafeteria; 2 classrooms; 3 computer laboratories; demonstration laboratory; food production kitchen; gourmet dining room; 2 laboratories; learning resource center; 2 lecture rooms; library; public restaurant; snack shop; student lounge; teaching kitchen.

Culinary Student Profile
300 total: 150 full-time; 150 part-time. 100 are under 25 years old; 175 are between 25 and 44 years old; 25 are over 44 years old.

Faculty
4 full-time; 12 part-time. 13 are industry professionals; 3 are culinary-accredited teachers. Prominent faculty: George Macht and Chris Thielman.

Expenses
Application fee: $10. Tuition: $31 per quarter hour. Program-related fees include: $70 for knives; $40 for uniforms.

Financial Aid
In 1996, 15 scholarships were awarded (average award was $200). Employment placement assistance is available. Employment opportunities within the program are available.

Application Information
Students are accepted for enrollment in January, March, June, and September. In 1996, 300 applied; 300 were accepted. Applicants must submit a formal application.

Contact
George Macht, Coordinator, Culinary Arts/Pastry Arts, 425 22nd Street, Glen Ellyn, IL 60137; 630-942-2315; Fax: 630-858-9399; E-mail: machtg@cdnet.cod.edu

College of Lake County

Grayslake, Illinois

General Information
Public, coeducational, two-year college. Suburban campus. Founded in 1967. Accredited by North Central Association of Colleges and Schools.

Program Information
Offered since 1987. Member of American Culinary Federation; Council on Hotel, Restaurant, and Institutional Education; Educational Foundation of the NRA; National Restaurant Association.
Program calendar is divided into semesters. 1-year Certificates in Food Service Management; Culinary Arts. 2-year Associate degree in Food Service Management.

Areas of Study
Baking; buffet catering; computer applications; controlling costs in food service; convenience cookery; culinary skill development; food preparation; food purchasing; food service communication; food service math; garde-manger; international cuisine; introduction to food service; kitchen management; management and human resources; meal planning; menu and facilities design; nutrition; restaurant opportunities; sanitation; saucier; soup, stock, sauce, and starch production.

Facilities
Bake shop; cafeteria; catering service; 4 classrooms; 12 computer laboratories; 2 demonstration laboratories; 2 food production kitchens; learning resource center; library; 2 public restaurants; snack shop.

Culinary Student Profile
150 total: 50 full-time; 100 part-time.

Faculty
1 full-time; 6 part-time. 4 are industry professionals; 3 are culinary-accredited teachers.

Expenses
Tuition: $51 per credit hour. Program-related fees include: $125 for tools and equipment; $50 for uniforms.

Financial Aid
In 1996, 2 scholarships were awarded (average award was $1000); 50 loans were granted (average loan was $500). Employment placement assistance is available. Employment opportunities within the program are available.

Application Information
Students are accepted for enrollment in January, June, and August. Applications are accepted continuously for fall, spring, and summer. In 1996, 40 applied; 30 were accepted. Applicants must submit a formal application.

Contact
Cliff Wener, Program Coordinator, Food Service Program, 19351 West Washington Street, Grayslake, IL 60030-1198; 847-343-2823; Fax: 847-223-7248.

Cooking Academy of Chicago

Chicago, Illinois

General Information
Private, coeducational, culinary institute. Urban campus. Founded in 1993.

Program Information
Offered since 1993. 6-month Certificates in Nutrition for Food Service; Baking and Pastry Career; Culinary Arts.

Areas of Study
Baking; buffet catering; culinary skill development; food preparation; garde-manger; international cuisine; meat cutting; meat fabrication; nutrition and food service; patisserie; sanitation; saucier; seafood processing; soup, stock, sauce, and starch production.

Facilities
Bake shop; cafeteria; catering service; 2 classrooms; 2 food production kitchens; library.

Culinary Student Profile
30 total: 15 full-time; 15 part-time.

Cooking Academy of Chicago *(continued)*

Faculty
2 full-time; 3 part-time. 3 are industry professionals; 2 are master chefs.

Expenses
Application fee: $100. Tuition: $4000 per program.

Financial Aid
Employment placement assistance is available. Employment opportunities within the program are available.

Application Information
Students are accepted for enrollment in January, February, May, August, and September. In 1996, 15 applied; 15 were accepted. Applicants must submit a formal application.

Contact
Nora Christensen, Director, 2500 West Bradley Place, Chicago, IL 60618; 773-478-9840; Fax: 773-478-3146.

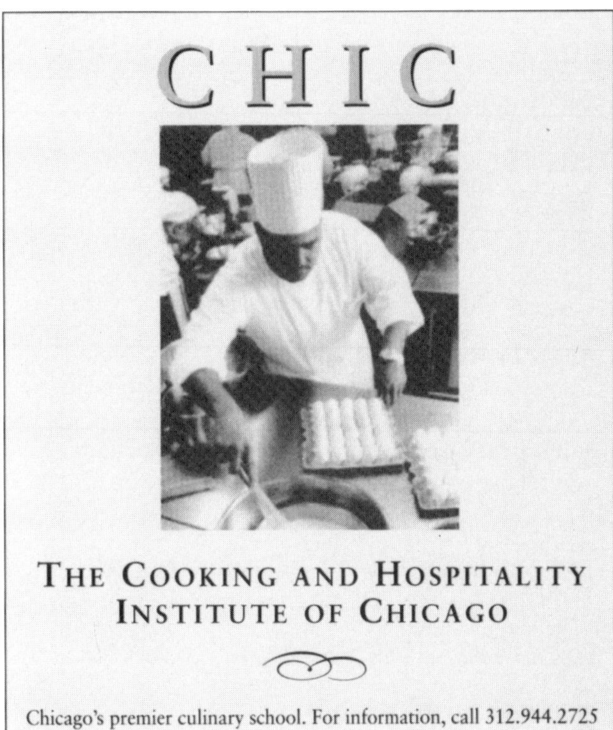

CHIC
THE COOKING AND HOSPITALITY INSTITUTE OF CHICAGO
Chicago's premier culinary school. For information, call 312.944.2725

The Cooking and Hospitality Institute of Chicago

Chicago, Illinois

General Information
Private, coeducational, two-year college. Urban campus. Founded in 1983.

Program Information
Offered since 1991. Accredited by American Culinary Federation Education Institute. Member of American Culinary Federation; American Institute of Wine & Food; Council on Hotel, Restaurant, and Institutional Education; Educational Foundation of the NRA; International Association of Culinary Professionals; James Beard Foundation, Inc.; National Restaurant Association. Program calendar is divided into trimesters. 2-year Associate degree in Culinary Arts. 8-month Certificates in Restaurant Management; Baking and Pastry; Professional Cooking.

Areas of Study
Baking; beverage management; computer applications; controlling costs in food service; culinary French; culinary skill development; food preparation; food purchasing; garde-manger; international cuisine; introduction to food service; kitchen management; management and human resources; meal planning; meat fabrication; menu and facilities design; nutrition; patisserie; sanitation; saucier; seafood processing; soup, stock, sauce, and starch production.

Facilities
Classroom; computer laboratory; demonstration laboratory; food production kitchen; gourmet dining room; laboratory; learning resource center; lecture room; library; public restaurant; student lounge; teaching kitchen.

Culinary Student Profile
930 total: 370 full-time; 560 part-time. 245 are under 25 years old; 620 are between 25 and 44 years old; 65 are over 44 years old.

Faculty
23 full-time; 3 part-time. 5 are industry professionals; 3 are master chefs; 15 are culinary-accredited teachers. Prominent faculty: Mark Facklam and Brent Holten.

Expenses
Application fee: $100. Tuition: $3565 per semester full-time, $1783 per semester part-time.

Financial Aid
In 1996, 12 scholarships were awarded (average award was $1000). Employment placement assistance is available. Employment opportunities within the program are available.

Application Information
Students are accepted for enrollment in January, March, May, July, September, and November. In 1996, 1,020 applied; 930 were accepted. Applicants must submit a formal application.

Contact
Jim Simpson, Director, 361 West Chestnut Street, Chicago, IL 60610-3050; 312-944-0882 Ext. 15; Fax: 312-944-8557; E-mail: chic@chicnet.org; World Wide Web: http://www.chicnet.org

Elgin Community College

Hospitality Department

Elgin, Illinois

General Information
Public, coeducational, two-year college. Suburban campus. Founded in 1949. Accredited by North Central Association of Colleges and Schools.

Program Information
Offered since 1971. Accredited by American Culinary Federation Education Institute. Member of American Culinary Federation; American Culinary Federation Educational Institute; American Institute of Baking; Council on Hotel, Restaurant, and Institutional Education; Educational Foundation of the NRA; National Restaurant Association. Program calendar is divided into semesters. 1-year Certificates in Baking; Cooking. 2-year Associate degrees in Restaurant Management; Culinary Arts.

Areas of Study
Baking; beverage management; controlling costs in food service; culinary French; culinary skill development; food preparation; food purchasing; garde-manger; introduction to food service; meat cutting; meat fabrication; menu and facilities design; nutrition; patisserie; restaurant opportunities; sanitation; saucier; seafood processing; soup, stock, sauce, and starch production.

Facilities
Bakery; cafeteria; catering service; 3 classrooms; computer laboratory; 2 demonstration laboratories; 3 food production kitchens; gourmet dining room; 4 laboratories; learning resource center; 5 lecture rooms; library; 2 public restaurants; snack shop; student lounge.

Culinary Student Profile
310 total: 140 full-time; 170 part-time.

Elgin Community College *(continued)*

Faculty
4 full-time; 9 part-time. 13 are culinary-accredited teachers. Prominent faculty: Michael Zema.

Special Programs
International exchange with Hotelfascule in Semmering, Austria.

Expenses
Application fee: $15. Tuition: $3000 per degree full-time, $33 per credit hour part-time. Program-related fees include: $50 for lab fees; $125 for smallwares and toolbox.

Financial Aid
In 1996, 15 scholarships were awarded (average award was $400); 12 loans were granted. Program-specific awards include 5 National Restaurant Association scholarships. Employment placement assistance is available. Employment opportunities within the program are available.

Housing
Average off-campus housing cost per month: $300.

Application Information
Students are accepted for enrollment in January, May, and August. In 1996, 40 applied; 35 were accepted. Applicants must submit a formal application and have a high school diploma or GED.

Contact
Michael Zema, Director, Hospitality Department, 1700 Spartan Drive, Elgin, IL 60123-7193; 847-697-1000; Fax: 847-931-3911.

Kendall College

Culinary School

Evanston, Illinois

General Information
Private, coeducational, four-year college. Suburban campus. Founded in 1934. Accredited by North Central Association of Colleges and Schools.

Program Information
Offered since 1985. Accredited by American Culinary Federation Education Institute. Program calendar is divided into quarters. 1-year Certificates in Pastry Arts; Culinary Arts. 2-year Associate degree in Culinary Arts.

Areas of Study
Baking; buffet catering; computer applications; confectionery show pieces; controlling costs in food service; culinary skill development; food preparation; food purchasing; garde-manger; international cuisine; introduction to food service; kitchen management; management and human resources; meal planning; meat fabrication; menu and facilities design; nutrition; patisserie; sanitation; saucier; seafood processing; soup, stock, sauce, and starch production; wines and spirits; techniques of healthy cooking.

Culinary Student Profile
280 total: 200 full-time; 80 part-time.

Faculty
13 full-time; 8 part-time.

Expenses
Tuition: $13,566 per year full-time, $324 per credit hour part-time.

Housing
Coed housing available.

Application Information
Students are accepted for enrollment in January, March, June, and September.

Contact
Admissions Office, Culinary School, 2408 Orrington Avenue, Evanston, IL 60201-2899; 847-866-1300; Fax: 847-866-1320.

Lexington College

Chicago, Illinois

General Information
Private, two-year college. Urban campus. Founded in 1977. Accredited by North Central Association of Colleges and Schools.

Program Information
Offered since 1977. Member of Council on Hotel, Restaurant, and Institutional Education; National Restaurant Association; Illinois Restaurant Association. Program calendar is divided into semesters. 2-year Associate degree in Food Service and Lodging.

Areas of Study
Baking; beverage management; culinary skill development; food purchasing; food service math; garde-manger; management and human resources; nutrition; sanitation.

Facilities
3 classrooms; computer laboratory; library; teaching kitchen.

Culinary Student Profile
55 total: 50 full-time; 5 part-time. 45 are under 25 years old; 7 are between 25 and 44 years old; 3 are over 44 years old.

Faculty
2 full-time; 15 part-time. 8 are industry professionals.

Expenses
Application fee: $25. Tuition: $3200 per semester full-time, $175 per credit hour part-time.

Financial Aid
In 1996, 5 scholarships were awarded (average award was $500). Employment placement assistance is available. Employment opportunities within the program are available.

Housing
12 culinary students housed on campus. Single-sex housing available. Average on-campus housing cost per month: $375.

Application Information
Students are accepted for enrollment in January and September. Application deadline for fall is continuous with a recommended date of August 25. Application deadline for spring is continuous with a recommended date of January 2. In 1996, 34 applied; 34 were accepted. Applicants must submit formal application, and international students must provide letters of reference.

Lexington College ***(continued)***

CONTACT
Mary Jane Markel, Director of Admissions, 10840 South Western Avenue, Chicago, IL 60643-3294; 773-779-3800; Fax: 773-779-7450.

LINCOLN LAND COMMUNITY COLLEGE

Hospitality Management

Springfield, Illinois

GENERAL INFORMATION
Public, coeducational, two-year college. Rural campus. Founded in 1967. Accredited by North Central Association of Colleges and Schools.

PROGRAM INFORMATION
Offered since 1994. Member of American Dietetic Association; Council on Hotel, Restaurant, and Institutional Education; Educational Foundation of the NRA; National Restaurant Association; Illinois Restaurant Association. Program calendar is divided into semesters. 1-year Certificate in Food and Beverage. 2-year Associate degree in Hospitality Management.

AREAS OF STUDY
Baking; buffet catering; controlling costs in food service; culinary skill development; food preparation; food purchasing; food service communication; food service math; international cuisine; introduction to food service; kitchen management; management and human resources; meal planning; nutrition; nutrition and food service; restaurant opportunities; sanitation; soup, stock, sauce, and starch production.

FACILITIES
Bake shop; bakery; cafeteria; catering service; classroom; computer laboratory; demonstration laboratory; food production kitchen; laboratory; learning resource center; lecture room; library; public restaurant; snack shop; student lounge; teaching kitchen.

CULINARY STUDENT PROFILE
75 total: 25 full-time; 50 part-time. 20 are under 25 years old; 40 are between 25 and 44 years old; 15 are over 44 years old.

FACULTY
2 full-time; 3 part-time. 2 are industry professionals; 1 is a master chef; 2 are culinary-accredited teachers. Prominent faculty: Charlyn Fargo.

EXPENSES
Tuition: $475 per semester full-time, $117 per semester part-time. Program-related fees include: $20 for materials.

FINANCIAL AID
In 1996, 4 scholarships were awarded (average award was $400); 10 loans were granted. Employment placement assistance is available. Employment opportunities within the program are available.

APPLICATION INFORMATION
Students are accepted for enrollment in January and August. Application deadline for fall is August 1. Application deadline for spring is January 1. In 1996, 20 applied; 20 were accepted. Applicants must submit a formal application.

CONTACT
Jay Kitterman, Director, Hospitality Management, Shepherd Road, Springfield, IL 62794-9256; 217-786-2772; Fax: 217-786-2495; E-mail: jkitterm@cabin.llcc.cc.il.us

MORAINE VALLEY COMMUNITY COLLEGE

Restaurant Hotel Management Program/Culinary Arts Management Program

Palos Hills, Illinois

GENERAL INFORMATION
Public, coeducational, two-year college. Suburban campus. Founded in 1967. Accredited by North Central Association of Colleges and Schools.

Program Information
Offered since 1984. Member of American Culinary Federation; American Culinary Federation Educational Institute; Council on Hotel, Restaurant, and Institutional Education; Educational Foundation of the NRA; Interntional Food Service Executives Association; National Restaurant Association. Program calendar is divided into semesters. 1-year Certificate in Beverage Management. 12-month Certificates in Restaurant/Hotel Management; Baking/Pastry Arts; Culinary Arts Management. 2-year Associate degrees in Culinary Arts Management; Restaurant/ Hotel Management.

Areas of Study
Baking; beverage management; buffet catering; controlling costs in food service; culinary French; culinary skill development; food preparation; food purchasing; food service math; garde-manger; international cuisine; introduction to food service; kitchen management; meal planning; meat cutting; meat fabrication; menu and facilities design; nutrition; nutrition and food service; patisserie; sanitation; saucier; seafood processing; soup, stock, sauce, and starch production; wines and spirits; culture and cuisine.

Facilities
Bake shop; catering service; computer laboratory; food production kitchen; gourmet dining room; laboratory; learning resource center; library; student lounge; teaching kitchen.

Culinary Student Profile
252 total: 98 full-time; 154 part-time.

Faculty
2 full-time; 8 part-time.

Special Programs
Internships.

Expenses
Tuition: $49 per credit hour.

Financial Aid
In 1996, 1 scholarship was awarded (award was $1000). Employment placement assistance is available.

Application Information
Students are accepted for enrollment in January and August. Application deadline for spring is January 1. Application deadline for fall is August 1. Application deadline for summer is June 1. Applicants must submit a formal application.

Contact
Anne L. Jachim, Assistant Professor, Culinary Arts Management, Restaurant Hotel Management Program/Culinary Arts Management Program, 10900 South 88th Avenue, Palos Hills, IL 60465-0937; 708-974-5320; Fax: 708-974-1184; E-mail: jachim@moraine.cc.il.us; World Wide Web: http://www.moraine.cc.il.us

Triton College

Hospitality Industry Administration

River Grove, Illinois

General Information
Public, coeducational, two-year college. Suburban campus. Founded in 1964. Accredited by North Central Association of Colleges and Schools.

Program Information
Offered since 1972. Accredited by American Culinary Federation Education Institute. Member of American Culinary Federation; American Culinary Federation Educational Institute; American Institute of Baking; American Institute of Wine & Food; Council on Hotel, Restaurant, and Institutional Education; Educational Foundation of the NRA; National Restaurant Association. Program calendar is divided into semesters. 1-year Certificate in Culinary Arts. 2-year Associate degree in Culinary Arts.

Areas of Study
Baking; beverage management; buffet catering; computer applications; controlling costs in food service; convenience cookery; culinary French; culinary skill development; food preparation; food purchasing; food service communication; garde-manger; international cuisine; introduction to

Triton College *(continued)*

food service; kitchen management; management and human resources; meal planning; meat cutting; meat fabrication; nutrition; nutrition and food service; patisserie; sanitation; saucier; seafood processing; soup, stock, sauce, and starch production; wines and spirits.

Facilities
Bake shop; bakery; 2 cafeterias; catering service; 4 classrooms; coffee shop; computer laboratory; 2 demonstration laboratories; 2 food production kitchens; 2 gardens; gourmet dining room; 2 laboratories; learning resource center; 3 lecture rooms; library; public restaurant; snack shop; 2 student lounges; teaching kitchen.

Culinary Student Profile
500 total: 400 full-time; 100 part-time. 200 are under 25 years old; 150 are between 25 and 44 years old; 150 are over 44 years old.

Faculty
4 full-time; 10 part-time. 4 are industry professionals; 2 are culinary-accredited teachers. Prominent faculty: Jerome Drosos and Jens Nielson.

Expenses
Tuition: $45 per semester hour.

Financial Aid
Employment placement assistance is available. Employment opportunities within the program are available.

Housing
Average off-campus housing cost per month: $400.

Application Information
Students are accepted for enrollment in January and August. Application deadline for fall is August 30. Application deadline for spring is January 30. In 1996, 150 applied; 125 were accepted. Applicants must submit a formal application.

Contact
Jerome J. Drosos, Coordinator/Chef Instructor, Hospitality Industry Administration, 2000 5th Avenue, River Grove, IL 60171-9983; 708-456-0300 Ext. 3624; Fax: 708-583-3108.

Washburne Trade School

Washburne Culinary Institute

Chicago, Illinois

General Information
Public, coeducational, two-year college. Urban campus. Founded in 1937.

Program Information
Offered since 1937. Member of American Culinary Federation; Educational Foundation of the NRA; Interntional Food Service Executives Association. Program calendar is year-round. 2-year Certificate in Culinary Arts.

Areas of Study
Baking; buffet catering; confectionery show pieces; controlling costs in food service; culinary French; culinary skill development; food preparation; food purchasing; food service communication; food service math; garde-manger; international cuisine; introduction to food service; kitchen management; management and human resources; meat cutting; meat fabrication; patisserie; restaurant opportunities; sanitation; saucier; seafood processing; soup, stock, sauce, and starch production; ice carving.

Facilities
Bake shop; bakery; cafeteria; catering service; 4 classrooms; demonstration laboratory; 6 food production kitchens; lecture room.

Culinary Student Profile
135 full-time.

Faculty
5 full-time; 3 part-time. 1 is an industry professional; 2 are culinary-accredited teachers. Prominent faculty: Dean Jaramillo and John Claybrooke.

Expenses
Application fee: $10. In-state tuition: $4185 per certificate. Out-of-state tuition: $14,421 per certificate.

Financial Aid
In 1996, 5 scholarships were awarded (average award was $4600). Employment placement assistance is available.

Application Information
Students are accepted for enrollment in January, May, and September. In 1996, 250 applied. Applicants must submit a formal application and have a high school diploma or GED.

Contact
Culinary Chefs Department, Washburne Culinary Institute, 3233 West 31st Street, Chicago, IL 60623; 773-579-6100; Fax: 773-376-5940.

William Rainey Harper College

Hospitality Management

Palatine, Illinois

General Information
Public, coeducational, two-year college. Suburban campus. Founded in 1965. Accredited by North Central Association of Colleges and Schools.

Program Information
Member of American Culinary Federation; Council on Hotel, Restaurant, and Institutional Education; Educational Foundation of the NRA; International Association of Culinary Professionals; Interntional Food Service Executives Association; National Restaurant Association; Retailer's Bakery Association. Program calendar is divided into semesters. 12-month Certificates in Hospitality Management; Hotel Management; Culinary Arts; Bread and Pastry Arts. 2-year Associate degree in Hospitality Management.

Areas of Study
Baking; buffet catering; confectionery show pieces; controlling costs in food service; convenience cookery; food preparation; food purchasing; garde-manger; introduction to food service; management and human resources; meal planning; meat cutting; menu and facilities design; nutrition; patisserie; restaurant opportunities; sanitation; saucier; seafood processing; soup, stock, sauce, and starch production; wines and spirits.

Facilities
Bakery; cafeteria; catering service; 6 classrooms; demonstration laboratory; food production kitchen; laboratory; learning resource center; lecture room; library; public restaurant; teaching kitchen.

Culinary Student Profile
160 total: 40 full-time; 120 part-time. 50 are under 25 years old; 60 are between 25 and 44 years old; 50 are over 44 years old.

Faculty
2 full-time; 8 part-time. 10 are industry professionals.

Expenses
Application fee: $20. Tuition: $3036 per degree full-time, $46 per credit hour part-time. Program-related fees include: $50 for lab fees per course.

Financial Aid
In 1996, 2 scholarships were awarded (average award was $500). Program-specific awards include professional management and "Pro-Start" scholarships. Employment placement assistance is available.

Application Information
Students are accepted for enrollment in January and August. In 1996, 40 applied; 40 were accepted. Applicants must submit a formal application.

Contact
Bruce Bohrer, Director of Admissions, Hospitality Management, 1200 West Algonquin Road, Palatine, IL 60067-7398; 847-925-6206; Fax: 847-925-6044; E-mail: info@harper.cc.il.us; World Wide Web: http://www.harper.cc.il.us

Wilton School of Cake Decorating

Darien, Illinois

General Information

Private, coeducational, culinary institute. Suburban campus. Founded in 1929.

Program Information

Offered since 1929. Member of American Institute of Wine & Food; International Association of Culinary Professionals. Certificates in Candy Making; Cake Decorating.

Areas of Study

Candy making; cake decorating.

Facilities

Classroom; demonstration laboratory; food production kitchen.

Culinary Student Profile

320 full-time.

Faculty

2 full-time; 4 part-time. 6 are industry professionals. Prominent faculty: Elain Gonzalez and Colette Peters.

Expenses

Application fee: $100. Tuition: $675 per 2 weeks.

Application Information

Students are accepted for enrollment year-round. In 1996, 350 applied; 350 were accepted. Applicants must submit a formal application.

Contact

Nancy Pakenham, School Secretary, 2240 West 75th Street, Woodbridge, IL 60517; 630-963-7100 Ext. 211; Fax: 630-963-7299; E-mail: npakenha@wilton.com

Ball State University

Muncie, Indiana

General Information

Public, coeducational, university. Urban campus. Founded in 1918. Accredited by North Central Association of Colleges and Schools.

Program Information

Offered since 1975. Program calendar is divided into semesters. 2-year Associate degree in Food Management. 4-year Bachelor's degree in Food Management.

Areas of Study

Beverage management; buffet catering; computer applications; controlling costs in food service; convenience cookery; culinary skill development; food preparation; food purchasing; introduction to food service; kitchen management; management and human resources; meal planning; nutrition; nutrition and food service; restaurant opportunities; sanitation; customer relations.

Culinary Student Profile

60 total: 40 full-time; 20 part-time.

Faculty

3 full-time; 4 part-time.

Expenses

Tuition: $3000 per semester.

Housing

Coed, apartment-style, and single-sex housing available.

Application Information

Students are accepted for enrollment in January, May, and August. Applicants must submit a formal application and letters of reference.

Contact

Lois A. Altman, Program Director, Food Management Program, Department of Family and Consumer Sciences, 2000 University Avenue,

Muncie, IN 47306-1099; 765-285-5931; Fax: 765-285-2314; E-mail: 00laaltman@bsu.edu

Indiana University-Purdue University Fort Wayne

Hospitality Management

Fort Wayne, Indiana

General Information
Public, coeducational, comprehensive institution. Suburban campus. Founded in 1917. Accredited by North Central Association of Colleges and Schools.

Program Information
Offered since 1976. Member of Confrerie de la Chaine des Rotisseurs; Council on Hotel, Restaurant, and Institutional Education; Educational Foundation of the NRA. Program calendar is divided into semesters. 2-year Associate degree in Hotel, Restaurant, and Tourism Management.

Areas of Study
Beverage management; buffet catering; computer applications; controlling costs in food service; convenience cookery; culinary French; culinary skill development; food preparation; food purchasing; introduction to food service; kitchen management; management and human resources; meal planning; menu and facilities design; nutrition; nutrition and food service; restaurant opportunities; sanitation; soup, stock, sauce, and starch production; wines and spirits; tourism.

Facilities
Bake shop; cafeteria; catering service; 10 classrooms; 5 computer laboratories; demonstration laboratory; food production kitchen; 2 gardens; gourmet dining room; laboratory; learning resource center; 10 lecture rooms; library; public restaurant; 4 snack shops; student lounge; teaching kitchen; ballroom.

Culinary Student Profile
60 total: 30 full-time; 30 part-time. 30 are under 25 years old; 20 are between 25 and 44 years old; 10 are over 44 years old.

Faculty
2 full-time; 5 part-time. 3 are industry professionals; 2 are culinary-accredited teachers. Prominent faculty: John B. Knight.

Special Programs
Annual visits to New York, Las Vegas, and Chicago.

Expenses
Application fee: $30. Tuition: $1700 per semester full-time, $100 per credit part-time.

Financial Aid
In 1996, 4 scholarships were awarded (average award was $500). Employment placement assistance is available.

Housing
Average off-campus housing cost per month: $600.

Application Information
Students are accepted for enrollment in January and August. In 1996, 20 applied; 20 were accepted. Applicants must submit a formal application.

Contact
John B. Knight, Director, Hospitality Management, Neff Hall, Room 330B, Fort Wayne, IN 46805-1499; 219-481-6562; Fax: 219-481-5472.

Ivy Tech State College-Central Indiana

Hospitality Administration Program

Indianapolis, Indiana

General Information
Public, coeducational, two-year college. Urban campus. Founded in 1963. Accredited by North Central Association of Colleges and Schools.

Ivy Tech State College-Central Indiana *(continued)*

Program Information
Accredited by American Culinary Federation Education Institute. Program calendar is divided into semesters. 2-year Associate degrees in Baking; Culinary Arts; Hospitality Administration.

Areas of Study
Baking; computer applications; food preparation; food purchasing; food service communication; garde-manger; international cuisine; introduction to food service; management and human resources; meat cutting; menu and facilities design; nutrition; sanitation; seafood processing; soup, stock, sauce, and starch production; wines and spirits.

Facilities
Bake shop; cafeteria; catering service; 20 classrooms; 3 computer laboratories; 2 food production kitchens; 2 laboratories; 3 lecture rooms; library; snack shop; student lounge; 2 teaching kitchens.

Culinary Student Profile
173 total: 100 full-time; 73 part-time. 62 are under 25 years old; 94 are between 25 and 44 years old; 17 are over 44 years old.

Faculty
2 full-time; 15 part-time. Prominent faculty: Vincent Kinkade and Alix Vandivier.

Expenses
Tuition: $6000 per degree.

Financial Aid
Employment placement assistance is available. Employment opportunities within the program are available.

Application Information
Students are accepted for enrollment in January, May, and August. Application deadline for summer is May 1. Application deadline for fall is July 1. Application deadline for spring is November 1. Applicants must submit a formal application.

Contact
Vincent Kinkade, Chairman, Hospitality Administration Program, One West 26th Street, PO Box 1763, Indianapolis, IN 46206-1763; 317-921-4619; Fax: 317-921-4753.

Ivy Tech State College-Northwest

Culinary Arts Program

Gary, Indiana

General Information
Public, coeducational, two-year college. Urban campus. Founded in 1963. Accredited by North Central Association of Colleges and Schools.

Program Information
Offered since 1985. Accredited by American Culinary Federation Education Institute. Member of American Culinary Federation; Educational Foundation of the NRA; National Restaurant Association. Program calendar is divided into semesters. 2-year Associate degree in Culinary Arts.

Areas of Study
Baking; buffet catering; computer applications; controlling costs in food service; culinary French; culinary skill development; food preparation; food purchasing; garde-manger; international cuisine; introduction to food service; management and human resources; meal planning; meat cutting; meat fabrication; menu and facilities design; nutrition; patisserie; sanitation; saucier; seafood processing; soup, stock, sauce, and starch production; wines and spirits.

Facilities
Bake shop; catering service; 2 computer laboratories; 2 demonstration laboratories; food production kitchen; gourmet dining room; learning resource center; library; student lounge; teaching kitchen.

Culinary Student Profile
120 total: 70 full-time; 50 part-time. 40 are under 25 years old; 60 are between 25 and 44 years old; 20 are over 44 years old.

Faculty
3 full-time; 8 part-time. 5 are industry professionals; 6 are culinary-accredited teachers.

Special Programs
National Restaurant Association shows, 2-week trip to France.

Expenses
In-state tuition: $64.55 per credit. Out-of-state tuition: $118.70 per credit. Program-related fees include: $175 for knife set; $57 for lab uniform; $63 for baking kit.

Financial Aid
In 1996, 8 scholarships were awarded (average award was $751.80). Employment placement assistance is available.

Application Information
Students are accepted for enrollment in January, May, and August. Applicants must submit a formal application and have a high school diploma or GED.

Contact
Sharon Purdy, Program Chair, Culinary Arts Program, 1440 East 35th Avenue, Gary, IN 46409-1499; 219-981-4400; Fax: 219-981-4415.

Vincennes University

Culinary Arts Department

Vincennes, Indiana

General Information
Public, coeducational, two-year college. Rural campus. Founded in 1801. Accredited by North Central Association of Colleges and Schools.

Program Information
Offered since 1993. Member of American Culinary Federation; American Culinary Federation Educational Institute; American Institute of Baking; Cooking Together Foundation; Council on Hotel, Restaurant, and Institutional Education; Educational Foundation of the NRA; International Association of Culinary Professionals; Les Amis d'Escoffier Society. Program calendar is divided into semesters. 2-year Associate degree.

Areas of Study
Baking; beverage management; buffet catering; computer applications; controlling costs in food service; convenience cookery; culinary French; culinary skill development; food preparation; food purchasing; food service communication; food service math; garde-manger; international cuisine; introduction to food service; kitchen management; management and human resources; meal planning; meat fabrication; menu and facilities design; nutrition and food service; restaurant opportunities; sanitation; saucier; seafood processing; soup, stock, sauce, and starch production.

Facilities
Bake shop; catering service; 2 classrooms; demonstration laboratory; food production kitchen; gourmet dining room; 2 laboratories; learning resource center; 2 lecture rooms; public restaurant; teaching kitchen.

Culinary Student Profile
60 total: 50 full-time; 10 part-time.

Faculty
4 full-time; 2 part-time. 1 is an industry professional; 1 is a culinary-accredited teacher. Prominent faculty: Carol Keusch and Robert Bird.

Expenses
Application fee: $20. Tuition: $1248 per semester full-time, $78 per credit hour part-time. Program-related fees include: $260 for knife set and pastry set.

Financial Aid
In 1996, 9 scholarships were awarded (average award was $6000). Employment placement assistance is available. Employment opportunities within the program are available.

Housing
Coed, apartment-style, and single-sex housing available. Average off-campus housing cost per month: $350.

Vincennes University *(continued)*

Application Information
Students are accepted for enrollment in May and December. In 1996, 42 applied; 42 were accepted. Applicants must submit a formal application.

Contact
Admissions Office, Culinary Arts Department, 1002 North First Street, Vincennes, IN 47591-5202; 812-888-4313; Fax: 812-888-5868.

Des Moines Area Community College

Culinary Arts Department

Ankeny, Iowa

General Information
Public, coeducational, two-year college. Urban campus. Founded in 1966. Accredited by North Central Association of Colleges and Schools.

Program Information
Offered since 1975. Accredited by American Culinary Federation Education Institute. Member of American Culinary Federation; American Culinary Federation Educational Institute; National Restaurant Association. Program calendar is divided into semesters. 2-year Associate degree in Culinary Arts.

Areas of Study
Baking; beverage management; buffet catering; computer applications; culinary French; culinary skill development; food preparation; food purchasing; garde-manger; international cuisine; introduction to food service; menu and facilities design; nutrition; sanitation.

Facilities
Bake shop; cafeteria; 3 classrooms; computer laboratory; demonstration laboratory; 2 food production kitchens; gourmet dining room; learning resource center; lecture room; library.

Culinary Student Profile
120 full-time. 80 are under 25 years old; 40 are between 25 and 44 years old.

Faculty
3 full-time; 5 part-time. 1 is an industry professional; 2 are culinary-accredited teachers.

Special Programs
French culinary exchange.

Expenses
Tuition: $55 per credit.

Financial Aid
Employment placement assistance is available.

Housing
Apartment-style housing available.

Application Information
Students are accepted for enrollment in January and September. In 1996, 60 applied; 60 were accepted. Applicants must submit a formal application.

Contact
Robert L. Anderson, Program Chair, Culinary Arts Department, 2006 South Ankeny Boulevard, Building #7, Ankeny, IA 50021-8995; 515-964-6532; Fax: 515-964-6486.

Iowa Lakes Community College

Culinary Arts Department

Emmetsburg, Iowa

General Information
Public, coeducational, two-year college. Rural campus. Founded in 1967.

Program Information
Offered since 1973. Member of Council on Hotel, Restaurant, and Institutional Education; Educational Foundation of the NRA; National Restaurant Association; American Hotel/Motel Association. Program calendar is divided into semesters. 2-year Associate degree in Hotel/Motel Restaurant Management.

Areas of Study
Beverage management; buffet catering; computer applications; controlling costs in food service; convenience cookery; culinary skill development; food preparation; food purchasing; food service communication; food service math; garde-manger; introduction to food service; kitchen management; management and human resources; meal planning; menu and facilities design; nutrition; restaurant opportunities; sanitation; saucier; soup, stock, sauce, and starch production; wines and spirits; hospitality law; marketing.

Facilities
Bakery; 2 cafeterias; 3 catering services; 3 classrooms; coffee shop; 5 computer laboratories; demonstration laboratory; 2 food production kitchens; gourmet dining room; 2 laboratories; 3 learning resource centers; 2 lecture rooms; 3 libraries; public restaurant; snack shop; student lounge; teaching kitchen.

Culinary Student Profile
45 total: 40 full-time; 5 part-time.

Faculty
1 full-time; 3 part-time. 2 are industry professionals; 2 are culinary-accredited teachers. Prominent faculty: Robert Halverson.

Special Programs
Iowa Hospitality Show in Des Moines, Midwest Hospitality Show in Minneapolis, Marriott-Tan-Tar-A Hospitality Show.

Expenses
Tuition: $1500 per semester full-time, $54 per credit part-time. Program-related fees include: $200 for knives; $130 for trip fees.

Financial Aid
In 1996, 5 scholarships were awarded (average award was $500); 35 loans were granted (average loan was $2600). Program-specific awards include National Restaurant Association scholarships, scholarships for freshmen ($150). Employment placement assistance is available. Employment opportunities within the program are available.

Housing
Coed and apartment-style housing available. Average on-campus housing cost per month: $675. Average off-campus housing cost per month: $200.

Application Information
Students are accepted for enrollment in January, May, and August. Application deadline for fall is August 1. Application deadline for spring is January 1. Application deadline for summer is May 1. Applicants must submit a formal application.

Contact
Robert Halverson, Coordinator, Culinary Arts Department, 3200 College Drive, Emmetsburg, IA 50536-1098; 712-852-5256; Fax: 712-852-2152.

Kirkwood Community College

Restaurant Management/Culinary Arts

Cedar Rapids, Iowa

General Information
Public, coeducational, two-year college. Urban campus. Founded in 1966. Accredited by North Central Association of Colleges and Schools.

Program Information
Offered since 1968. Accredited by American Culinary Federation Education Institute. Member of American Culinary Federation; American Culinary Federation Educational Institute; American Dietetic Association; Council on Hotel, Restaurant, and Institutional Education; Educational Foundation of the NRA; National Restaurant Association. Program calendar is divided into semesters. 1-year Certificate in Bakery. 1-year Diploma in Food Service Training. 2-year Associate degrees in Dietetic Technician; Culinary Arts; Restaurant Management.

Kirkwood Community College *(continued)*

Areas of Study
Baking; beverage management; computer applications; controlling costs in food service; culinary skill development; food preparation; food purchasing; food service communication; food service math; garde-manger; international cuisine; kitchen management; management and human resources; meal planning; menu and facilities design; nutrition; sanitation; soup, stock, sauce, and starch production; wines and spirits; conversational Spanish; culinary competition.

Facilities
Bakery; cafeteria; catering service; 2 classrooms; computer laboratory; demonstration laboratory; food production kitchen; gourmet dining room; learning resource center; library; public restaurant.

Culinary Student Profile
110 total: 100 full-time; 10 part-time.

Faculty
3 full-time; 6 part-time. 7 are industry professionals; 2 are dietitians. Prominent faculty: Mary Jane German and Mary Rhiner.

Special Programs
Professional meetings and conventions (local, state, national).

Expenses
Tuition: $57 per credit hour. Program-related fees include: $275 for knives; $150 for uniforms.

Financial Aid
Employment placement assistance is available. Employment opportunities within the program are available.

Application Information
Students are accepted for enrollment in January, May, and August. In 1996, 45 applied; 45 were accepted. Applicants must submit a formal application.

Contact
Mary Jane German, Coordinator, Restaurant Management/Culinary Arts, 6301 Kirkwood Boulevard, SW, Cedar Rapids, IA 52406; 319-398-4981; Fax: 319-398-5667; E-mail: mgerman@kirkwood.cc.ia.us

American Institute of Baking

Baking Science and Technology

Manhattan, Kansas

General Information
Private, coeducational, culinary institute. Rural campus. Founded in 1919.

Program Information
Offered since 1919. Member of American Institute of Baking. 4-month Certificate in Baking Science and Technology.

Areas of Study
Baking; maintenance engineering.

Facilities
4 bake shops; bakery; 4 classrooms; computer laboratory; 4 demonstration laboratories; laboratory; library; student lounge; cookie-cracker production line.

Culinary Student Profile
72 full-time. 10 are under 25 years old; 52 are between 25 and 44 years old; 10 are over 44 years old.

Faculty
8 full-time. 8 are industry professionals.

Special Programs
Half-day tours of grain elevator and flour mill, half-day tours of commercial wholesale bakeries, half-day tours of Kansas wheat farm.

Expenses
Application fee: $45. Tuition: $4000 per certificate.

Financial Aid
In 1996, 40 scholarships were awarded (average award was $2500); 15 loans were granted (average loan was $2000). Employment placement assistance is available.

Housing
Average off-campus housing cost per month: $600.

Application Information
Students are accepted for enrollment in February and September. In 1996, 160 applied; 155 were accepted. Applicants must submit a formal

application and letters of reference and have a college degree or 2 years of work experience.

Contact
Ken Embers, Director of Admissions, Baking Science and Technology, 1213 Bakers Way, Manhattan, KS 66502; 800-633-5137; Fax: 785-537-1493; E-mail: kembers@aibonline.org; World Wide Web: http://www.aibonline.org

Kansas City Kansas Area Vocational Technical School

Professional Cooking

Kansas City, Kansas

General Information
Public, coeducational, adult vocational school. Urban campus. Founded in 1972.

Program Information
Offered since 1972. Program calendar is year-round. 7-month Certificate in Food Service.

Areas of Study
Baking; buffet catering; computer applications; controlling costs in food service; convenience cookery; culinary skill development; food preparation; food purchasing; food service communication; food service math; garde-manger; international cuisine; introduction to food service; kitchen management; meal planning; meat cutting; menu and facilities design; nutrition; nutrition and food service; patisserie; restaurant opportunities; sanitation; saucier; soup, stock, sauce, and starch production; child-care cookery.

Facilities
Bake shop; bakery; cafeteria; classroom; coffee shop; computer laboratory; demonstration laboratory; food production kitchen; gourmet dining room; learning resource center; lecture room; library; public restaurant; snack shop; student lounge.

Culinary Student Profile
25 total: 15 full-time; 10 part-time. 4 are under 25 years old; 18 are between 25 and 44 years old; 3 are over 44 years old.

Faculty
2 full-time; 3 part-time.

Expenses
Application fee: $25. In-state tuition: $720 per certificate. Out-of-state tuition: $5400 per certificate. Program-related fees include: $150 for textbooks; $50 for uniforms.

Financial Aid
Program-specific awards include 10 Missouri Restaurant Association scholarships, 5 Kansas City Hotel/Motel Association scholarships. Employment placement assistance is available. Employment opportunities within the program are available.

Application Information
Students are accepted for enrollment in January, February, March, April, May, June, August, September, October, November, and December. Applicants must take an entrance exam and complete an interview.

Contact
Deborah Reynolds, Enrollment Counselor, Professional Cooking, 2220 North 59th Street, Kansas City, KS 66104; 913-596-5500; Fax: 913-596-5509.

Wichita Area Technical College

Food Service Education

Wichita, Kansas

General Information
Public, coeducational, technical college. Urban campus.

Program Information
Offered since 1979. Member of American Culinary Federation; National Restaurant Association.

Wichita Area Technical College *(continued)*

2-year Associate degree in Food Service Management. 9-month Diploma in Food Service Production.

Areas of Study
Baking; buffet catering; computer applications; controlling costs in food service; culinary skill development; food preparation; food purchasing; food service math; garde-manger; international cuisine; introduction to food service; kitchen management; management and human resources; meal planning; nutrition and food service; restaurant opportunities; sanitation; soup, stock, sauce, and starch production.

Facilities
Bake shop; cafeteria; classroom; computer laboratory; food production kitchen; learning resource center; library.

Culinary Student Profile
21 full-time. 11 are under 25 years old; 8 are between 25 and 44 years old; 2 are over 44 years old.

Faculty
3 full-time. 1 is an industry professional; 2 are culinary-accredited teachers. Prominent faculty: Colette Baptista and Dan Hypse.

Expenses
Tuition: $1863 per 9 months. Program-related fees include: $100 for uniforms.

Financial Aid
Employment placement assistance is available.

Application Information
Students are accepted for enrollment in August. In 1996, 30 applied; 21 were accepted. Applicants must submit a formal application.

Contact
Colette Baptista, Coordinating Instructor, Food Service Education, 324 North Emporia, Wichita, KS 67217; 316-833-4360; Fax: 316-833-4341.

Kentucky Tech Elizabethtown

Food Service Technology

Elizabethtown, Kentucky

General Information
Public, coeducational, technical institute. Rural campus. Founded in 1967.

Program Information
Offered since 1967. Program calendar is divided into semesters. 18-month Diplomas in Food Service Worker (Hospital) studies; Head Baker studies. 22-month Diploma in Kitchen Supervisor studies.

Areas of Study
Baking; beverage management; buffet catering; computer applications; culinary skill development; food preparation; food purchasing; food service communication; food service math; introduction to food service; kitchen management; meal planning; meat cutting; meat fabrication; menu and facilities design; nutrition; nutrition and food service; restaurant opportunities; sanitation.

Facilities
Cafeteria; classroom; computer laboratory; learning resource center; lecture room; teaching kitchen.

Culinary Student Profile
23 total: 18 full-time; 5 part-time. 5 are under 25 years old; 13 are between 25 and 44 years old; 5 are over 44 years old.

Faculty
2 full-time; 1 part-time. 1 is a culinary-accredited teacher; 1 is a dining room supervisor. Prominent faculty: Brenda Harrington.

Special Programs
Field trips.

Expenses
Application fee: $20. Tuition: $310 per semester full-time, $210 per semester part-time. Program-related fees include: $150 for uniforms; $150 for books; $25 for general supplies.

Financial Aid
Program-specific awards include scholarships available through local organizations. Employment placement assistance is available. Employment opportunities within the program are available.

Application Information
Students are accepted for enrollment in January and July. Application deadline for fall is continuous with a recommended date of July 15. Application deadline for spring is continuous with a recommended date of December 15. In 1996, 30 applied; 30 were accepted. Applicants must submit a formal application and academic transcripts and take the Test of Adult Basic Education.

Contact
Rene J. Emond, Registrar, Food Service Technology, 505 University Drive, Elizabethtown, KY 42701; 502-766-5133; Fax: 502-737-0505.

Kentucky Tech-Jefferson Campus

Food Service Technology Program

Louisville, Kentucky

General Information
Public, coeducational, two-year college. Rural campus. Founded in 1972. Accredited by Southern Association of Colleges and Schools.

Program Information
Offered since 1972. Member of National Restaurant Association. Program calendar is divided into quarters. 18-month Diploma in Food Service Technology.

Areas of Study
Baking; computer applications; food preparation; food purchasing; introduction to food service; kitchen management; nutrition; nutrition and food service; sanitation; soup, stock, sauce, and starch production.

Facilities
Bake shop; cafeteria; catering service; classroom; computer laboratory; demonstration laboratory; food production kitchen; laboratory; learning resource center; lecture room; student lounge; teaching kitchen.

Culinary Student Profile
17 total: 15 full-time; 2 part-time. 6 are under 25 years old; 10 are between 25 and 44 years old; 1 is over 44 years old.

Faculty
1 full-time; 1 part-time. 1 is an industry professional; 1 is a culinary-accredited teacher.

Expenses
Application fee: $20. Tuition: $1050 per diploma.

Financial Aid
Employment placement assistance is available. Employment opportunities within the program are available.

Application Information
Students are accepted for enrollment in January, April, July, and October. In 1996, 5 applied; 4 were accepted. Applicants must submit a formal application.

Contact
Tara Parker, Registrar, Food Service Technology Program, 727 West Chestnut Street, Louisville, KY 40203; 502-595-4136; Fax: 502-595-4399.

Sullivan College

National Center for Hospitality Studies

Louisville, Kentucky

General Information
Private, coeducational, comprehensive institution. Suburban campus. Founded in 1962. Accredited by Southern Association of Colleges and Schools.

Program Information
Offered since 1987. Accredited by American Culinary Federation Education Institute. Member of American Culinary Federation; American Culinary Federation Educational Institute; American Dietetic Association; Council on Hotel, Restaurant, and Institutional Education; Educational Foundation of the NRA; National

Sullivan College ***(continued)***

Restaurant Association; Kentucky Restaurant Association. Program calendar is divided into quarters. 12-month Diploma in Professional Cook studies. 18-month Associate degrees in Travel and Tourism; Professional Catering; Hotel and Restaurant Management; Culinary Arts; Baking and Pastry Arts. 9-month Diplomas in Travel and Tourism; Professional Baker studies.

Areas of Study

Baking; beverage management; computer applications; confectionery show pieces; controlling costs in food service; culinary skill development; food preparation; food purchasing; food service math; garde-manger; international cuisine; kitchen management; management and human resources; meat cutting; menu and facilities design; nutrition; patisserie; restaurant opportunities; sanitation; saucier; seafood processing; soup, stock, sauce, and starch production; wines and spirits; hotel restaurant management; professional catering.

Facilities

3 bake shops; bakery; cafeteria; catering service; 35 classrooms; 6 computer laboratories; demonstration laboratory; 8 food production kitchens; gourmet dining room; 8 laboratories; library; public restaurant; 8 teaching kitchens.

Faculty

14 full-time; 3 part-time. 11 are industry professionals; 1 is a master chef; 3 are culinary-accredited teachers. Prominent faculty: Thomas Hickey and Walter Rhea.

Special Programs

3-month restaurant practicum in 3½-star fine dining restaurant located on campus.

Expenses

Application fee: $75. Tuition: $19,260 per 18 months full-time, $3210 per quarter part-time. Program-related fees include: $4250 for comprehensive lab fees (includes uniforms and kniv.

Financial Aid

In 1996, 8 scholarships were awarded (average award was $5150); 104 loans were granted (average loan was $1940). Employment placement assistance is available. Employment opportunities within the program are available.

Housing

176 culinary students housed on campus. Apartment-style housing available. Average on-campus housing cost per month: $300.

Application Information

Students are accepted for enrollment in January, March, June, and September. Application deadline for winter is continuous with a recommended date of January 5. Application deadline for spring is continuous with a recommended date of March 30. Application deadline for fall is continuous with a recommended date of September 28. In 1996, 618 applied; 587 were accepted. Applicants must submit a formal application.

Contact

Greg Cawthon, Director of Admissions, National Center for Hospitality Studies, 3101 Bardstown Road, Louisville, KY 40205; 800-844-1354; Fax: 502-454-4880; E-mail: admissions@sullivan.edu

See color display following page 140.

University of Kentucky, Jefferson Community College

Business Division

Louisville, Kentucky

General Information

Public, coeducational, two-year college. Urban campus. Founded in 1968. Accredited by Southern Association of Colleges and Schools.

Program Information

Offered since 1974. Accredited by American Culinary Federation Education Institute. Member of American Culinary Federation; American Culinary Federation Educational Institute; Educational Foundation of the NRA; National

Restaurant Association. Program calendar is divided into semesters. 2-year Associate degree in Culinary Arts.

Areas of Study
Baking; beverage management; buffet catering; computer applications; confectionery show pieces; controlling costs in food service; convenience cookery; culinary French; culinary skill development; food preparation; food purchasing; food service communication; garde-manger; international cuisine; kitchen management; meal planning; meat cutting; meat fabrication; menu and facilities design; nutrition; nutrition and food service; patisserie; restaurant opportunities; sanitation; saucier; seafood processing; soup, stock, sauce, and starch production.

Facilities
Bake shop; cafeteria; classroom; computer laboratory; demonstration laboratory; food production kitchen; gourmet dining room; laboratory; learning resource center; 3 lecture rooms; library; student lounge; teaching kitchen.

Culinary Student Profile
64 total: 44 full-time; 20 part-time. 30 are under 25 years old; 34 are between 25 and 44 years old.

Faculty
2 full-time; 2 part-time. 1 is an industry professional; 2 are culinary-accredited teachers; 1 is a registered dietitian. Prominent faculty: Nancy Russman and Gail Crawford.

Expenses
In-state tuition: $510 per semester full-time, $42.50 per credit hour part-time. Out-of-state tuition: $1530 per semester full-time, $127.50 per credit hour part-time. Program-related fees include: $40 for technology fees for updating equipment.

Financial Aid
Program-specific awards include Kentucky Restaurant Association scholarship. Employment opportunities within the program are available.

Housing
Average off-campus housing cost per month: $450.

Application Information
Students are accepted for enrollment in January and August. In 1996, 17 applied; 17 were accepted. Applicants must submit a formal application and interview with the program coordinator.

Contact
Gail Crawford, Program Coordinator, Business Division, 109 East Broadway, Louisville, KY 40202-2005; 502-584-0181 Ext. 2317; Fax: 502-584-0181 Ext. 2414.

West Kentucky State Vocational Technical School

Food Service Technology

Paducah, Kentucky

General Information
Public, coeducational, technical institute. Urban campus. Founded in 1911.

Program Information
Program calendar is divided into quarters. Certificates in School Cafeteria Cook studies; Dietary Aide studies; Cook's Helper studies; Cafeteria Attendant studies; Baker's Helper studies. Diplomas in Head Baker studies; Head School Cook studies; Kitchen Supervisor studies; Restaurant Cook studies.

Areas of Study
Baking; beverage management; buffet catering; controlling costs in food service; convenience cookery; culinary skill development; food preparation; food purchasing; food service communication; food service math; introduction to food service; kitchen management; management and human resources; meal planning; nutrition; sanitation; soup, stock, sauce, and starch production.

Culinary Student Profile
25 total: 20 full-time; 5 part-time.

West Kentucky State Vocational Technical School *(continued)*

Faculty
2 full-time.

Expenses
In-state tuition: $155 per quarter. Out-of-state tuition: $310 per quarter.

Application Information
Students are accepted for enrollment in January, March, July, and October. Applicants must submit a formal application.

Contact
Mary Sanderson, Food Service Director, Food Service Technology, Blandville Road, Paducah, KY 42002; 502-554-4991; Fax: 502-554-9754.

Culinary Arts Institute of Louisiana

Baton Rouge, Louisiana

General Information
Private, coeducational, culinary institute. Urban campus. Founded in 1988.

Program Information
Offered since 1988. Member of American Culinary Federation; American Institute of Wine & Food; International Association of Culinary Professionals; James Beard Foundation, Inc.; Women Chefs and Restaurateurs; Louisiana Restaurant Association. Program calendar is divided into months. 15-month Associate degree in Professional Cooking and Restaurant Management.

Areas of Study
Baking; beverage management; buffet catering; confectionery show pieces; controlling costs in food service; culinary French; culinary skill development; food preparation; food purchasing; food service math; garde-manger; international cuisine; introduction to food service; kitchen management; management and human resources; meal planning; meat cutting; meat fabrication; menu and facilities design; nutrition; nutrition and food service; sanitation; saucier; seafood processing; soup, stock, sauce, and starch production; wines and spirits.

Facilities
Catering service; 3 classrooms; demonstration laboratory; food production kitchen; gourmet dining room; laboratory; library; public restaurant; teaching kitchen.

Culinary Student Profile
60 full-time.

Faculty
4 full-time.

Expenses
Tuition: $17,314 per degree.

Financial Aid
In 1996, 1 scholarship was awarded (award was $5000); 78 loans were granted (average loan was $6520). Employment placement assistance is available. Employment opportunities within the program are available.

Housing
Average off-campus housing cost per month: $500.

Application Information
Students are accepted for enrollment in January, February, March, April, May, June, July, August, September, and October. In 1996, 76 applied; 54 were accepted. Applicants must submit a formal application and letters of reference.

Contact
Sharon B. Burke, Director, 427 Lafayette Street, Baton Rouge, LA 70802; 504-343-6233; Fax: 504-338-4880; E-mail: school@caila.com

Delgado Community College

Culinary Arts Program

New Orleans, Louisiana

General Information
Public, coeducational, two-year college. Urban campus. Founded in 1921. Accredited by Southern Association of Colleges and Schools.

Program Information
Accredited by American Culinary Federation Education Institute. Member of American Culinary Federation; American Culinary Federation Educational Institute; American Institute of Wine & Food; Council on Hotel, Restaurant, and Institutional Education; International Association of Culinary Professionals; National Restaurant Association; Women Chefs and Restaurateurs; Louisiana Restaurant Association. Program calendar is divided into semesters. 1-year Certificate in Culinary Arts. 3-year Associate degree in Culinary Arts.

Areas of Study
Baking; beverage management; buffet catering; computer applications; controlling costs in food service; culinary French; culinary skill development; food preparation; food purchasing; food service communication; food service math; garde-manger; international cuisine; introduction to food service; kitchen management; management and human resources; meal planning; meat cutting; menu and facilities design; nutrition; patisserie; restaurant opportunities; sanitation; saucier; seafood processing; soup, stock, sauce, and starch production; wines and spirits.

Culinary Student Profile
175 total: 150 full-time; 25 part-time.

Faculty
5 full-time; 3 part-time.

Expenses
Tuition: $600 per semester full-time, $200 per semester part-time.

Application Information
Students are accepted for enrollment in March. In 1996, 60 applied; 60 were accepted. Applicants must submit a formal application and letters of reference.

Contact
Iva Bergeron, Director, Culinary Arts Program, 615 City Park Avenue, New Orleans, LA 70119-4399; 504-483-4208; Fax: 504-483-4893.

Louisiana Technical College, Lafayette Campus

Lafayette, Louisiana

General Information
Public, coeducational, technical institute. Urban campus.

Program Information
Accredited by American Culinary Federation Education Institute. Program calendar is divided into quarters. 18-month Associate degree in Hotel Hospitality Management. 18-month Diploma in Culinary Arts.

Areas of Study
Baking; beverage management; buffet catering; computer applications; controlling costs in food service; convenience cookery; culinary French; culinary skill development; food preparation; food purchasing; food service communication; food service math; garde-manger; international cuisine; introduction to food service; kitchen management; management and human resources; meal planning; meat fabrication; menu and facilities design; nutrition; patisserie; restaurant opportunities; sanitation; saucier; seafood processing; soup, stock, sauce, and starch production.

Culinary Student Profile
44 total: 43 full-time; 1 part-time.

Faculty
6 full-time; 1 part-time.

Expenses
Application fee: $5. In-state tuition: $105 per quarter full-time, $15 per credit part-time. Out-of-state tuition: $210 per quarter full-time, $30 per credit part-time.

Application Information
Students are accepted for enrollment in February, July, August, and November. Applicants must submit a formal application and complete an entrance exam.

Louisiana Technical College, Lafayette Campus *(continued)*

Contact
Jerry Fonnier, Program Coordinator, 1101 Bertrand Drive, Lafayette, LA 70506; 318-262-5962 Ext. 232; Fax: 318-262-5122.

Louisiana Technical College-Baton Rouge Campus

Culinary Arts and Occupations

Baton Rouge, Louisiana

General Information
Public, coeducational, technical institute. Urban campus. Founded in 1974.

Program Information
Offered since 1974. Member of American Culinary Federation; Educational Foundation of the NRA. Program calendar is divided into quarters. 18-month Diploma in Culinary Arts and Occupations. 3-month Certificates in Supervision; Nutrition; Sanitation.

Areas of Study
Baking; computer applications; controlling costs in food service; culinary skill development; food preparation; food purchasing; food service communication; food service math; kitchen management; management and human resources; meat cutting; nutrition and food service; sanitation; soup, stock, sauce, and starch production.

Facilities
Bake shop; cafeteria; classroom; computer laboratory; food production kitchen.

Culinary Student Profile
12 full-time.

Faculty
1 full-time. 1 is an industry professional. Prominent faculty: Michael Travasos.

Expenses
Application fee: $5. Tuition: $630 per diploma full-time, $75 per half credit part-time. Program-related fees include: $50 for uniforms; $300 for textbooks.

Financial Aid
Employment placement assistance is available. Employment opportunities within the program are available.

Application Information
Students are accepted for enrollment in March, May, August, and November. Applicants must submit a formal application.

Contact
Michael Travasos, Instructor, Culinary Arts and Occupations, 3250 North Acadian Highway, E, Baton Rouge, LA 70805; 504-359-9226.

Louisiana Technical College-Sidney N. Collier Campus

Culinary Arts

New Orleans, Louisiana

General Information
Public, coeducational, two-year college. Suburban campus. Founded in 1952.

Program Information
Program calendar is divided into quarters. 1,800-hour Certificate in Culinary Arts.

Areas of Study
Baking; beverage management; buffet catering; computer applications; controlling costs in food service; culinary skill development; food preparation; food purchasing; food service communication; food service math; garde-manger; international cuisine; introduction to food service; kitchen management; management and human resources; meal planning; meat cutting; nutrition; patisserie; sanitation; seafood processing; soup, stock, sauce, and starch production.

Culinary Student Profile
15 full-time.

Faculty
2 full-time; 1 part-time.

Expenses
Tuition: $105 per quarter.

Application Information
Students are accepted for enrollment year-round. Applicants must submit a formal application and complete an entrance exam.

Contact
Levi Lewis, Sr., Director, Culinary Arts, 3727 Louisa Street, New Orleans, LA 70126; 504-942-8333 Ext. 132; Fax: 504-942-8337.

Louisiana Technical College-Sowela Campus

Culinary Arts and Occupations

Lake Charles, Louisiana

General Information
Public, coeducational, two-year college. Suburban campus. Founded in 1940.

Program Information
Offered since 1977. Member of Educational Foundation of the NRA. Program calendar is year-round. 18-month Diploma in Culinary Arts and Occupations.

Areas of Study
Baking; buffet catering; computer applications; controlling costs in food service; convenience cookery; culinary skill development; food preparation; food purchasing; food service communication; food service math; garde-manger; international cuisine; introduction to food service; kitchen management; management and human resources; meal planning; meat fabrication; nutrition; nutrition and food service; sanitation; saucier; soup, stock, sauce, and starch production; food and beverage service; basic accounting.

Facilities
Bakery; cafeteria; classroom; computer laboratory; demonstration laboratory; food production kitchen; garden; gourmet dining room; laboratory; learning resource center; lecture room; library; snack shop; student lounge; teaching kitchen.

Culinary Student Profile
54 total: 30 full-time; 24 part-time.

Faculty
2 full-time; 1 part-time. 2 are culinary-accredited teachers. Prominent faculty: Lee Thibodeaux and Strranley Leger.

Expenses
Tuition: $650 per diploma full-time, $76 per quarter part-time. Program-related fees include: $426 for books; $104 for tools; $209 for uniforms.

Financial Aid
Program-specific awards include Rotary Club scholarship. Employment placement assistance is available.

Application Information
Students are accepted for enrollment year-round. In 1996, 20 applied; 18 were accepted. Applicants must submit a formal application.

Contact
Susan Simmons, Student Services Director, Culinary Arts and Occupations, 3820 J. Bennett Johnston Avenue, Lake Charles, LA 70615; 318-491-2687; Fax: 318-491-2135.

Nicholls State University

Chef John Folse Culinary Institute

Thibodaux, Louisiana

General Information
Public, coeducational, comprehensive institution. Rural campus. Founded in 1948. Accredited by Southern Association of Colleges and Schools.

Program Information
Offered since 1994. Member of American Culinary Federation; Council on Hotel, Restaurant, and Institutional Education; National Restaurant

Nicholls State University *(continued)*

Association; Research Chefs Association Society for the Advancement of Foodservice Research. Program calendar is divided into semesters. 2-year Associate degree in Culinary Arts. 4-year Bachelor's degree in Culinary Arts.

Areas of Study
Baking; beverage management; buffet catering; computer applications; confectionery show pieces; controlling costs in food service; convenience cookery; culinary French; culinary skill development; food preparation; food purchasing; garde-manger; international cuisine; introduction to food service; kitchen management; management and human resources; meal planning; meat cutting; meat fabrication; menu and facilities design; nutrition; nutrition and food service; patisserie; restaurant opportunities; sanitation; saucier; seafood processing; soup, stock, sauce, and starch production; wines and spirits.

Facilities
Bakery; cafeteria; catering service; 3 classrooms; 2 computer laboratories; demonstration laboratory; food production kitchen; gourmet dining room; 3 laboratories; 2 learning resource centers; 3 lecture rooms; library; public restaurant; snack shop; student lounge; 2 teaching kitchens.

Culinary Student Profile
120 total: 105 full-time; 15 part-time. 37 are under 25 years old; 80 are between 25 and 44 years old; 3 are over 44 years old.

Faculty
3 full-time; 2 part-time. Prominent faculty: Jerald W. Chesser and John Folse.

Expenses
Application fee: $10. In-state tuition: $1009 per semester full-time, $89 per credit hour part-time. Out-of-state tuition: $2304 per semester full-time, $194 per credit hour part-time. Program-related fees include: $330 for knives; $300 for uniforms; $50 for lab fees.

Financial Aid
In 1996, 3 scholarships were awarded (average award was $1000). Employment placement assistance is available.

Housing
Single-sex housing available. Average on-campus housing cost per month: $338.

Application Information
Students are accepted for enrollment in January and August. Application deadline for fall is July 15. Application deadline for spring is November 15. In 1996, 80 applied; 63 were accepted. Applicants must submit a formal application.

Contact
Jerald W. Chesser, Director, Chef John Folse Culinary Institute, Chef John Folse Culinary Institute, PO Box 2099, Thibodaux, LA 70310; 504-449-7100; Fax: 504-449-7089; E-mail: jfci-jwc@nich-nsunet.nich.edu

Sclafani's Cooking School, Inc.

Commercial Cook/Baker

Metaire, Louisiana

General Information
Private, coeducational, culinary institute. Suburban campus. Founded in 1987.

Program Information
Offered since 1987. Member of American Culinary Federation; Louisiana Restaurant Association. Program calendar is divided into months. 4-week Certificates in Sautéing; Sauces; Lamb Preparation; Commercial Cooking/Baking.

Areas of Study
Baking; culinary skill development; food preparation; food service math; garde-manger; introduction to food service; kitchen management; management and human resources; meal planning; restaurant opportunities; sanitation; saucier; soup, stock, sauce, and starch production.

Facilities
Bake shop; bakery; classroom; demonstration laboratory; food production kitchen; library; teaching kitchen.

Culinary Student Profile
110 full-time. 23 are under 25 years old; 70 are between 25 and 44 years old; 17 are over 44 years old.

Faculty
3 full-time. 2 are industry professionals. Prominent faculty: Frank P. Sclafani, Sr.

Expenses
Tuition: $2145 per 4 weeks.

Financial Aid
Employment placement assistance is available.

Housing
Average off-campus housing cost per month: $450.

Application Information
Students are accepted for enrollment year-round. In 1996, 110 applied; 108 were accepted. Applicants must interview and take the Wonderlic aptitude test.

Contact
Frank P. Sclafani, Sr., President, Commercial Cook/Baker, 107 Gennaro Place, Metaire, LA 70001; 504-833-7861; Fax: 504-833-7872; E-mail: sclafani@gnofn.org

Southern Maine Technical College

Hospitality Management

South Portland, Maine

General Information
Public, coeducational, two-year college. Suburban campus. Founded in 1946. Accredited by New England Association of Schools and Colleges.

Program Information
Offered since 1958. Member of American Culinary Federation Educational Institute; American Dietetic Association; American Institute of Baking; Council on Hotel, Restaurant, and Institutional Education; Educational Foundation of the NRA; National Restaurant Association; Maine Restaurant Association. Program calendar is divided into semesters. 2-year Associate degrees in Hotel, Motel, and Restaurant Management; Culinary Arts. 2-year Diploma in Culinary Arts.

Areas of Study
Baking; beverage management; buffet catering; computer applications; controlling costs in food service; culinary skill development; food preparation; food purchasing; food service communication; garde-manger; international cuisine; introduction to food service; kitchen management; management and human resources; meal planning; meat fabrication; nutrition; nutrition and food service; restaurant opportunities; sanitation; saucier; soup, stock, sauce, and starch production.

Facilities
Bake shop; bakery; 7 classrooms; computer laboratory; demonstration laboratory; 4 food production kitchens; gourmet dining room; 3 laboratories; learning resource center; lecture room; library; public restaurant; 2 teaching kitchens.

Culinary Student Profile
120 total: 100 full-time; 20 part-time.

Faculty
8 full-time; 2 part-time. 4 are culinary arts instructors; 2 are dietetic instructors; 2 are hospitality management instructors. Prominent faculty: Robert Latham and Robert Lyna.

Special Programs
Tours of food service establishments, guest lectures.

Expenses
Application fee: $20. In-state tuition: $1188 per semester full-time, $66 per credit part-time. Out-of-state tuition: $2610 per semester full-time, $145 per credit part-time. Program-related fees include: $238 for technology fees; $200 for uniforms, knives, and equipment.

Financial Aid
In 1996, 12 scholarships were awarded (average award was $500). Employment placement assistance is available. Employment opportunities within the program are available.

Southern Maine Technical College *(continued)*

HOUSING
Coed and single-sex housing available. Average on-campus housing cost per month: $190. Average off-campus housing cost per month: $300.

APPLICATION INFORMATION
Students are accepted for enrollment in January and August. Application deadline for fall is continuous with a recommended date of April 1. Application deadline for spring is continuous with a recommended date of November 1. In 1996, 160 applied; 90 were accepted. Applicants must submit a formal application and have a high school diploma.

CONTACT
Robert Weimont, Director of Admissions, Hospitality Management, Fort Road, South Portland, ME 04106; 207-767-9520; Fax: 207-767-9671.

The Culinary Arts program features an ocean-side restaurant (80 seats), 4 large commercial kitchens, and 4 full-time faculty members with many years of real-world experience in the region's finest restaurants. The Hotel, Motel and Restaurant Management program features the McKernan Hospitality Center, the new world-class ocean-side inn. The Center features 2 kitchens, 2 dining rooms, 2 conference rooms, and 8 guest rooms. Both programs offer high-quality rigorous programs in tremendous facilities on the beautiful 70-acre campus only 4 miles from Portland, Maine's largest and most livable city. Tuition costs are very reasonable. Prospective students are encouraged to visit this comprehensive facility.

BALTIMORE INTERNATIONAL COLLEGE

School of Culinary Arts

Baltimore, Maryland

GENERAL INFORMATION
Private, coeducational, two-year college. Urban campus. Founded in 1972. Accredited by Middle States Association of Colleges and Schools.

PROGRAM INFORMATION
Offered since 1972. Member of American Culinary Federation; International Association of Culinary Professionals. Program calendar is divided into semesters. 12-month Certificate in Professional Cooking. 15-week Diploma in Epicurean Cooking and Baking. 18-month Associate degrees in Institutional Food Service Management; Hotel/Motel/Innkeeping Management; Food and Beverage Management; Professional Cooking and Baking; Professional Baking and Pastry; Professional Cooking. 22-month Certificate in Culinary Arts. 9-month Certificate in Professional Baking and Pastry.

AREAS OF STUDY
Baking; beverage management; buffet catering; computer applications; confectionery show pieces; controlling costs in food service; convenience cookery; culinary French; culinary skill development; food preparation; food purchasing; food service communication; food service math; garde-manger; international cuisine; introduction to food service; kitchen management; management and human resources; meal planning; meat cutting; meat fabrication; menu and facilities design; nutrition; nutrition and food service; patisserie; restaurant opportunities; sanitation; saucier; seafood processing; soup, stock, sauce, and starch production; wines and spirits.

FACILITIES
3 bake shops; cafeteria; 10 classrooms; coffee shop; computer laboratory; 2 demonstration laboratories; 6 food production kitchens; 4 gardens; gourmet dining room; 6 laboratories; learning resource center; 10 lecture rooms; library; public restaurant; snack shop; student lounge; 7 teaching kitchens; pub.

CULINARY STUDENT PROFILE
850 full-time.

FACULTY
16 full-time; 8 part-time. 4 are industry professionals; 2 are master chefs; 11 are culinary-accredited teachers.

SPECIAL PROGRAMS
Students required to take 1-3 courses at college campus in Ireland.

Expenses

Application fee: $35. Tuition: $3360 per semester.

Financial Aid

In 1996, 452 scholarships were awarded (average award was $1500); 100 loans were granted (average loan was $4500). Employment placement assistance is available. Employment opportunities within the program are available.

Housing

150 culinary students housed on campus. Coed and apartment-style housing available. Average on-campus housing cost per month: $560.

Application Information

Students are accepted for enrollment in January, April, May, July, September, and October. In 1996, 1,149 applied; 1,122 were accepted. Applicants must submit a formal application.

Contact

Lisa Rice, Director of Admissions, School of Culinary Arts, Commerce Exchange, 17 Commerce Street, Baltimore, MD 21202-3230; 410-752-4710; Fax: 410-752-3730.

Baltimore International College, America's international hospitality college, offers degree, certificate, and diploma programs in the US and Europe. Students choose from 6 associate degree programs within the School of Culinary Arts and the School of Business and Management. These include professional cooking, professional cooking and baking, professional baking and pastry, food and beverage management, hotel/motel/innkeeping management, and institutional food service management. The College has a campus in downtown Baltimore and at Virginia Park, an eighteenth-century estate outside of Dublin, Ireland. The College's mission is to provide qualified students with the education and experience necessary to pursue careers in the hospitality industry.

L'Academie de Cuisine

Gaithersburg, Maryland

General Information

Private, coeducational, culinary institute. Suburban campus. Founded in 1976.

Program Information

Offered since 1976. Member of American Institute of Wine & Food; Confrerie de la Chaine des Rotisseurs; International Association of Culinary Professionals; National Restaurant Association. Program calendar is year-round. 26-week Certificate in Pastry Arts. 48-week Certificate in Culinary Arts.

Areas of Study

Buffet catering; controlling costs in food service; culinary French; culinary skill development; food preparation; food purchasing; food service math; garde-manger; international cuisine; kitchen management; meal planning; meat cutting; meat fabrication; menu and facilities design; nutrition; nutrition and food service; patisserie; sanitation; saucier; seafood processing; soup, stock, sauce, and starch production; wines and spirits.

Facilities

Demonstration laboratory; 2 food production kitchens; library; student lounge.

Culinary Student Profile

122 total: 72 full-time; 50 part-time.

Faculty

7 full-time; 7 part-time. Prominent faculty: Francois Diont and Mark Ramsdell.

Special Programs

1-week culinary tour in Gascony, France.

Expenses

Application fee: $75. Tuition: $15,000 per year full-time, $8200 per 26 weeks part-time. Program-related fees include: $500 for books, knives, and uniforms.

Financial Aid

Employment placement assistance is available.

L'Academie de Cuisine *(continued)*

Application Information
Students are accepted for enrollment in January, March, July, and September. Applicants must submit a formal application and letters of reference.

Contact
Carol McClure, Assistant Director, 16006 Industrial Drive, Gaithersburg, MD 20877; 301-670-8670; Fax: 301-670-0450; E-mail: lacademe@erols.com

Berkshire Community College

Business Division

Pittsfield, Massachusetts

General Information
Public, coeducational, two-year college. Rural campus. Founded in 1960. Accredited by New England Association of Schools and Colleges.

Program Information
Program calendar is divided into semesters. 1-year Certificate in Culinary Arts. 2-year Associate degree in Hotel and Restaurant Management.

Areas of Study
Baking; beverage management; computer applications; garde-manger; kitchen management; management and human resources; nutrition; sanitation; wines and spirits; food service management; quantity foods.

Culinary Student Profile
45 total: 30 full-time; 15 part-time.

Faculty
2 full-time; 2 part-time.

Special Programs
Required 400 hours of internship/cooperative work.

Expenses
Application fee: $10. In-state tuition: $2552 per year full-time, $88 per credit part-time. Out-of-state tuition: $6032 per year full-time, $208 per credit part-time.

Financial Aid
Employment placement assistance is available. Employment opportunities within the program are available.

Application Information
Students are accepted for enrollment in January and September. Applicants must submit a formal application and have a high school diploma or GED.

Contact
Nancy Simonds-Ruderman, Professor, Business Division, 1350 West Street, Pittsfield, MA 01201-5786; 413-499-4660 Ext. 229; Fax: 413-447-7840.

Boston University

Culinary Arts

Boston, Massachusetts

General Information
Private, coeducational, university. Urban campus. Founded in 1839. Accredited by New England Association of Schools and Colleges.

Program Information
Offered since 1989. Member of American Institute of Wine & Food; International Association of Culinary Professionals; Oldways Preservation and Exchange Trust; Women Chefs and Restaurateurs. Program calendar is divided into semesters. 4-month Certificate in Culinary Arts.

Areas of Study
Baking; buffet catering; confectionery show pieces; controlling costs in food service; convenience cookery; culinary French; culinary skill development; food preparation; food purchasing; garde-manger; international cuisine; kitchen management; meal planning; meat cutting; menu and facilities design; nutrition; patisserie;

restaurant opportunities; sanitation; saucier; seafood processing; soup, stock, sauce, and starch production; wines and spirits.

Facilities
Cafeteria; catering service; 4 classrooms; coffee shop; demonstration laboratory; 8 food production kitchens; lecture room; library; public restaurant; snack shop; student lounge; teaching kitchen.

Culinary Student Profile
12 full-time. 3 are under 25 years old; 7 are between 25 and 44 years old; 2 are over 44 years old.

Faculty
2 full-time. Prominent faculty: Jacques Pejrin and Jasper White.

Expenses
Application fee: $35. Tuition: $6400 per certificate.

Financial Aid
Employment placement assistance is available. Employment opportunities within the program are available.

Application Information
Students are accepted for enrollment in January and September. Application deadline for fall is July 16. Application deadline for spring is December 16. In 1996, 50 applied; 24 were accepted. Applicants must submit a formal application and have prior food industry employment.

Contact
Mary Donlon, Assistant Director, Culinary Arts, 808 Commonwealth Avenue, Boston, MA 02215; 617-353-9852; Fax: 617-353-4130.

Bristol Community College

Culinary Arts Department

Fall River, Massachusetts

General Information
Public, coeducational, two-year college. Suburban campus. Founded in 1965. Accredited by New England Association of Schools and Colleges.

Program Information
Offered since 1985. Program calendar is divided into semesters. 1-year Certificate in Culinary Arts.

Areas of Study
Baking; controlling costs in food service; culinary skill development; food preparation; food service math; introduction to food service; patisserie; sanitation; saucier; soup, stock, sauce, and starch production.

Facilities
Bake shop; cafeteria; catering service; classroom; food production kitchen; lecture room; library; teaching kitchen.

Culinary Student Profile
19 full-time. 8 are under 25 years old; 6 are between 25 and 44 years old; 5 are over 44 years old.

Faculty
1 full-time; 3 part-time. 3 are industry professionals; 1 is a culinary-accredited teacher. Prominent faculty: John Caressimo.

Expenses
Application fee: $10. Tuition: $78 per credit hour.

Financial Aid
Employment placement assistance is available.

Application Information
Students are accepted for enrollment in September. In 1996, 43 applied; 39 were accepted. Applicants must submit a formal application.

Contact
John Caressimo, Director, Culinary Arts Department, 777 Elsbree Street, Fall River, MA 02720-7395; 508-678-2811 Ext. 2111; Fax: 508-676-7146.

Bunker Hill Community College

Culinary Arts Program

Boston, Massachusetts

General Information
Public, coeducational, two-year college. Urban campus. Founded in 1973. Accredited by New England Association of Schools and Colleges.

Bunker Hill Community College *(continued)*

Program Information

Offered since 1978. Member of American Culinary Federation; Council on Hotel, Restaurant, and Institutional Education; Educational Foundation of the NRA; National Restaurant Association; Food Service Consultants International. Program calendar is divided into semesters. 1-year Certificate in Culinary Arts. 2-year Associate degree in Culinary Arts.

Areas of Study

Baking; beverage management; buffet catering; computer applications; controlling costs in food service; convenience cookery; culinary skill development; food preparation; food purchasing; garde-manger; international cuisine; introduction to food service; kitchen management; meal planning; meat cutting; menu and facilities design; nutrition; sanitation; saucier; seafood processing; soup, stock, sauce, and starch production; wines and spirits.

Facilities

Bake shop; bakery; catering service; classroom; demonstration laboratory; food production kitchen; gourmet dining room; learning resource center; lecture room; library; public restaurant; teaching kitchen.

Culinary Student Profile

240 total: 200 full-time; 40 part-time. 160 are under 25 years old; 75 are between 25 and 44 years old; 5 are over 44 years old.

Faculty

4 full-time; 4 part-time. 2 are industry professionals; 6 are culinary-accredited teachers.

Expenses

Application fee: $15. Tuition: $1200 per semester full-time, $78 per credit part-time. Program-related fees include: $150 for knives; $90 for uniforms.

Financial Aid

In 1996, 15 scholarships were awarded (average award was $500). Employment placement assistance is available. Employment opportunities within the program are available.

Application Information

Students are accepted for enrollment in January and September. In 1996, 125 applied; 100 were accepted. Applicants must submit a formal application.

Contact

Doug Clifford, Director: Enrollment Services, Culinary Arts Program, 250 New Rutherford Avenue, Boston, MA 02129; 617-228-2417; Fax: 617-228-2082.

The Cambridge School of Culinary Arts

Professional Chef's Program

Cambridge, Massachusetts

General Information

Private, coeducational, culinary institute. Urban campus. Founded in 1974.

Program Information

Offered since 1974. Member of American Culinary Federation Educational Institute; American Institute of Wine & Food; James Beard Foundation, Inc.; National Restaurant Association; Oldways Preservation and Exchange Trust; Women Chefs and Restaurateurs. Program calendar is divided into trimesters. 37-week Diploma in Professional Chef's Program.

Areas of Study

Baking; buffet catering; confectionery show pieces; controlling costs in food service; convenience cookery; culinary French; culinary skill development; food preparation; food purchasing; food service math; garde-manger; international cuisine; introduction to food service; kitchen management; management and human resources; meal planning; meat cutting; meat fabrication; nutrition; nutrition and food service; patisserie; restaurant opportunities; sanitation; saucier; soup, stock, sauce, and starch production; wines and spirits.

Facilities
3 lecture rooms; library; student lounge; 3 teaching kitchens.

Culinary Student Profile
150 full-time.

Faculty
12 full-time; 12 part-time.

Special Programs
Culinary excursions.

Expenses
Application fee: $45. Tuition: $10,800 per diploma. Program-related fees include: $175 for books; $100 for class materials; $430 for supplies.

Financial Aid
Employment placement assistance is available. Employment opportunities within the program are available.

Application Information
Students are accepted for enrollment in January and September. Applicants must submit a formal application and a resume, a statement of purpose, and academic transcripts.

Contact
Admissions, Professional Chef's Program, 2020 Massachusetts Avenue, Cambridge, MA 02140-2124; 617-354-2020; Fax: 617-576-1963.

Endicott College

International Programme for Hospitality Studies

Beverly, Massachusetts

General Information
Private, coeducational, comprehensive institution. Suburban campus. Founded in 1939. Accredited by New England Association of Schools and Colleges.

Program Information
Offered since 1994. Member of Club Managers Association. Program calendar is divided into semesters. 4-year Bachelor's degrees in Hotel, Restaurant, and Travel Administration; International Hospitality.

Areas of Study
Beverage management; buffet catering; computer applications; controlling costs in food service; culinary French; culinary skill development; food preparation; food purchasing; food service communication; garde-manger; international cuisine; introduction to food service; kitchen management; management and human resources; meal planning; meat cutting; menu and facilities design; nutrition; patisserie; restaurant opportunities; sanitation; saucier; soup, stock, sauce, and starch production; wines and spirits.

Facilities
Bake shop; 2 cafeterias; 2 catering services; 50 classrooms; 4 computer laboratories; demonstration laboratory; food production kitchen; gourmet dining room; 2 learning resource centers; public restaurant; student lounge; teaching kitchen.

Culinary Student Profile
106 total: 100 full-time; 6 part-time. 80 are under 25 years old; 20 are between 25 and 44 years old; 6 are over 44 years old.

Faculty
6 full-time; 3 part-time. 6 are industry professionals; 2 are master chefs; 1 is a culinary-accredited teacher. Prominent faculty: Brendan Cronin.

Special Programs
14-day European wine trip, 2-semester internship, study abroad opportunities.

Expenses
Application fee: $25. Tuition: $12,970 per year full-time, $397 per credit part-time.

Financial Aid
Employment placement assistance is available.

Housing
Coed, apartment-style, and single-sex housing available.

Endicott College *(continued)*

Application Information
Students are accepted for enrollment in January and September. In 1996, 80 applied; 64 were accepted. Applicants must submit a formal application, letters of reference, an essay, and academic transcripts.

Contact
Sharon Woodbury, Director of Admissions, International Programme for Hospitality Studies, 376 Hale Street, Beverly, MA 01915-2096; 978-921-1000; Fax: 978-232-2520; E-mail: s.woodbar@endicott.edu

Essex Agricultural and Technical Institute

Culinary Arts and Food Service

Hathorne, Massachusetts

General Information
Public, coeducational, two-year college. Suburban campus. Founded in 1913. Accredited by New England Association of Schools and Colleges.

Program Information
Offered since 1965. Member of American Dietetic Association; American Institute of Baking; Institute of Food Technologists. Program calendar is divided into semesters. 12-month Certificates in Cooking; Baking. 2-year Associate degree in Baking and Cooking. 2-year Certificates in Cooking; Baking.

Areas of Study
Baking; beverage management; buffet catering; confectionery show pieces; controlling costs in food service; culinary skill development; food preparation; food purchasing; food service communication; food service math; garde-manger; international cuisine; introduction to food service; management and human resources; menu and facilities design; nutrition; sanitation; wines and spirits.

Facilities
Bake shop; cafeteria; 5 classrooms; computer laboratory; 3 demonstration laboratories; food production kitchen; 4 laboratories; learning resource center; 3 lecture rooms; library; public restaurant; student lounge; teaching kitchen.

Culinary Student Profile
72 total: 60 full-time; 12 part-time. 55 are under 25 years old; 12 are between 25 and 44 years old; 5 are over 44 years old.

Faculty
4 full-time; 2 part-time. 2 are industry professionals; 4 are culinary-accredited teachers. Prominent faculty: James Cristello.

Expenses
Tuition: $1050 per year full-time, $48 per credit hour part-time. Program-related fees include: $150 for food consumption by students; $450 for knives and program-related activities.

Financial Aid
In 1996, 10 scholarships were awarded (average award was $800); 7 loans were granted (average loan was $2625). Employment placement assistance is available.

Application Information
Students are accepted for enrollment in September. Application deadline for fall is continuous with a recommended date of March 15. In 1996, 80 applied; 79 were accepted. Applicants must submit a formal application and letters of reference and be a resident of Massachusetts.

Contact
G. Don Glazer, Jr., Director of Admissions, Culinary Arts and Food Service, 562 Maple Street, PO Box 562, Hathorne, MA 01937; 508-774-0050 Ext. 215; Fax: 508-774-6530; E-mail: aggie@eati.usal.com

Holyoke Community College

Culinary Arts Department

Holyoke, Massachusetts

General Information
Public, coeducational, two-year college. Suburban

campus. Founded in 1946. Accredited by New England Association of Schools and Colleges.

Program Information
Offered since 1991. Member of American Dietetic Association; Council on Hotel, Restaurant, and Institutional Education; National Restaurant Association. Program calendar is divided into semesters. 12-month Certificate in Culinary Arts.

Areas of Study
Baking; controlling costs in food service; culinary skill development; food preparation; food purchasing; food service math; international cuisine; management and human resources; nutrition; patisserie; restaurant opportunities; sanitation.

Facilities
Bake shop; 3 classrooms; computer laboratory; demonstration laboratory; food production kitchen; gourmet dining room; learning resource center; library; student lounge; teaching kitchen.

Culinary Student Profile
60 total: 50 full-time; 10 part-time.

Faculty
3 full-time; 2 part-time. 3 are industry professionals; 2 are master chefs.

Expenses
In-state tuition: $2320 per program full-time, $80 per credit hour part-time. Out-of-state tuition: $6670 per program full-time, $261 per credit hour part-time. Program-related fees include: $138 for knife kit and uniforms.

Financial Aid
Employment placement assistance is available.

Application Information
Students are accepted for enrollment in January and September. Application deadline for fall is September 1. Application deadline for spring is January 15. In 1996, 52 applied; 52 were accepted. Applicants must submit a formal application.

Contact
Hugh Robert, Chair, Culinary Arts Department, 303 Homestead Avenue, Holyoke, MA 01040-1099; 413-552-2229; Fax: 413-534-8975.

Newbury College

Brookline, Massachusetts

General Information
Private, coeducational, two-year college. Suburban campus. Founded in 1962. Accredited by New England Association of Schools and Colleges.

Program Information
Offered since 1983. Member of Retail Bakers of America. Program calendar is divided into semesters. 11-month Certificates in Buffet Catering; Pastry Arts. 2-year Associate degree in Culinary Arts.

Areas of Study
Baking; buffet catering; controlling costs in food service; culinary French; culinary skill development; food preparation; food purchasing; garde-manger; international cuisine; meat cutting; menu and facilities design; nutrition; patisserie; sanitation; saucier; wines and spirits.

Facilities
2 bake shops; cafeteria; computer laboratory; food production kitchen; gourmet dining room; learning resource center; library; student lounge; 3 teaching kitchens.

Culinary Student Profile
400 total: 200 full-time; 200 part-time.

Faculty
10 full-time; 8 part-time. 18 are industry professionals.

Expenses
Application fee: $30. Tuition: $11,780 per year full-time, $390 per credit part-time. Program-related fees include: $500 for knives and uniforms.

Financial Aid
In 1996, 18 scholarships were awarded (average award was $600). Employment placement assistance is available.

Housing
200 culinary students housed on campus. Coed housing available. Average on-campus housing cost per month: $775.

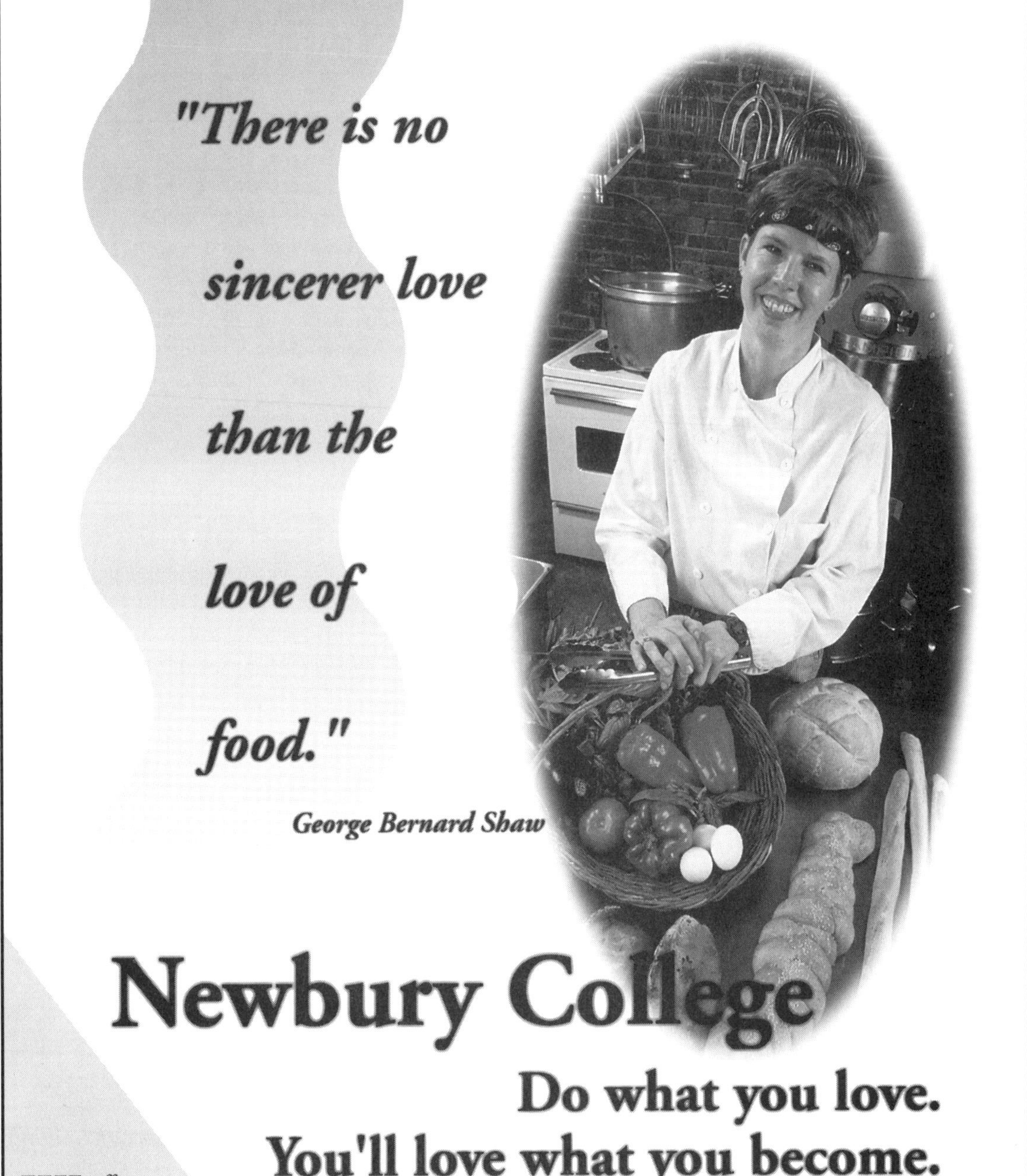

"There is no
sincerer love
than the
love of
food."
George Bernard Shaw
Newbury College
Do what you love.
You'll love what you become.
We offer Associate degree programs in Culinary Arts & Hospitality Management.
129 Fisher Avenue, Brookline, Massachusetts 02146
website: http://www.newbury.edu
e-mail: info@newbury.edu
1 (800) NEWBURY

Newbury College *(continued)*

Application Information
Students are accepted for enrollment in January and September. Applications are accepted continuously for fall and spring. Applicants must submit a formal application and letters of reference.

Contact
Admission Office, 129 Fisher Avenue, Brookline, MA 02146-5750; 617-730-7007; Fax: 617-731-9618; E-mail: info@newbury.edu; World Wide Web: http://www.newbury.edu

See display on page 136.

Charles Stewart Mott Community College

Culinary Arts Program

Flint, Michigan

General Information
Public, coeducational, two-year college. Urban campus. Founded in 1923. Accredited by North Central Association of Colleges and Schools.

Program Information
Offered since 1984. Program calendar is divided into semesters. 1-year Certificate in Baking and Pastry Arts. 2-year Associate degrees in Culinary Arts; Food Service Management.

Areas of Study
Baking; beverage management; culinary skill development; food preparation; food purchasing; food service math; garde-manger; introduction to food service; kitchen management; management and human resources; meal planning; meat cutting; menu and facilities design; nutrition; patisserie; sanitation; specialty desserts; à la carte dining.

Culinary Student Profile
97 total: 50 full-time; 47 part-time.

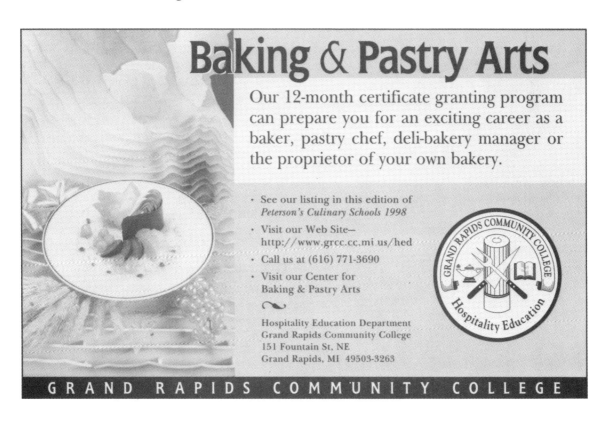

Charles Stewart Mott Community College *(continued)*

Faculty
3 full-time; 4 part-time.

Special Programs
Internships.

Expenses
In-state tuition: $56.50 per contact hour. Out-of-state tuition: $81.50 per contact hour.

Financial Aid
Employment placement assistance is available. Employment opportunities within the program are available.

Application Information
Students are accepted for enrollment in January and September. Applicants must submit a formal application.

Contact
Grace Alexander, Coordinator, Culinary Arts Program, 1401 East Court Street, Flint, MI 48503-2089; 810-232-7845; Fax: 810-232-9442.

Grand Rapids Community College

Hospitality Education Department

Grand Rapids, Michigan

General Information
Public, coeducational, two-year college. Urban campus. Founded in 1914. Accredited by North Central Association of Colleges and Schools.

Program Information
Offered since 1973. Accredited by American Culinary Federation Education Institute. Member of American Culinary Federation; American Culinary Federation Educational Institute; American Institute of Baking; Confrerie de la Chaine des Rotisseurs; Council on Hotel, Restaurant, and Institutional Education; Educational Foundation of the NRA; Interntional Food Service Executives Association; National Restaurant Association; Society of Wine Educators; Tasters Guild International; Retail Bakers of America. Program calendar is divided into semesters. 12-month Certificate in Baking and Pastry Arts. 21-month Associate degrees in Culinary Management; Culinary Arts.

Areas of Study
Baking; beverage management; buffet catering; computer applications; confectionery show pieces; controlling costs in food service; culinary skill development; food preparation; food purchasing; food service math; garde-manger; international cuisine; introduction to food service; kitchen management; management and human resources; meat fabrication; menu and facilities design; nutrition; nutrition and food service; patisserie; restaurant opportunities; sanitation; seafood processing; soup, stock, sauce, and starch production; wines and spirits; table service; deli-bakery operations.

Facilities
3 bake shops; cafeteria; catering service; 10 computer laboratories; demonstration laboratory; 8 food production kitchens; garden; 5 gourmet dining rooms; learning resource center; lecture room; 2 libraries; public restaurant; 2 snack shops; 3 student lounges; 6 banquet rooms; bakery/delicatessen.

Culinary Student Profile
305 total: 260 full-time; 45 part-time. 160 are under 25 years old; 110 are between 25 and 44 years old; 35 are over 44 years old.

Faculty
12 full-time; 4 part-time. 14 are industry professionals; 2 are master chefs. Prominent faculty: Robert Garlough and Gilles Rennsson.

Special Programs
International exchange program, international culinary study tours, California Wine Country study tours.

Expenses
Application fee: $20. In-state tuition: $5986 per degree full-time, $82 per credit hour part-time. Out-of-state tuition: $6789 per degree full-time,

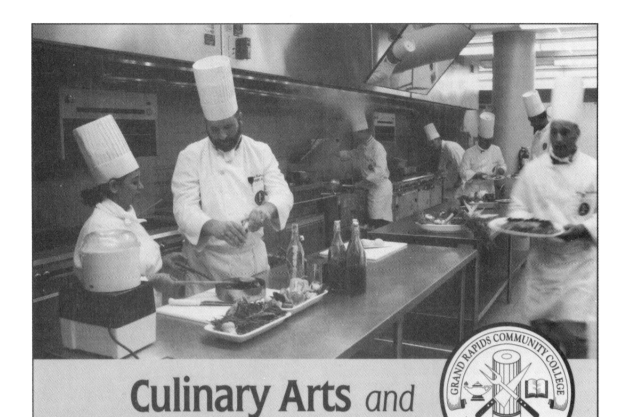

Grand Rapids Community College *(continued)*

$93 per credit hour part-time. Program-related fees include: $110 for knife kit; $250 for uniforms; $800 for textbooks.

Financial Aid
In 1996, 19 scholarships were awarded (average award was $750). Employment placement assistance is available. Employment opportunities within the program are available.

Application Information
Students are accepted for enrollment in January and August. Application deadline for fall is continuous with a recommended date of May 1. Application deadline for spring is continuous with a recommended date of October 1. Application deadline for summer is continuous with a recommended date of Febuary 1. In 1996, 170 applied; 160 were accepted. Applicants must submit a formal application.

Contact
Robert Garlough, Director, Hospitality Education Department, 151 Fountain Street, NE, Grand Rapids, MI 49503; 616-771-3690; Fax: 616-771-3698; E-mail: bgarloug@post.grcc.cc.mi.us; World Wide Web: http://www.grcc.cc.mi.us/hed

See display on page 139.

Henry Ford Community College

Culinary Arts/Hotel Restaurant Management

Dearborn, Michigan

General Information
Public, coeducational, two-year college. Suburban campus. Founded in 1938. Accredited by North Central Association of Colleges and Schools.

Program Information
Offered since 1972. Accredited by American Culinary Federation Education Institute. Program calendar is divided into semesters. 1-year Certificate in Food Management. 2-year Associate degrees in Hotel/Restaurant Management; Culinary Arts.

Areas of Study
Baking; beverage management; computer applications; confectionery show pieces; controlling costs in food service; culinary skill development; food preparation; food purchasing; food service communication; food service math; garde-manger; international cuisine; introduction to food service; kitchen management; management and human resources; meal planning; meat cutting; meat fabrication; menu and facilities design; nutrition; patisserie; sanitation; saucier; seafood processing; soup, stock, sauce, and starch production; wines and spirits.

Culinary Student Profile
300 total: 150 full-time; 150 part-time.

Faculty
5 full-time; 10 part-time.

Expenses
Application fee: $30. In-state tuition: $49 per credit. Out-of-state tuition: $87 per credit.

Application Information
Students are accepted for enrollment in January, May, July, and September. Applicants must submit a formal application and have a high school diploma or GED.

Contact
Dennis Konarski, Culinary Director, Culinary Arts/Hotel Restaurant Management, 5101 Evergreen Road, Dearborn, MI 48128-1495; 313-845-6390; Fax: 313-845-9784.

Macomb Community College

Culinary Arts/Hospitality Department

Clinton Township, Michigan

General Information
Public, coeducational, two-year college. Suburban campus. Founded in 1954. Accredited by North Central Association of Colleges and Schools.

Sullivan College Celebrates 10 Years of International Excellence

The Sullivan College Advantages

Sullivan College has earned a reputation as one of the most successful culinary schools in America by offering students a unique opportunity to prepare for a successful career in today's professional hospitality world.

Sullivan College's National Center for Hospitality Studies has been listed among the ***top-five culinary programs in the nation*** and has one of the few ***American Culinary Federation Educational Institute accredited*** Baking & Pastry Arts and Culinary Arts degree programs in the U.S. The College is regionally accredited by the Commission on Colleges of the Southern Association of Colleges and Schools to award Associate, Bachelor's and Master's degrees.

Located in Louisville, Kentucky, Sullivan provides its students with access to over 1,200 restaurants, as well as numerous other hospitality and travel related industries.

Modern, furnished apartments near the campus are available for out of town students. Each apartment has access to the laundry rooms, club house and the swimming pool. Daily transportation is provided by the College.

A Wide Range of Options

Sullivan offers students seven different career-in-a-year diploma and associate degree programs in:

- ***Professional Baker***
- ***Professional Cook***
- ***Baking & Pastry Arts***
- ***Culinary Arts***
- ***Professional Catering***
- ***Hotel/Restaurant Management***
- ***Travel & Tourism***

Sullivan will introduce a Bachelor of Science degree in Hospitality Management in the spring of 1998.

A World Class Faculty

Sullivan's hospitality division has grown to nearly 500 students from 38 states and from a number of foreign countries who choose to come to Sullivan College for their hospitality career training.

The College's hospitality staff includes ***four International Culinary Olympic Gold-Medal winners***, two chef-instructors are London Guild Master Chefs and one holds the Certified Master Pastry Chef designation. The faculty bring over two hundred years of culinary, hospitality and hotel experience to share with their students.

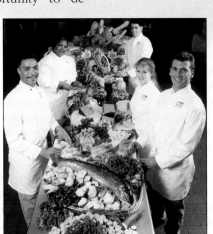

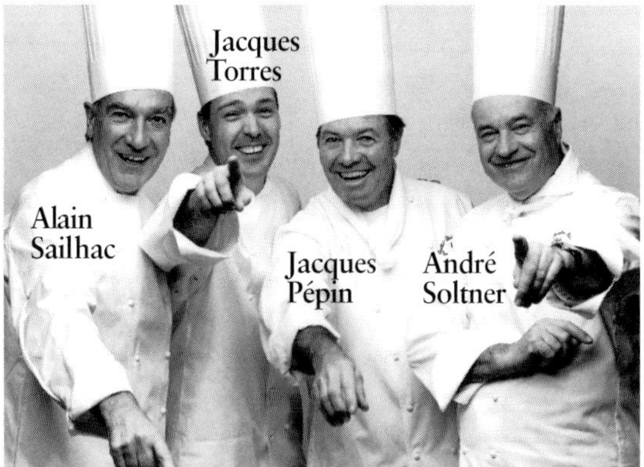

PROGRAM INFORMATION
Offered since 1969. Accredited by American Culinary Federation Education Institute. Member of American Culinary Federation; American Culinary Federation Educational Institute; American Institute of Baking; Council on Hotel, Restaurant, and Institutional Education; Educational Foundation of the NRA; National Restaurant Association; Michigan Restaurant Association. Program calendar is divided into semesters. 1-year Certificates in National Restaurant Association Diploma; Supervision; Food Production. 2-year Associate degree in Culinary Arts/Hospitality. 2-year Certificate in American Culinary Federation Certified Cook.

AREAS OF STUDY
Baking; beverage management; buffet catering; confectionery show pieces; controlling costs in food service; culinary skill development; food preparation; food purchasing; food service communication; food service math; garde-manger; international cuisine; kitchen management; management and human resources; meal planning; meat cutting; meat fabrication; menu and facilities design; nutrition; nutrition and food service; patisserie; restaurant opportunities; sanitation; saucier; seafood processing; soup, stock, sauce, and starch production; wines and spirits.

FACILITIES
Bake shop; bakery; cafeteria; 4 classrooms; coffee shop; computer laboratory; demonstration laboratory; food production kitchen; gourmet dining room; learning resource center; lecture room; library; public restaurant; snack shop; student lounge; 3 teaching kitchens.

CULINARY STUDENT PROFILE
165 total: 55 full-time; 110 part-time.

FACULTY
3 full-time; 6 part-time. 3 are industry professionals; 1 is a culinary-accredited teacher; 2 are Total Alcohol Management certified. Prominent faculty: David F. Schneider and Henry Anderson.

SPECIAL PROGRAMS
14-day tour of Paris and northern France.

EXPENSES
Application fee: $10. Tuition: $52 per credit hour. Program-related fees include: $50 for kitchen classes; $145 for cutlery.

FINANCIAL AID
In 1996, 6 scholarships were awarded (average award was $300). Employment placement assistance is available. Employment opportunities within the program are available.

APPLICATION INFORMATION
Students are accepted for enrollment in January and August. In 1996, 45 applied; 45 were accepted. Applicants must submit a formal application.

CONTACT
David F. Schneider, Department Coordinator, Culinary Arts/Hospitality Department, 44575 Garfield Road, K-124-1, Clinton Township, MI 48316; 810-286-2088; Fax: 810-286-2038.

MONROE COUNTY COMMUNITY COLLEGE

Culinary Arts Department

Monroe, Michigan

GENERAL INFORMATION
Public, coeducational, two-year college. Rural campus. Founded in 1964. Accredited by North Central Association of Colleges and Schools.

PROGRAM INFORMATION
Offered since 1981. Accredited by American Culinary Federation Education Institute. Member of American Culinary Federation; American Culinary Federation Educational Institute; Educational Foundation of the NRA; National Restaurant Association. Program calendar is divided into semesters. 2-year Associate degree in Culinary Skills and Management. 2-year Certificate in Culinary Skills and Management.

AREAS OF STUDY
Baking; beverage management; buffet catering; computer applications; controlling costs in food

Monroe County Community College *(continued)*

service; culinary skill development; food preparation; food purchasing; food service communication; food service math; garde-manger; international cuisine; introduction to food service; kitchen management; meal planning; meat cutting; menu and facilities design; nutrition; nutrition and food service; patisserie; sanitation; seafood processing; soup, stock, sauce, and starch production.

FACILITIES
Bake shop; cafeteria; catering service; classroom; computer laboratory; demonstration laboratory; food production kitchen; gourmet dining room; laboratory; learning resource center; lecture room; library; student lounge; teaching kitchen.

CULINARY STUDENT PROFILE
40 full-time.

FACULTY
2 full-time; 2 part-time. 1 is a culinary-accredited teacher; 1 is a lab technician. Prominent faculty: Kevin Thomas and Vicki Lyn LaValle.

SPECIAL PROGRAMS
2-day trip to NRA Food Show in Chicago.

EXPENSES
In-state tuition: $46 per credit hour. Out-of-state tuition: $80 per credit hour. Program-related fees include: $40 for lab fees per semester.

FINANCIAL AID
In 1996, 2 scholarships were awarded (average award was $500); 10 loans were granted (average loan was $1000). Employment placement assistance is available.

APPLICATION INFORMATION
Students are accepted for enrollment in January, May, and September. Application deadline for fall is continuous with a recommended date of May 1. In 1996, 40 applied; 20 were accepted. Applicants must submit a formal application and letters of reference and interview.

CONTACT
Kevin Thomas, Instructor, Culinary Arts Department, 1555 South Raisinville Road, Monroe, MI 48161-9047; 313-384-4150; Fax: 313-242-9711.

NORTHERN MICHIGAN UNIVERSITY

Department of Consumer and Family Studies, College of Technology and Applied Sciences

Marquette, Michigan

GENERAL INFORMATION
Public, coeducational, comprehensive institution. Rural campus. Founded in 1899. Accredited by North Central Association of Colleges and Schools.

PROGRAM INFORMATION
Offered since 1980. Member of American Culinary Federation; American Culinary Federation Educational Institute; Council on Hotel, Restaurant, and Institutional Education; Educational Foundation of the NRA; National Restaurant Association. Program calendar is divided into semesters. 1-year Certificate in Culinary Arts. 2-year Associate degree in Food Service Operations. 4-year Bachelor's degree in Restaurant and Institutional Management.

AREAS OF STUDY
Baking; beverage management; buffet catering; computer applications; controlling costs in food service; convenience cookery; culinary skill development; food preparation; food purchasing; food service math; international cuisine; introduction to food service; kitchen management; management and human resources; meal planning; menu and facilities design; nutrition and food service; restaurant opportunities; sanitation; soup, stock, sauce, and starch production; wines and spirits.

FACILITIES
Bake shop; cafeteria; catering service; 3 classrooms; computer laboratory; food production kitchen; garden; gourmet dining room; 3 laboratories; learning resource center; library; public restaurant; teaching kitchen.

CULINARY STUDENT PROFILE
103 total: 63 full-time; 40 part-time.

Faculty
3 full-time; 1 part-time. 2 are industry professionals; 1 is a culinary-accredited teacher. Prominent faculty: David Sonderschafer and Ted Bogdan.

Expenses
Tuition: $90 per credit hour.

Financial Aid
Employment placement assistance is available. Employment opportunities within the program are available.

Housing
Coed, apartment-style, and single-sex housing available.

Application Information
Students are accepted for enrollment in January and August. In 1996, 34 applied; 34 were accepted. Applicants must submit a formal application.

Contact
David Sonderschafer, Professor, Department of Consumer and Family Studies, College of Technology and Applied Sciences, 1401 Presque Isle Avenue, Marquette, MI 49855-5301; 906-227-2066; Fax: 906-227-1549.

Northwestern Michigan College

Culinary Arts Department

Traverse City, Michigan

General Information
Public, coeducational, two-year college. Suburban campus. Founded in 1951. Accredited by North Central Association of Colleges and Schools.

Program Information
Offered since 1992. Accredited by American Culinary Federation Education Institute. Member of American Culinary Federation; American Culinary Federation Educational Institute; Council on Hotel, Restaurant, and Institutional Education; Educational Foundation of the NRA; Tasters Guild International. Program calendar is divided into semesters. 2-year Associate degrees in Restaurant Management; Culinary Arts.

Areas of Study
Baking; buffet catering; computer applications; controlling costs in food service; culinary skill development; food preparation; food purchasing; food service communication; food service math; garde-manger; international cuisine; introduction to food service; kitchen management; management and human resources; meal planning; nutrition; nutrition and food service; patisserie; sanitation; soup, stock, sauce, and starch production.

Facilities
Bake shop; bakery; cafeteria; catering service; 3 classrooms; 2 computer laboratories; demonstration laboratory; food production kitchen; laboratory; 2 learning resource centers; 2 lecture rooms; library; public restaurant; student lounge; teaching kitchen; vineyard.

Culinary Student Profile
85 total: 45 full-time; 40 part-time. 25 are under 25 years old; 50 are between 25 and 44 years old; 10 are over 44 years old.

Faculty
4 full-time; 7 part-time. 9 are industry professionals; 2 are culinary-accredited teachers. Prominent faculty: Fred Laughlin and Lucy House.

Expenses
Application fee: $15. Tuition: $109 per credit hour. Program-related fees include: $275 for knives and uniforms.

Financial Aid
In 1996, 8 scholarships were awarded (average award was $750); 30 loans were granted (average loan was $1000). Employment placement assistance is available. Employment opportunities within the program are available.

Housing
20 culinary students housed on campus. Coed, apartment-style, and single-sex housing available. Average on-campus housing cost per month: $200. Average off-campus housing cost per month: $450.

Northwestern Michigan College *(continued)*

Application Information
Students are accepted for enrollment in January and August. Application deadline for fall is August 15. Application deadline for spring is December 15. In 1996, 30 applied; 30 were accepted. Applicants must submit a formal application.

Contact
Fred Laughlin, Department Chair, Culinary Arts Department, 1701 East Front Street, Traverse City, MI 49686-3061; 616-922-1197; Fax: 616-922-1134; E-mail: flaughli@nmc.edu

Oakland Community College

Hospitality/Culinary Arts

Farmington Hills, Michigan

General Information
Public, coeducational, two-year college. Suburban campus. Founded in 1964. Accredited by North Central Association of Colleges and Schools.

Program Information
Offered since 1965. Accredited by American Culinary Federation Education Institute. Member of American Culinary Federation; American Culinary Federation Educational Institute; American Institute of Baking; Council on Hotel, Restaurant, and Institutional Education; Educational Foundation of the NRA; National Restaurant Association; Tasters Guild International. Program calendar is divided into semesters. 2-year Associate degrees in Hotel Management; Restaurant Management; Culinary Arts. 3-year Certificate in Culinary Apprentice.

Areas of Study
Baking; beverage management; computer applications; confectionery show pieces; controlling costs in food service; culinary skill development; food preparation; food purchasing; garde-manger; international cuisine; management and human resources; meat cutting; meat fabrication; menu and facilities design; nutrition; patisserie; sanitation; seafood processing; soup, stock, sauce, and starch production; wines and spirits; operations management.

Facilities
2 bake shops; cafeteria; catering service; 3 classrooms; computer laboratory; demonstration laboratory; 2 food production kitchens; gourmet dining room; 4 laboratories; learning resource center; lecture room; library; public restaurant; teaching kitchen; bakery retail center.

Culinary Student Profile
150 total: 100 full-time; 50 part-time.

Faculty
9 full-time; 3 part-time. 5 are culinary-accredited teachers; 2 are food management professionals.

Expenses
Tuition: $552 per semester full-time, $46 per credit hour part-time. Program-related fees include: $330 for tool kits.

Financial Aid
In 1996, 5 scholarships were awarded (average award was $250). Employment placement assistance is available. Employment opportunities within the program are available.

Application Information
Students are accepted for enrollment in January, May, and September. Application deadline for fall is August 15. Application deadline for winter is December 15. Application deadline for spring is April 15. Applicants must submit a formal application.

Contact
Susan Baier, Department Chairperson, Hospitality/ Culinary Arts, 27055 Orchard Lake Road, Farmington Hills, MI 48334; 248-471-7786; Fax: 248-471-7553; E-mail: smbaier@occ.cc.mi.us

Schoolcraft College

Culinary Arts

Livonia, Michigan

General Information
Public, coeducational, two-year college. Suburban campus. Founded in 1961. Accredited by North Central Association of Colleges and Schools.

Program Information
Offered since 1964. Member of American Culinary Federation; Educational Foundation of the NRA; National Restaurant Association; Michigan Restaurant Association. Program calendar is divided into semesters. 1-year Certificates in Culinary Management; Culinary Arts. 2-year Associate degree in Culinary Arts.

Areas of Study
Baking; buffet catering; computer applications; confectionery show pieces; controlling costs in food service; culinary French; culinary skill development; food preparation; food purchasing; food service math; garde-manger; international cuisine; introduction to food service; kitchen management; management and human resources; meal planning; meat cutting; meat fabrication; menu and facilities design; nutrition; patisserie; restaurant opportunities; sanitation; saucier; seafood processing; soup, stock, sauce, and starch production; wines and spirits; ice carving; salon competition.

Facilities
Bake shop; bakery; 2 cafeterias; catering service; 2 computer laboratories; demonstration laboratory; 2 food production kitchens; garden; gourmet dining room; laboratory; learning resource center; lecture room; 2 libraries; public restaurant; snack shop; student lounge; 6 teaching kitchens.

Culinary Student Profile
96 total: 23 full-time; 73 part-time. 31 are under 25 years old; 48 are between 25 and 44 years old; 17 are over 44 years old.

Faculty
6 full-time; 7 part-time. 13 are industry professionals. Prominent faculty: Kevin Gawronski and Leopold Schaeli.

Special Programs
Salon competitions in Singapore and Switzerland.

Expenses
Tuition: $5025 per degree full-time, $75 per credit part-time. Program-related fees include: $1380 for lab fees.

Financial Aid
In 1996, 50 scholarships were awarded (average award was $400); 2 loans were granted (average loan was $2020). Employment placement assistance is available. Employment opportunities within the program are available.

Housing
Average off-campus housing cost per month: $580.

Application Information
Students are accepted for enrollment in January and August. Applications are accepted continuously for fall and winter. In 1996, 120 applied; 103 were accepted. Applicants must submit a formal application.

Contact
Kevin Gawronski, Culinary Manager, Culinary Arts, 18600 Haggerty Road; 313-462-4423; Fax: 313-462-4581; E-mail: kgawrons@schoolcraft.cc.mi.us

Washtenaw Community College

Culinary and Hospitality Management

Ann Arbor, Michigan

General Information
Public, coeducational, two-year college. Suburban campus. Founded in 1965. Accredited by North Central Association of Colleges and Schools.

Program Information
Offered since 1971. Member of Council on Hotel, Restaurant, and Institutional Education; Educational Foundation of the NRA; National Restaurant Association. Program calendar is divided into semesters. 1-year Certificate in Food

Washtenaw Community College *(continued)*

Production Specialty. 2-year Associate degrees in Culinary Arts Technology; Hotel/Restaurant Management.

Areas of Study

Baking; beverage management; buffet catering; computer applications; controlling costs in food service; culinary skill development; food preparation; food purchasing; food service math; garde-manger; international cuisine; introduction to food service; kitchen management; management and human resources; meal planning; menu and facilities design; nutrition and food service; patisserie; restaurant opportunities; sanitation; saucier; seafood processing; soup, stock, sauce, and starch production.

Facilities

Bake shop; cafeteria; catering service; 5 classrooms; coffee shop; 3 computer laboratories; 2 demonstration laboratories; gourmet dining room; 3 laboratories; learning resource center; 5 lecture rooms; library; public restaurant; snack shop; 5 student lounges; 3 teaching kitchens.

Culinary Student Profile

100 total: 60 full-time; 40 part-time. 85 are under 25 years old; 10 are between 25 and 44 years old; 5 are over 44 years old.

Faculty

4 full-time; 14 part-time. 8 are industry professionals; 6 are culinary-accredited teachers. Prominent faculty: Don L. Garrett and Jill Beauchamp.

Expenses

Application fee: $15. In-state tuition: $52 per credit hour full-time, $52 per credit hour part-time. Out-of-state tuition: $95 per credit hour full-time, $95 per credit hour part-time. Program-related fees include: $60 for uniforms; $60 for equipment.

Financial Aid

In 1996, 12 scholarships were awarded (average award was $500). Employment placement assistance is available.

Application Information

Students are accepted for enrollment in January, May, and September. In 1996, 150 applied; 150 were accepted. Applicants must submit a formal application.

Contact

Don L. Garrett, Department Chair, Culinary and Hospitality Management, 4800 East Huron River Drive, PO Box D-1, Ann Arbor, MI 48106; 313-973-3601; Fax: 313-677-5414; E-mail: dgarrett@orchard.washtenaw.cc.mi.us

Alexandria Technical College

Dietary Manager Program

Alexandria, Minnesota

General Information

Public, coeducational, two-year college. Rural campus. Founded in 1961. Accredited by North Central Association of Colleges and Schools.

Program Information

Offered since 1969. Member of Dietary Managers Association. Program calendar is divided into quarters. 1-quarter Certificate in ServSafe. 12-month Diploma in Dietary Manager.

Areas of Study

Computer applications; controlling costs in food service; food preparation; food purchasing; food service communication; food service math; introduction to food service; kitchen management; management and human resources; meal planning; nutrition; nutrition and food service; sanitation.

Facilities

Cafeteria; catering service; classroom; 4 computer laboratories; food production kitchen; learning resource center; 5 lecture rooms; snack shop; student lounge.

Culinary Student Profile

30 total: 24 full-time; 6 part-time.

Faculty
1 full-time; 2 part-time. 1 is a registered dietitian. Prominent faculty: Maryln Lehmkuhl.

Special Programs
Various field trips to health-care and other food service departments.

Expenses
Application fee: $35. Tuition: $2400 per 12 months. Program-related fees include: $65 for thermometer, malpractice insurance, and handouts.

Financial Aid
In 1996, 2 scholarships were awarded (average award was $150). Employment placement assistance is available. Employment opportunities within the program are available.

Housing
Average off-campus housing cost per month: $300.

Application Information
Students are accepted for enrollment in September. In 1996, 16 applied; 16 were accepted. Applicants must submit a formal application.

Contact
Admissions Office, Dietary Manager Program, 1601 Jefferson Street, Alexandria, MN 56308-3707; 320-762-0221; Fax: 320-762-4501.

Hennepin Technical College

Culinary Arts Department

Brooklyn Park, Minnesota

General Information
Public, coeducational, two-year college. Suburban campus. Founded in 1972.

Program Information
Offered since 1972. Accredited by American Culinary Federation Education Institute. Member of American Culinary Federation; American Culinary Federation Educational Institute. Program calendar is divided into semesters. 15-month Certificate in Culinary Arts.

Areas of Study
Baking; buffet catering; computer applications; confectionery show pieces; controlling costs in food service; convenience cookery; culinary French; culinary skill development; food preparation; food purchasing; food service communication; food service math; garde-manger; international cuisine; introduction to food service; kitchen management; meal planning; meat cutting; meat fabrication; menu and facilities design; nutrition; nutrition and food service; restaurant opportunities; sanitation; saucier; seafood processing; soup, stock, sauce, and starch production; wines and spirits.

Facilities
4 bake shops; cafeteria; 2 catering services; 3 classrooms; 5 computer laboratories; 3 demonstration laboratories; 2 food production kitchens; 2 gourmet dining rooms; 4 laboratories; learning resource center; 4 lecture rooms; library; 2 public restaurants; snack shop; student lounge; 4 teaching kitchens.

Culinary Student Profile
75 total: 60 full-time; 15 part-time.

Faculty
4 full-time; 3 part-time. 3 are industry professionals; 4 are culinary-accredited teachers. Prominent faculty: Carlo Castagneri and Rich Forphal.

Expenses
Tuition: $42.65 per credit. Program-related fees include: $100 for knives; $300 for books; $20 for food shows.

Financial Aid
In 1996, 4 scholarships were awarded (average award was $300); 2 loans were granted (average loan was $300). Program-specific awards include culinary arts scholarships ($400), 4 Geneva Club scholarships ($500), 3 Minneapolis ACF scholarships ($500). Employment placement assistance is available. Employment opportunities within the program are available.

Application Information
Students are accepted for enrollment in January, February, March, April, May, September, October,

Hennepin Technical College *(continued)*

November, and December. In 1996, 62 applied; 62 were accepted. Applicants must submit a formal application.

CONTACT
Diane Benson, Counselor, Culinary Arts Department, 9000 Brooklyn Boulevard, Brooklyn Park, MN 55445; 612-425-3800 Ext. 2463; Fax: 612-550-2119.

NATIONAL BAKING CENTER

Minneapolis, Minnesota

GENERAL INFORMATION
Private, coeducational, culinary institute. Urban campus. Founded in 1996.

PROGRAM INFORMATION
Offered since 1996. Member of The Bread Bakers Guild of America. Program calendar is year-round. Certificate in Baking.

AREAS OF STUDY
Baking; patisserie.

FACILITIES
3 bake shops; 2 lecture rooms; reception area.

FACULTY
3 full-time. 2 are industry professionals; 1 is a master baker. Prominent faculty: Didier Rosada and Philippe LeCorre.

SPECIAL PROGRAMS
26-week internships.

EXPENSES
Tuition: $535 per 3 days.

APPLICATION INFORMATION
Students are accepted for enrollment year-round.

CONTACT
Tom McMahon, 818 Dunwoody Boulevard, Minneapolis, MN 55403; 612-374-3303; Fax: 612-374-3332; E-mail: nbc@dunwoody.tec.mn.us

NORTHWEST TECHNICAL COLLEGE

Chef Training

Moorhead, Minnesota

GENERAL INFORMATION
Public, coeducational, two-year college. Urban campus. Founded in 1965. Accredited by North Central Association of Colleges and Schools.

PROGRAM INFORMATION
Offered since 1965. Program calendar is divided into semesters. 2-year Diploma in Chef Training.

AREAS OF STUDY
Baking; buffet catering; computer applications; controlling costs in food service; convenience cookery; food preparation; food purchasing; food service communication; garde-manger; kitchen management; meal planning; meat cutting; meat fabrication; menu and facilities design; nutrition; nutrition and food service; restaurant opportunities; seafood processing; soup, stock, sauce, and starch production; salads, sandwiches, and appetizers; production supervision.

FACILITIES
Bakery; cafeteria; catering service; 2 classrooms; 2 computer laboratories; demonstration laboratory; food production kitchen; laboratory; learning resource center; library.

CULINARY STUDENT PROFILE
50 total: 45 full-time; 5 part-time.

FACULTY
2 full-time. Prominent faculty: Kim Brewster and Colleen Kraft.

EXPENSES
Application fee: $20. Tuition: $716 per quarter full-time, $44.75 per credit part-time. Program-related fees include: $120 for knives.

FINANCIAL AID
Employment placement assistance is available.

APPLICATION INFORMATION
Students are accepted for enrollment in March, June, July, September, and December. Application

deadline for fall is continuous with a recommended date of August 1. Application deadline for winter is continuous with a recommended date of November 1. Application deadline for spring is continuous with a recommended date of Febuary 1. Applicants must submit a formal application.

Contact

Janet Hohenstein, Admissions, Chef Training, 1900 28th Avenue South, Moorhead, MN 56560; 218-299-6512; Fax: 218-236-0342; E-mail: janeth@mail.ntc.mnscu.edu; World Wide Web: http://www.ntc-online.com

South Central Technical College

Culinary Arts

North Mankato, Minnesota

General Information

Public, coeducational, two-year college. Urban campus. Founded in 1968.

Program Information

Offered since 1968. Member of American Culinary Federation; National Restaurant Association. Program calendar is divided into quarters. 18-month Diploma in Culinary Arts. 2-year Associate degree in Culinary Arts.

Areas of Study

Baking; buffet catering; computer applications; controlling costs in food service; culinary skill development; food preparation; food purchasing; food service math; garde-manger; introduction to food service; kitchen management; meal planning; meat cutting; meat fabrication; menu and facilities design; nutrition; sanitation; saucier; seafood processing; soup, stock, sauce, and starch production; ice carving; classical pastries.

Facilities

Bake shop; cafeteria; 5 catering services; 2 classrooms; 2 computer laboratories; demonstration laboratory; 2 food production kitchens; learning resource center; lecture room; library; public restaurant; snack shop; student lounge; teaching kitchen.

Culinary Student Profile

32 total: 22 full-time; 10 part-time.

Faculty

1 full-time; 1 part-time. 1 is a culinary-accredited teacher. Prominent faculty: James R. Hanson.

Expenses

Application fee: $20. Tuition: $650 per quarter. Program-related fees include: $100 for lab fees; $120 for knives; $140 for uniforms.

Financial Aid

In 1996, 4 scholarships were awarded (average award was $400); 15 loans were granted (average loan was $1200). Employment placement assistance is available. Employment opportunities within the program are available.

Housing

Average off-campus housing cost per month: $300.

Application Information

Students are accepted for enrollment in January, March, and September. Applications are accepted continuously for fall, winter, and spring. In 1996, 20 applied; 20 were accepted. Applicants must submit a formal application.

Contact

Jim Hanson, Instructor, Culinary Arts, 1920 Lee Boulevard, North Mankato, MN 56003; 507-389-7229; Fax: 507-388-9951.

Mississippi University for Women

Division of Interdisciplinary Studies

Columbus, Mississippi

General Information

Public, coeducational, comprehensive institution. Rural campus. Founded in 1884. Accredited by Southern Association of Colleges and Schools.

Mississippi University for Women *(continued)*

Program Information
Offered since 1997. Program calendar is divided into semesters. 1-year Certificate in Culinary Arts. 4-year Bachelor's degree in Culinary Arts.

Areas of Study
Baking; beverage management; buffet catering; computer applications; controlling costs in food service; culinary French; culinary skill development; food preparation; food purchasing; food service communication; food service math; garde-manger; international cuisine; introduction to food service; kitchen management; management and human resources; meal planning; meat cutting; meat fabrication; nutrition; patisserie; sanitation; saucier; seafood processing; soup, stock, sauce, and starch production; chareuterie; food for special diets.

Culinary Student Profile
46 total: 25 full-time; 21 part-time.

Faculty
2 full-time; 3 part-time.

Expenses
In-state tuition: $2284 per semester full-time, $93.50 per credit hour part-time. Out-of-state tuition: $4786 per semester full-time, $104.25 per credit hour part-time.

Housing
Single-sex housing available.

Application Information
Students are accepted for enrollment in January and September. Applicants must submit a formal application, SAT or ACT scores, and academic transcripts.

Contact
Division of Interdisciplinary Studies, Division of Interdisciplinary Studies, Box W1639, Columbus, MS 39701; 601-241-7472; Fax: 601-241-7627.

Missouri Culinary Institute

Lexington, Missouri

General Information
Private, coeducational, culinary institute. Rural campus. Founded in 1995.

Program Information
Offered since 1995. Member of American Culinary Federation; American Institute of Wine & Food; National Restaurant Association; Greater Kansas City Chef's Association. Program calendar is divided into quarters. 1-day Certificate in Specialty Cooking Classes. 1-week Certificate in Specialty Cooking Seminars. 12-week Certificate in Culinary Arts.

Areas of Study
Baking; buffet catering; controlling costs in food service; convenience cookery; culinary French; culinary skill development; food preparation; food purchasing; food service communication; food service math; garde-manger; international cuisine; introduction to food service; kitchen management; meal planning; meat cutting; menu and facilities design; nutrition; nutrition and food service; patisserie; restaurant opportunities; sanitation; saucier; soup, stock, sauce, and starch production; barbecue (smoking); barbecue rubs and sauces.

Facilities
Catering service; classroom; food production kitchen; lecture room; library; public restaurant; teaching kitchen.

Culinary Student Profile
24 total: 12 full-time; 12 part-time.

Faculty
2 full-time; 1 part-time. 2 are industry professionals; 1 is a master chef. Prominent faculty: Terry Kopp and Dorothy Kopp.

Special Programs
1-day tours of Kansas City brewing companies, coffee companies, and organic gardens.

Expenses
Application fee: $50. Tuition: $2500 per course full-time, $10 per hour part-time. Program-related fees include: $300 for books, cutlery, and uniforms; $100 for registration; $700 for lab fees.

Financial Aid
Program-specific awards include 7 scholarships of $1000 each.

Housing
Average off-campus housing cost per month: $250.

Application Information
Students are accepted for enrollment in January, February, March, April, May, June, July, August, September, October, and November. In 1996, 4 applied; 4 were accepted. Applicants must submit a formal application and letters of reference.

Contact
Terry Kopp, Chef/Instructor, 13 & 24 Highway Junction, Route 1, Box 224 F, Lexington, MO 64067; 816-259-6464.

St. Louis Community College at Forest Park

Hotel, Restaurant, and Culinary Management Department

St. Louis, Missouri

General Information
Public, coeducational, two-year college. Suburban campus. Founded in 1962. Accredited by North Central Association of Colleges and Schools.

Program Information
Offered since 1964. Program calendar is divided into semesters. 2-year Associate degree in Hospitality Studies.

Areas of Study
Baking; beverage management; buffet catering; confectionery show pieces; controlling costs in food service; culinary skill development; food preparation; food purchasing; food service math; garde-manger; international cuisine; introduction to food service; kitchen management; management and human resources; meal planning; menu and facilities design; nutrition and food service; sanitation.

Facilities
Bake shop; computer laboratory; demonstration laboratory; food production kitchen; garden; gourmet dining room; learning resource center; library; public restaurant; student lounge; teaching kitchen.

Culinary Student Profile
175 total: 75 full-time; 100 part-time. 50 are under 25 years old; 100 are between 25 and 44 years old; 25 are over 44 years old.

Faculty
4 full-time; 20 part-time. 18 are industry professionals; 2 are master chefs; 4 are culinary-accredited teachers. Prominent faculty: Kathy Schiffman and Mike Downey.

Expenses
Tuition: $42 per credit hour.

Financial Aid
In 1996, 30 scholarships were awarded (average award was $250). Employment placement assistance is available.

Application Information
Students are accepted for enrollment in January and August. Applications are accepted continuously for fall and spring. Applicants must submit a formal application.

Contact
Kathy Schiffman, Department Chair, Hotel, Restaurant, and Culinary Management Department, 560 Oakland Avenue, St. Louis, MO 63110; 314-644-9767; Fax: 314-951-9405; E-mail: Kschiffman@fpmail.stlcc.cc.mo.us

The University of Montana-Missoula

Culinary Arts Department

Missoula, Montana

General Information
Public, coeducational, university. Rural campus. Founded in 1893. Accredited by Northwest Association of Schools and Colleges.

Program Information
Accredited by American Culinary Federation Education Institute. Member of American Culinary Federation; American Culinary Federation Educational Institute; Educational Foundation of the NRA; National Restaurant

The University of Montana-Missoula *(continued)*

Association. Program calendar is divided into semesters. 2-year Associate degrees in Food and Beverage Management; Food Service Management. 9-month Certificate in Culinary Arts.

Areas of Study
Baking; beverage management; buffet catering; computer applications; confectionery show pieces; controlling costs in food service; convenience cookery; culinary skill development; food preparation; food purchasing; food service communication; food service math; garde-manger; international cuisine; introduction to food service; kitchen management; management and human resources; meal planning; meat fabrication; menu and facilities design; nutrition; nutrition and food service; patisserie; restaurant opportunities; sanitation; saucier; seafood processing; soup, stock, sauce, and starch production; wines and spirits.

Facilities
Bakery; cafeteria; catering service; 3 classrooms; coffee shop; computer laboratory; food production kitchen; garden; gourmet dining room; laboratory; learning resource center; 11 lecture rooms; library; public restaurant; snack shop; student lounge; teaching kitchen.

Culinary Student Profile
67 total: 62 full-time; 5 part-time. 46 are under 25 years old; 10 are between 25 and 44 years old; 11 are over 44 years old.

Faculty
3 full-time; 2 part-time. 3 are industry professionals. Prominent faculty: Frank L. Sonnenberg and Ross Lodahl.

Expenses
Application fee: $40. Tuition: $1114.65 per semester full-time, $802.20 per semester part-time. Program-related fees include: $435 for food supplies.

Financial Aid
In 1996, 5 scholarships were awarded (average award was $550). Employment placement assistance is available. Employment opportunities within the program are available.

Housing
7 culinary students housed on campus. Coed, apartment-style, and single-sex housing available. Average on-campus housing cost per month: $326. Average off-campus housing cost per month: $424.

Application Information
Students are accepted for enrollment in January and August. In 1996, 35 applied; 24 were accepted. Applicants must submit a formal application.

Contact
Wendy Wyatt, Associate Director of Admissions, Culinary Arts Department, 909 South Avenue West, Missoula, MT 59801; 406-243-7811; Fax: 406-243-7899; E-mail: wwyatt@selway.umt.edu

Central Community College-Hastings Campus

Hotel, Motel, and Restaurant Management

Hastings, Nebraska

General Information
Public, coeducational, two-year college. Rural campus. Founded in 1966. Accredited by North Central Association of Colleges and Schools.

Program Information
Offered since 1970. Member of Council on Hotel, Restaurant, and Institutional Education; National Restaurant Association. Program calendar is divided into semesters. 2-year Associate degrees in Culinary Arts; Restaurant Management; Hotel Management.

Areas of Study
Baking; beverage management; computer applications; controlling costs in food service; culinary skill development; food preparation; food purchasing; food service math; garde-manger; international cuisine; introduction to food service; kitchen management; management and human resources; meal planning; menu and facilities design; nutrition and food service; patisserie; sanitation; saucier; soup, stock, sauce, and starch production.

Facilities
Bake shop; cafeteria; catering service; 3 classrooms; 2 computer laboratories; food production kitchen; learning resource center; 2 lecture rooms; library; public restaurant; teaching kitchen.

Culinary Student Profile
38 total: 30 full-time; 8 part-time. 25 are under 25 years old; 9 are between 25 and 44 years old; 4 are over 44 years old.

Faculty
2 full-time; 1 part-time. 2 are industry professionals.

Expenses
Tuition: $662.30 per semester full-time, $42.60 per credit hour part-time. Program-related fees include: $60 for uniforms and technical manuals.

Financial Aid
In 1996, 4 scholarships were awarded (average award was $200). Employment placement assistance is available.

Housing
Coed and single-sex housing available.

Application Information
Students are accepted for enrollment in January, February, March, April, May, June, August, September, October, November, and December. In 1996, 35 applied; 35 were accepted. Applicants must submit a formal application.

Contact
Bob Glenn, Admissions Officer, Hotel, Motel, and Restaurant Management, PO Box 1024, Hastings, NE 68902-1024; 402-463-9811; Fax: 402-461-2454; E-mail: glehsts@cccadm.gi.cccneb.edu

Metropolitan Community College

Food Arts and Management

Omaha, Nebraska

General Information
Public, coeducational, two-year college. Urban campus. Founded in 1974. Accredited by North Central Association of Colleges and Schools.

Program Information
Offered since 1976. Accredited by American Culinary Federation Education Institute, Council on Hotel Restaurant Industry Educators. Member of American Culinary Federation; American Culinary Federation Educational Institute; Council on Hotel, Restaurant, and Institutional Education; Educational Foundation of the NRA; International Wine & Food Society; National Restaurant Association. Program calendar is divided into quarters. 2-year Associate degrees in Food Management; Food/Culinary Arts. 3-year Associate degree in Chef Apprentice.

Areas of Study
Baking; buffet catering; computer applications; controlling costs in food service; convenience cookery; culinary skill development; food preparation; food purchasing; food service communication; food service math; garde-manger; international cuisine; introduction to food service; kitchen management; management and human resources; meal planning; menu and facilities design; nutrition; nutrition and food service; restaurant opportunities; sanitation; saucier; seafood processing; soup, stock, sauce, and starch production.

Facilities
Bake shop; cafeteria; 4 classrooms; computer laboratory; food production kitchen; garden; 2 gourmet dining rooms; 2 laboratories; learning resource center; 4 lecture rooms; library; public restaurant; teaching kitchen.

Culinary Student Profile
125 total: 75 full-time; 50 part-time. 50 are under 25 years old; 70 are between 25 and 44 years old; 5 are over 44 years old.

Faculty
2 full-time; 12 part-time. 4 are culinary-accredited teachers. Prominent faculty: James E. Trebbien.

Expenses
Tuition: $27.50 per credit hour.

Financial Aid
In 1996, 4 scholarships were awarded (average award was $500). Program-specific awards include Omaha Restaurant Association scholarships.

Metropolitan Community College *(continued)*

Employment placement assistance is available. Employment opportunities within the program are available.

Application Information

Students are accepted for enrollment in March, June, September, and December. In 1996, 70 applied; 68 were accepted.

Contact

Jim Trebbien, Division Representative, Food Arts and Management, PO Box 3777, Omaha, NE 68103-0777; 402-457-2510; Fax: 402-457-2515; E-mail: jtrebbien@metropo.mccneb.edu; World Wide Web: http://www.mcc.neb.edu

Southeast Community College, Lincoln Campus

Lincoln, Nebraska

General Information

Public, coeducational, two-year college. Urban campus. Founded in 1973. Accredited by North Central Association of Colleges and Schools.

Program Information

Offered since 1973. Member of American Culinary Federation Educational Institute; American Dietetic Association; Council on Hotel, Restaurant, and Institutional Education; Educational Foundation of the NRA; National Restaurant Association. Program calendar is divided into quarters. 18-month Associate degrees in Dietetic Technology; Food Service Management; Culinary Arts.

Areas of Study

Baking; beverage management; buffet catering; computer applications; controlling costs in food service; culinary French; culinary skill development; food preparation; food purchasing; food service math; garde-manger; introduction to food service; kitchen management; management and human resources; meal planning; meat fabrication; menu and facilities design; nutrition; sanitation; saucier; soup, stock, sauce, and starch production.

Facilities

Bakery; cafeteria; catering service; 2 classrooms; computer laboratory; food production kitchen; learning resource center; library; student lounge.

Culinary Student Profile

65 total: 50 full-time; 15 part-time. 45 are under 25 years old; 18 are between 25 and 44 years old; 2 are over 44 years old.

Faculty

1 full-time; 5 part-time. 3 are industry professionals; 1 is a culinary-accredited teacher.

Special Programs

Trip to the National Restaurant Association show.

Expenses

Application fee: $25. Tuition: $26 per credit hour.

Financial Aid

In 1996, 4 scholarships were awarded (average award was $300). Employment placement assistance is available. Employment opportunities within the program are available.

Housing

Average off-campus housing cost per month: $350.

Application Information

Students are accepted for enrollment in January, March, July, and October. Application deadline for fall is continuous with a recommended date of August 15. Application deadline for winter is continuous with a recommended date of October 15. Application deadline for spring is continuous with a recommended date of January 15. In 1996, 50 applied; 48 were accepted. Applicants must submit a formal application.

Contact

Jo Taylor, Program Chair, Food Service Program, 8800 O Street, Lincoln, NE 68520-1299; 402-437-2465; Fax: 402-437-2404.

Community College of Southern Nevada

Culinary Arts Department

North Las Vegas, Nevada

General Information
Public, coeducational, two-year college. Urban campus. Founded in 1971. Accredited by Northwest Association of Schools and Colleges.

Program Information
Offered since 1989. Accredited by American Culinary Federation Education Institute. Member of American Culinary Federation; American Culinary Federation Educational Institute; Council on Hotel, Restaurant, and Institutional Education; Educational Foundation of the NRA; National Restaurant Association. Program calendar is divided into semesters. 1-year Certificate in Culinary Arts. 2-year Associate degree in Culinary Arts.

Areas of Study
Baking; beverage management; buffet catering; computer applications; controlling costs in food service; culinary French; culinary skill development; food preparation; food purchasing; food service math; garde-manger; international cuisine; introduction to food service; management and human resources; meat cutting; meat fabrication; menu and facilities design; nutrition; patisserie; sanitation; saucier; seafood processing; soup, stock, sauce, and starch production; wines and spirits.

Facilities
2 bake shops; cafeteria; catering service; 6 classrooms; coffee shop; computer laboratory; demonstration laboratory; 2 food production kitchens; garden; learning resource center; library; public restaurant; snack shop; 2 student lounges; 4 teaching kitchens.

Culinary Student Profile
845 total: 230 full-time; 615 part-time.

Faculty
5 full-time; 18 part-time. 19 are industry professionals; 4 are culinary-accredited teachers. Prominent faculty: Giovanni Joe Delrosario.

Special Programs
Culinary tour of France, Middle-East, and China. Aromatics tour of southern California.

Expenses
Application fee: $10. Tuition: $716 per semester full-time, $39 per credit part-time. Program-related fees include: $80 for knives; $70 for uniforms.

Financial Aid
In 1996, 97 scholarships were awarded (average award was $700). Employment placement assistance is available. Employment opportunities within the program are available.

Application Information
Students are accepted for enrollment in January, May, and August. In 1996, 240 applied; 240 were accepted. Applicants must submit a formal application.

Contact
Giovanni Joe Delrosario, Director, Culinary Arts Department, 3200 East Cheyene, North Las Vegas, NV 89030; 702-651-4192; Fax: 702-651-4743; E-mail: delrosar@ccsn.nevada.edu

Truckee Meadows Community College

Culinary Arts Department

Reno, Nevada

General Information
Public, coeducational, two-year college. Urban campus. Founded in 1971. Accredited by Northwest Association of Schools and Colleges.

Program Information
Offered since 1979. Member of American Culinary Federation; American Culinary Federation Educational Institute. Program calendar is divided

Truckee Meadows Community College (continued)

into semesters. 1-year Certificate in Food Service Technology. 2-year Associate degree in Food Service Technology.

Areas of Study
Buffet catering; computer applications; controlling costs in food service; culinary skill development; food preparation; food purchasing; garde-manger; introduction to food service; kitchen management; management and human resources; nutrition; nutrition and food service; sanitation; saucier; soup, stock, sauce, and starch production.

Facilities
Bake shop; bakery; cafeteria; catering service; 3 classrooms; 4 computer laboratories; food production kitchen; gourmet dining room; 8 laboratories; learning resource center; 3 lecture rooms; 2 libraries; snack shop; student lounge; teaching kitchen.

Culinary Student Profile
84 total: 24 full-time; 60 part-time. 19 are under 25 years old; 55 are between 25 and 44 years old; 10 are over 44 years old.

Faculty
2 full-time; 7 part-time. 7 are industry professionals. Prominent faculty: George Skivofilakas and Edward Chow.

Expenses
Application fee: $5. Tuition: $2000 per 2 years full-time, $38 per credit part-time.

Financial Aid
In 1996, 4 scholarships were awarded (average award was $300). Employment placement assistance is available. Employment opportunities within the program are available.

Housing
Average off-campus housing cost per month: $650.

Application Information
Students are accepted for enrollment in January and August. Application deadline for fall is August 1. Application deadline for spring is January 1. In 1996, 26 applied; 26 were accepted. Applicants must submit a formal application.

Contact
George Skirofilakas, Professor, Culinary Arts Department, 7000 Dandini Boulevard, Reno, NV 89512-3901; 702-673-7096; Fax: 702-673-7018.

University of Nevada, Las Vegas

Department of Food and Beverage Management

Las Vegas, Nevada

General Information
Public, coeducational, university. Urban campus. Founded in 1957. Accredited by Northwest Association of Schools and Colleges.

Program Information
Offered since 1967. Program calendar is divided into semesters. 4-year Bachelor's degree in Culinary Arts Management.

Areas of Study
Beverage management; buffet catering; computer applications; controlling costs in food service; food purchasing; introduction to food service; kitchen management; management and human resources; meal planning; menu and facilities design; nutrition; nutrition and food service; restaurant opportunities; sanitation; wines and spirits.

Facilities
Bake shop; catering service; 25 classrooms; computer laboratory; demonstration laboratory; food production kitchen; 3 gourmet dining rooms; laboratory; 25 lecture rooms; library; 2 teaching kitchens; bar/lounge.

Culinary Student Profile
1,725 total: 965 full-time; 760 part-time.

Faculty
38 full-time; 20 part-time. 12 are industry professionals; 2 are culinary-accredited teachers. Prominent faculty: Claude Lambertz and John Stefanelli.

Expenses
Application fee: $40. Tuition: $1536 per semester full-time, $66.50 per semester credit part-time. Program-related fees include: $500 for supplies.

Financial Aid
Employment placement assistance is available.

Housing
Coed housing available. Average off-campus housing cost per month: $400.

Application Information
Students are accepted for enrollment in January, May, and August. Applications are accepted continuously for summer. Application deadline for spring is continuous with a recommended date of December 15. Application deadline for fall is continuous with a recommended date of July 15. In 1996, 600 applied; 485 were accepted. Applicants must submit a formal application.

Contact
Rhonda Montgomery, Assistant Dean, Department of Food and Beverage Management, Harrah Hotel College; Office for Student Affairs, Las Vegas, NV 89154-6039; 702-895-3616; Fax: 702-895-3127; E-mail: hoaadviz@nevada.edu

New Hampshire College

The Culinary Institute of New Hampshire College

Manchester, New Hampshire

General Information
Private, coeducational, comprehensive institution. Suburban campus. Founded in 1932. Accredited by New England Association of Schools and Colleges.

Program Information
Offered since 1983. Accredited by American Culinary Federation Education Institute. Member of American Culinary Federation; American Culinary Federation Educational Institute. Program calendar is divided into semesters. 2-year Associate degree in Culinary Arts. 4-year Bachelor's degrees in Restaurant Management; Hotel Management; Hotel Management/Restaurant Management.

Areas of Study
Baking; beverage management; buffet catering; computer applications; confectionery show pieces; controlling costs in food service; convenience cookery; culinary skill development; food preparation; food purchasing; food service math; garde-manger; international cuisine; introduction to food service; kitchen management; management and human resources; meal planning; meat cutting; meat fabrication; menu and facilities design; nutrition and food service; restaurant opportunities; sanitation; saucier; seafood processing; soup, stock, sauce, and starch production; wines and spirits.

Facilities
2 bake shops; cafeteria; 5 classrooms; coffee shop; computer laboratory; 4 food production kitchens; gourmet dining room; learning resource center; lecture room; library; public restaurant; student lounge.

Culinary Student Profile
185 total: 150 full-time; 35 part-time. 150 are under 25 years old; 25 are between 25 and 44 years old; 10 are over 44 years old.

Faculty
6 full-time; 12 part-time.

Special Programs
Internships at various locations in the United States.

Expenses
Tuition: $11,200 per year full-time, $1200 per 3-credit course part-time. Program-related fees include: $100 for uniforms; $200 for knives.

Financial Aid
Employment placement assistance is available. Employment opportunities within the program are available.

Housing
Coed, apartment-style, and single-sex housing available. Average off-campus housing cost per month: $600.

New Hampshire College *(continued)*

Application Information
Students are accepted for enrollment in January and September. Application deadline for fall is continuous with a recommended date of March 15. Application deadline for spring is continuous with a recommended date of December 1. In 1996, 220 applied; 178 were accepted. Applicants must submit a formal application, letters of reference, an essay, and academic transcripts.

Contact
Brad Poznanski, Director of Admission, The Culinary Institute of New Hampshire College, 2500 North River Road, Manchester, NH 03106-1045; 800-642-4968; Fax: 603-645-9693; E-mail: admision@nhc.edu; World Wide Web: http://www.nhc.edu

See color display following page 140.

New Hampshire Community Technical College

Culinary Arts Department

Berlin, New Hampshire

General Information
Public, coeducational, two-year college. Rural campus. Founded in 1966. Accredited by New England Association of Schools and Colleges.

Program Information
Member of American Culinary Federation; Institute of Food Technologists; National Restaurant Association. Program calendar is divided into semesters. 1-year Certificate in Culinary Arts. 2-year Associate degree in Culinary Arts.

Areas of Study
Baking; beverage management; buffet catering; computer applications; confectionery show pieces; controlling costs in food service; culinary skill development; food preparation; food purchasing; food service communication; food service math; garde-manger; international cuisine; introduction to food service; kitchen management; management and human resources; meal planning; meat cutting; meat fabrication; menu and facilities design; nutrition; nutrition and food service; patisserie; restaurant opportunities; sanitation; saucier; seafood processing; soup, stock, sauce, and starch production; wines and spirits.

Facilities
Bake shop; bakery; cafeteria; catering service; 3 classrooms; 5 computer laboratories; 2 demonstration laboratories; 3 food production kitchens; gourmet dining room; 2 laboratories; learning resource center; library; public restaurant; student lounge; 3 teaching kitchens.

Culinary Student Profile
40 total: 35 full-time; 5 part-time.

Faculty
2 full-time; 2 part-time. 2 are industry professionals; 2 are culinary-accredited teachers. Prominent faculty: Kurt Hohmeister and Steve Griffiths.

Expenses
Application fee: $10. Tuition: $1440 per semester full-time, $110 per credit part-time. Program-related fees include: $160 for knives; $150 for uniforms; $250 for books.

Financial Aid
In 1996, 1 scholarship was awarded (award was $250). Employment placement assistance is available.

Housing
Coed and apartment-style housing available. Average off-campus housing cost per month: $150.

Application Information
Students are accepted for enrollment in January and September. In 1996, 37 applied; 28 were accepted. Applicants must submit a formal application and interview.

Contact
Kurt Hohmeister, Professor, Culinary Arts Department, 2020 Riverside Drive, Berlin, NH 03570-3717; 603-752-1113; Fax: 603-752-6335.

University of New Hampshire

Food Services Management, Thompson School of Applied Sciences

Durham, New Hampshire

General Information
Public, coeducational, university. Suburban campus. Founded in 1866. Accredited by New England Association of Schools and Colleges.

Program Information
Offered since 1964. Program calendar is divided into semesters. 2-year Associate degrees in Dietetic Technician studies; Restaurant Management.

Areas of Study
Beverage management; buffet catering; computer applications; controlling costs in food service; convenience cookery; culinary skill development; food preparation; food purchasing; food service math; introduction to food service; kitchen management; management and human resources; meal planning; meat cutting; meat fabrication; menu and facilities design; nutrition; restaurant opportunities; sanitation; wines and spirits.

Culinary Student Profile
65 total: 45 full-time; 20 part-time.

Faculty
4 full-time; 2 part-time.

Expenses
In-state tuition: $5889 per year full-time, $160 per credit hour part-time. Out-of-state tuition: $14,749 per year full-time, $176 per credit hour part-time.

Housing
Coed, apartment-style, and single-sex housing available.

Application Information
Students are accepted for enrollment in January and September. Applicants must submit a formal application and SAT scores and have a high school diploma or GED.

Contact
Emily Creighton, Admissions Officer, Food Services Management, Thompson School of Applied Sciences, Cole Hall, Mast Road, Durham, NH 03824; 603-862-1025; Fax: 603-862-2915.

Atlantic Community College

Academy of Culinary Arts

Mays Landing, New Jersey

General Information
Public, coeducational, two-year college. Rural campus. Founded in 1966. Accredited by Middle States Association of Colleges and Schools.

Program Information
Offered since 1981. Member of American Culinary Federation; American Culinary Federation Educational Institute; Council on Hotel, Restaurant, and Institutional Education; Educational Foundation of the NRA; National Restaurant Association. Program calendar is divided into semesters. 1-year Certificate in Pastry/Baking. 2-year Associate degrees in Food Service Management; Culinary Arts. 6-month Certificate in Culinary Arts and Management.

Areas of Study
Baking; beverage management; buffet catering; controlling costs in food service; culinary skill development; food preparation; food purchasing; food service math; garde-manger; international cuisine; introduction to food service; kitchen management; meal planning; meat cutting; menu and facilities design; nutrition; patisserie; sanitation; saucier; soup, stock, sauce, and starch production; wines and spirits; restaurant operations.

Facilities
2 bake shops; bakery; catering service; computer laboratory; food production kitchen; garden; gourmet dining room; learning resource center; 4 lecture rooms; library; snack shop; student lounge;

Atlantic Community College ***(continued)***

8 teaching kitchens; perishable/non-perishable storeroom; banquet room.

Culinary Student Profile
420 total: 380 full-time; 40 part-time.

Faculty
15 full-time; 5 part-time. 15 are industry professionals; 1 is a culinary-accredited teacher; 1 is a licensed dietitian.

Expenses
Application fee: $30. Tuition: $14,000 per 2 years full-time, $580 per course part-time. Program-related fees include: $300 for knives and specialty tools; $350 for textbooks; $200 for uniforms.

Financial Aid
In 1996, 50 scholarships were awarded (average award was $1140). Employment placement assistance is available. Employment opportunities within the program are available.

Housing
Average off-campus housing cost per month: $425.

Application Information
Students are accepted for enrollment in January and September. Application deadline for fall is continuous with a recommended date of July 1. Application deadline for spring is continuous with a recommended date of November 1. In 1996, 463 applied.

Contact
Linda McLeod, Admissions Assistant, Academy of Culinary Arts, 5100 Black Horse Pike, Mays Landing, NJ 08330-2699; 609-343-5009; Fax: 609-343-4921; E-mail: mcleod@nsvm.atlantic.edu; World Wide Web: http://www.atlantic.edu/

New Jersey's largest cooking school offers students hands-on training with an international faculty and a $4.6-million facility with 8 teaching kitchens, classrooms, a gourmet restaurant, a banquet room, a computer lab, and a retail store. A 2-year associate degree or a 6-month training program prepares students to enter a field with a strong job outlook. Because the Academy is a division of a public institution, students' costs are roughly half those at comparable 2-year cooking schools. Financial aid, scholarships, work-study, a payment plan, housing, and placement are offered. Campus tours or a videotape are available. Telephone: 609-343-5000, 800-654-CHEF (toll-free); e-mail: accadmit@nsvm.atlantic.edu; World Wide Web: http://www.atlantic.edu/.

Bergen Community College

Hospitality Management

Paramus, New Jersey

General Information
Public, coeducational, two-year college. Suburban campus. Founded in 1965. Accredited by Middle States Association of Colleges and Schools.

Program Information
Offered since 1974. Member of Council on Hotel, Restaurant, and Institutional Education; Educational Foundation of the NRA; National Restaurant Association. Program calendar is divided into semesters. 1-year Certificate in Culinary Arts. 2-year Associate degree in Hospitality Management.

Areas of Study
Baking; beverage management; computer applications; controlling costs in food service; food preparation; food purchasing; garde-manger; international cuisine; introduction to food service; menu and facilities design; nutrition; sanitation.

Facilities
Cafeteria; 6 classrooms; 2 food production kitchens; gourmet dining room; learning resource center; library; public restaurant; teaching kitchen.

Culinary Student Profile
160 total: 120 full-time; 40 part-time. 70 are under 25 years old; 60 are between 25 and 44 years old; 30 are over 44 years old.

Faculty
4 full-time; 3 part-time. 3 are industry professionals; 2 are master chefs; 2 are culinary-accredited teachers.

Expenses
Application fee: $40. Tuition: $1000 per semester full-time, $65 per credit hour part-time.

Financial Aid
Employment placement assistance is available. Employment opportunities within the program are available.

Application Information
Students are accepted for enrollment in January and September. In 1996, 90 applied; 90 were accepted. Applicants must submit a formal application.

Contact
David Cohen, Professor, Hospitality Management, 400 Paramus Road, Paramus, NJ 07652-1595; 201-447-7192; Fax: 201-612-5240.

Burlington County College

Food Service Management

Pemberton, New Jersey

General Information
Public, coeducational, two-year college. Suburban campus. Founded in 1966. Accredited by Middle States Association of Colleges and Schools.

Program Information
Offered since 1997. Member of Educational Foundation of the NRA. Program calendar is divided into semesters. 2-year Associate degree in Food Service Management.

Areas of Study
Controlling costs in food service; food preparation; food purchasing; menu and facilities design; nutrition and food service; sanitation; marketing for hospitality; managing quantity food service.

Facilities
Cafeteria; classroom; food production kitchen; lecture room; library.

Culinary Student Profile
25 total: 5 full-time; 20 part-time. 5 are under 25 years old; 20 are between 25 and 44 years old.

Faculty
1 full-time. 1 is a registered dietitian. Prominent faculty: Steven F. Bergonzoni.

Expenses
Application fee: $16.50. Tuition: $741 per semester full-time, $57 per credit hour part-time.

Application Information
Students are accepted for enrollment in January, May, July, and September. Application deadline for fall is September 9. Application deadline for spring is January 30. Application deadline for summer is May 26. In 1996, 15 applied; 15 were accepted. Applicants must submit a formal application.

Contact
Steven F. Bergonzoni, Director, Food Service Management, Route 530, Pemberton, NJ 08068-1599; 609-894-9311 Ext. 7780; Fax: 609-726-0442.

Hudson County Community College

Culinary Arts Institute

Jersey City, New Jersey

General Information
Public, coeducational, two-year college. Urban campus. Founded in 1974. Accredited by Middle States Association of Colleges and Schools.

Program Information
Offered since 1983. Member of American Culinary Federation; Confrerie de la Chaine des Rotisseurs; Educational Foundation of the NRA; National Restaurant Association; Les Amis d'Escoffier Society. Program calendar is divided into semesters. 1-semester Certificates in Baking; Garde

Hudson County Community College *(continued)*

Manger; Hot Food. 1-year Certificate in Culinary Arts. 2-year Associate degree in Culinary Arts.

AREAS OF STUDY
Baking; buffet catering; computer applications; confectionery show pieces; controlling costs in food service; convenience cookery; culinary French; culinary skill development; food preparation; food purchasing; garde-manger; international cuisine; introduction to food service; menu and facilities design; nutrition; patisserie; sanitation; saucier; soup, stock, sauce, and starch production; wines and spirits; table service.

FACILITIES
Bake shop; bakery; cafeteria; 3 classrooms; coffee shop; 8 computer laboratories; 2 food production kitchens; gourmet dining room; learning resource center; 2 lecture rooms; library; 2 student lounges; 4 teaching kitchens.

CULINARY STUDENT PROFILE
238 total: 181 full-time; 57 part-time. 139 are under 25 years old; 88 are between 25 and 44 years old; 11 are over 44 years old.

FACULTY
7 full-time; 6 part-time. 7 are culinary-accredited teachers.

SPECIAL PROGRAMS
Externship in Italy co-sponsored by International Italian Guild of Professional Restaurateurs.

EXPENSES
Application fee: $15. In-state tuition: $4250 per 2 years full-time, $62.50 per credit part-time. Out-of-state tuition: $187.50 per credit part-time. Program-related fees include: $1000 for lab fees per semester; $200 for knife kit; $130 for uniforms.

FINANCIAL AID
In 1996, 5 scholarships were awarded (average award was $400); 3 loans were granted (average loan was $500). Employment placement assistance is available. Employment opportunities within the program are available.

APPLICATION INFORMATION
Students are accepted for enrollment in January, May, and September. In 1996, 222 applied; 195 were accepted. Applicants must submit a formal application.

CONTACT
Siroun Meguerditchian, Culinary Arts Institute, 161 Newkirk Street, Jersey City, NJ 07306; 201-714-2193; Fax: 201-656-1522.

MIDDLESEX COUNTY COLLEGE

Hotel Restaurant and Institution Management

Edison, New Jersey

GENERAL INFORMATION
Public, coeducational, two-year college. Suburban campus. Founded in 1964. Accredited by Middle States Association of Colleges and Schools.

PROGRAM INFORMATION
Offered since 1964. Member of American Dietetic Association; Council on Hotel, Restaurant, and Institutional Education; Educational Foundation of the NRA; Institute of Food Technologists; National Association for the Specialty Food Trade, Inc.; National Restaurant Association; American Hotel and Motel Association. Program calendar is divided into semesters. 2-year Associate degree in Hotel, Restaurant, and Institution Management. 3-semester Certificate in Culinary Arts.

AREAS OF STUDY
Baking; beverage management; computer applications; controlling costs in food service; culinary skill development; food preparation; food purchasing; food service communication; food service math; garde-manger; international cuisine; introduction to food service; kitchen management; management and human resources; meal planning; menu and facilities design; nutrition; nutrition and food service; restaurant opportunities; sanitation; soup, stock, sauce, and starch production; wines and spirits; facilities layout and design.

Facilities

Bake shop; cafeteria; catering service; 2 classrooms; computer laboratory; demonstration laboratory; food production kitchen; laboratory; learning resource center; 2 lecture rooms; library; public restaurant; 3 student lounges; teaching kitchen.

Culinary Student Profile

100 total: 50 full-time; 50 part-time. 25 are under 25 years old; 50 are between 25 and 44 years old; 25 are over 44 years old.

Faculty

5 full-time; 4 part-time. 7 are industry professionals; 2 are culinary-accredited teachers.

Special Programs

Externships, cooperative work experiences, Walt Disney World College Program (culinary practicum).

Expenses

Application fee: $25. Tuition: $5080 per degree. Program-related fees include: $150 for equipment and uniforms; $200 for course fees.

Financial Aid

Employment placement assistance is available. Employment opportunities within the program are available.

Application Information

Students are accepted for enrollment in January, May, and September. Application deadline for fall is August 4. Application deadline for spring is December 31. Application deadline for summer is May 1. In 1996, 112 applied; 112 were accepted. Applicants must submit a formal application.

Contact

Marilyn Laskowski-Sachnoff, Chairman, Hotel Restaurant and Institution Management, 2600 Woodbridge Avenue, PO Box 3050, Edison, NJ 08818-3050; 732-906-2538; Fax: 732-561-1885; E-mail: mlsachnoff@aol.com

Morris County School of Technology

Culinary Arts Program

Denville, New Jersey

General Information

Public, coeducational, technical institute. Suburban campus. Founded in 1970.

Program Information

Offered since 1970. Member of American Culinary Federation; American Culinary Federation Educational Institute; Educational Foundation of the NRA; National Restaurant Association. 1-year Certificates in Hospitality Management; Food Preparation. 30-hour Certificate in ServSafe Sanitation.

Areas of Study

Baking; buffet catering; computer applications; controlling costs in food service; culinary skill development; food preparation; food purchasing; food service math; garde-manger; international cuisine; introduction to food service; kitchen management; management and human resources; menu and facilities design; restaurant opportunities; sanitation; soup, stock, sauce, and starch production; hospitality management.

Facilities

Bake shop; bakery; cafeteria; 4 classrooms; computer laboratory; 2 demonstration laboratories; 2 food production kitchens; gourmet dining room; learning resource center; 4 lecture rooms; student lounge; 2 teaching kitchens.

Culinary Student Profile

120 total: 100 full-time; 20 part-time.

Faculty

6 full-time. 2 are industry professionals; 4 are culinary-accredited teachers.

Special Programs

Cross training for supermarkets industry for take-home, chef express, and baking departments.

Morris County School of Technology ***(continued)***

Expenses

Application fee: $50. In-state tuition: $2300 per 10 months full-time, $1150 per 10 months part-time. Out-of-state tuition: $3000 per 10 months full-time, $1800 per 10 months part-time. Program-related fees include: $80 for uniforms.

Financial Aid

In 1996, 4 scholarships were awarded (average award was $100); 5 loans were granted (average loan was $2500). Employment placement assistance is available. Employment opportunities within the program are available.

Application Information

Students are accepted for enrollment in September. Application deadline for fall is continuous with a recommended date of July 31. In 1996, 120 applied; 92 were accepted. Applicants must submit a formal application and academic transcripts.

Contact

Diane Gironda, Guidance Counselor, Culinary Arts Program, 400 East Main Street, Denville, NJ 07834; 201-627-4600 Ext. 220; Fax: 201-627-6979; E-mail: girondad@mcuts.org

Santa Fe Community College

Culinary Arts/Hospitality Department

Santa Fe, New Mexico

General Information

Public, coeducational, two-year college. Rural campus. Founded in 1983. Accredited by North Central Association of Colleges and Schools.

Program Information

Offered since 1985. Member of American Culinary Federation; American Culinary Federation Educational Institute. Program calendar is divided into semesters. 1-year Certificate in Culinary Arts. 2-year Associate degree in Culinary Arts.

Areas of Study

Baking; beverage management; buffet catering; computer applications; confectionery show pieces; controlling costs in food service; convenience cookery; culinary French; culinary skill development; food preparation; food purchasing; food service communication; food service math; garde-manger; international cuisine; introduction to food service; kitchen management; management and human resources; meal planning; meat cutting; meat fabrication; menu and facilities design; nutrition; nutrition and food service; patisserie; restaurant opportunities; sanitation; saucier; seafood processing; soup, stock, sauce, and starch production; wines and spirits; Southwest cuisine.

Facilities

Cafeteria; catering service; classroom; demonstration laboratory; food production kitchen; laboratory; learning resource center; lecture room; library; public restaurant; teaching kitchen.

Culinary Student Profile

200 total: 100 full-time; 100 part-time.

Faculty

1 full-time; 8 part-time. 4 are industry professionals; 4 are culinary-accredited teachers. Prominent faculty: Bill Weiland and Maurice Zeck.

Expenses

Tuition: $250 per semester full-time, $20 per credit hour part-time. Program-related fees include: $12 for linen and cleaning supplies.

Financial Aid

In 1996, 10 scholarships were awarded (average award was $250). Employment placement assistance is available. Employment opportunities within the program are available.

Housing

Average off-campus housing cost per month: $500.

Application Information

Students are accepted for enrollment in January, May, and August. In 1996, 40 applied; 40 were accepted. Applicants must submit a formal application.

Contact
Bill Weiland, Director, Culinary Arts/Hospitality Department, PO Box 4187, 6401 Richards Avenue, Santa Fe, NM 87502-4187; 505-438-1600; Fax: 505-438-1237; E-mail: bweiland@santa-fe.cc.nm.us

Southwestern Indian Polytechnic Institute

Culinary Arts Program

Albuquerque, New Mexico

General Information
Public, coeducational, two-year college. Urban campus. Founded in 1971. Accredited by North Central Association of Colleges and Schools.

Program Information
Offered since 1971. Member of American Culinary Federation; American Culinary Federation Educational Institute; Educational Foundation of the NRA; Interntional Food Service Executives Association; National Restaurant Association. Program calendar is divided into trimesters. 3-trimester Certificate in Culinary Arts. 5-trimester Associate degree in Culinary Arts.

Areas of Study
Baking; convenience cookery; culinary skill development; food preparation; food service math; kitchen management; meal planning; nutrition; sanitation.

Facilities
Bake shop; cafeteria; 2 classrooms; computer laboratory; food production kitchen; library; snack shop.

Culinary Student Profile
25 full-time.

Faculty
1 full-time; 1 part-time. 1 is an industry professional; 1 is a culinary-accredited teacher. Prominent faculty: Danny R. Gomez.

Expenses
Program-related fees include: $250 for books and uniforms.

Financial Aid
Employment placement assistance is available. Employment opportunities within the program are available.

Housing
Coed housing available.

Application Information
Students are accepted for enrollment in January, May, and August. Applicants must submit a formal application.

Contact
Alfred Green, Department Chair, Culinary Arts Program, 9169 Coors, NW, Box 10146, Albuquerque, NM 87184-0146; 505-897-5359; Fax: 505-897-5343.

Technical Vocational Institute Community College

Culinary Arts Program

Albuquerque, New Mexico

General Information
Public, coeducational, two-year college. Urban campus. Founded in 1965. Accredited by North Central Association of Colleges and Schools.

Program Information
Offered since 1973. Accredited by American Culinary Federation Education Institute. Member of American Culinary Federation; American Culinary Federation Educational Institute; American Institute of Baking; New Mexico Restaurant Association. Program calendar is divided into trimesters. 2-year Associate degree in Culinary Arts. 4-month Certificate in Food Service Management. 8-month Certificates in Baking; Quantity Food Preparation.

Areas of Study
Baking; beverage management; buffet catering; computer applications; controlling costs in food service; culinary skill development; food

Technical Vocational Institute Community College *(continued)*

preparation; food purchasing; food service math; garde-manger; international cuisine; introduction to food service; kitchen management; meal planning; meat cutting; meat fabrication; nutrition; sanitation; saucier; seafood processing; soup, stock, sauce, and starch production.

Facilities
Bake shop; bakery; 2 cafeterias; catering service; 10 classrooms; computer laboratory; demonstration laboratory; 2 food production kitchens; garden; gourmet dining room; 3 laboratories; learning resource center; 2 libraries; 3 public restaurants; 3 student lounges; 2 teaching kitchens.

Culinary Student Profile
210 total: 160 full-time; 50 part-time.

Faculty
5 full-time; 2 part-time. Prominent faculty: Carmine J. Russo.

Special Programs
Cooperative education, field trips.

Expenses
Tuition: $966 per trimester full-time, $80.50 per credit part-time.

Financial Aid
Program-specific awards include tuition waiver for New Mexico residents. Employment placement assistance is available. Employment opportunities within the program are available.

Application Information
Students are accepted for enrollment in January, May, and September. Applicants must submit a formal application.

Contact
Admissions Office, Culinary Arts Program, Student Services Building, 900 University Boulevard, SE, Albuquerque, NM 87106; 505-224-3194.

Adirondack Community College

Commercial Cooking/Occupational Education

Queensbury, New York

General Information
Public, coeducational, two-year college. Urban campus. Founded in 1960. Accredited by Middle States Association of Colleges and Schools.

Program Information
Program calendar is divided into semesters. 1-year Certificate in Commercial Cooking. 2-year Associate degree in Food Service.

Areas of Study
Baking; computer applications; culinary skill development; food preparation; food service math; meat fabrication; nutrition; sanitation; saucier; seafood processing; soup, stock, sauce, and starch production; wines and spirits; hospitality law; American regional cuisine.

Facilities
Bake shop; cafeteria; catering service; 3 classrooms; 2 coffee shops; computer laboratory; demonstration laboratory; food production kitchen; gourmet dining room; 2 learning resource centers; library; public restaurant; 6 student lounges.

Culinary Student Profile
48 total: 25 full-time; 23 part-time.

Faculty
1 full-time; 4 part-time. 4 are industry professionals; 1 is a culinary-accredited teacher.

Special Programs
Articulation agreement with Johnson & Wales University, tours of coffee, brewery, ice cream, and cheese production sites.

Expenses
Application fee: $50. Tuition: $75 per credit hour. Program-related fees include: $250 for knives; $130 for uniforms.

Financial Aid
In 1996, 15 scholarships were awarded (average award was $1000). Employment placement assistance is available. Employment opportunities within the program are available.

Application Information
Students are accepted for enrollment in January and September. In 1996, 30 applied; 30 were accepted. Applicants must submit a formal application.

Contact
Lee Brown, Admissions, Commercial Cooking/Occupational Education, 640 Bay Road, Queensbury, NY 12804; 518-743-2200; Fax: 518-743-2264.

Culinary Institute of America

Hyde Park, New York

General Information
Private, coeducational, culinary institute. Rural campus. Founded in 1946.

Program Information
Offered since 1946. Member of American Culinary Federation; American Dietetic Association; American Institute of Baking; American Institute of Wine & Food; Confrerie de la Chaine des Rotisseurs; Council on Hotel, Restaurant, and Institutional Education; International Association of Culinary Professionals; International Foodservice Editorial Council; James Beard Foundation, Inc.; National Association for the Specialty Food Trade, Inc.; Oldways Preservation and Exchange Trust; Society of Wine Educators; Sommelier Society of America; The Bread Bakers Guild of America; Women Chefs and Restaurateurs. Program calendar is year-round. 21-month Associate degrees in Baking and Pastry Arts; Culinary Arts. 30-week Certificate in Baking and Pastry Arts. 38-month Bachelor's degrees in Baking and Pastry Arts Management; Culinary Arts Management.

Areas of Study
Baking; beverage management; buffet catering; computer applications; confectionery show pieces; controlling costs in food service; culinary French; culinary skill development; food preparation; food purchasing; food service communication; food service math; garde-manger; international cuisine; introduction to food service; kitchen management; management and human resources; meal planning; meat cutting; meat fabrication; menu and facilities design; nutrition; patisserie; restaurant opportunities; sanitation; saucier; seafood processing; soup, stock, sauce, and starch production; wines and spirits; charcuterie.

Facilities
7 bake shops; 3 cafeterias; 24 classrooms; 5 computer laboratories; 2 demonstration laboratories; 2 food production kitchens; 4 gourmet dining rooms; laboratory; learning resource center; 4 lecture rooms; library; 4 public restaurants; snack shop; 2 student lounges; 36 teaching kitchens; amphitheatre.

Culinary Student Profile
2,019 full-time.

Faculty
105 full-time; 3 part-time. 52 are industry professionals; 13 are master chefs; 43 are culinary-accredited teachers. Prominent faculty: Markus Farbinger and Dan Budd.

Special Programs
4-week tour of wine country and restaurants in California.

Expenses
Application fee: $30. Tuition: $6995 per semester full-time, $4900 per semester part-time. Program-related fees include: $820 for tool kit, textbooks, uniforms, and accident insurance; $140 for practical exams; $345 for externship.

Financial Aid
In 1996, 1,002 scholarships were awarded (average award was $1700); 1,743 loans were granted (average loan was $7000). Employment placement assistance is available. Employment opportunities within the program are available.

Culinary Institute of America *(continued)*

Housing
1,200 culinary students housed on campus. Coed housing available. Average on-campus housing cost per month: $400.

Application Information
Students are accepted for enrollment year-round. In 1996, 2,025 applied; 1,573 were accepted. Applicants must submit a formal application, letters of reference, and academic transcripts.

Contact
Susan Weatherly, Director of Enrollment Services, 433 Albany Post Road, Hyde Park, NY 12538-1499; 800-CULINARY; World Wide Web: http://www.ciachef.edu

See affiliated program in St. Helena, California.

See display on page 168.

Erie Community College, City Campus

Hotel Technology/Culinary Arts

Buffalo, New York

General Information
Public, coeducational, two-year college. Urban campus. Founded in 1946. Accredited by Middle States Association of Colleges and Schools.

Program Information
Offered since 1984. Member of American Culinary Federation; Council on Hotel, Restaurant, and Institutional Education; Educational Foundation of the NRA; National Restaurant Association. Program calendar is divided into semesters. 2-year Associate degree in Hotel/Culinary Arts.

Areas of Study
Baking; beverage management; buffet catering; computer applications; controlling costs in food service; culinary skill development; food preparation; food purchasing; food service math; garde-manger; international cuisine; introduction to food service; management and human resources; meal planning; menu and facilities design; nutrition; sanitation; soup, stock, sauce, and starch production; wines and spirits.

Facilities
Bake shop; cafeteria; 3 classrooms; computer laboratory; demonstration laboratory; food production kitchen; gourmet dining room; learning resource center; library; teaching kitchen.

Culinary Student Profile
65 total: 50 full-time; 15 part-time.

Faculty
5 full-time; 3 part-time. 2 are industry professionals; 5 are culinary-accredited teachers.

Expenses
Application fee: $50. Tuition: $2500 per year full-time, $80 per credit hour part-time. Program-related fees include: $150 for knives; $150 for books.

Financial Aid
Employment placement assistance is available. Employment opportunities within the program are available.

Application Information
Students are accepted for enrollment in January and September. In 1996, 70 applied; 50 were accepted. Applicants must submit a formal application.

Contact
Paul J. Cannamela, Department Chair, Hotel Technology/Culinary Arts, 121 Ellicott Street, Buffalo, NY 14203-2601; 716-851-1035; Fax: 716-851-1129.

The French Culinary Institute

Classic Culinary Arts

New York, New York

General Information
Private, coeducational, culinary institute. Urban campus. Founded in 1984.

The French Culinary Institute *(continued)*

Program Information
Offered since 1984. Member of American Institute of Baking; American Institute of Wine & Food; Council on Hotel, Restaurant, and Institutional Education; Educational Foundation of the NRA; International Association of Culinary Professionals; James Beard Foundation, Inc.; National Restaurant Association; Women Chefs and Restaurateurs. Program calendar is divided into 6-week cycles. 6-month Certificates in Classic Culinary Arts; Classic Pastry Arts.

Areas of Study
Culinary French; culinary skill development; patisserie.

Facilities
6 classrooms; computer laboratory; demonstration laboratory; gourmet dining room; lecture room; library; public restaurant.

Culinary Student Profile
575 total: 300 full-time; 275 part-time.

Faculty
15 full-time; 3 part-time. 18 are industry professionals. Prominent faculty: Jacques Pépin and Andre Sóltner.

Expenses
Application fee: $100. Tuition: $18,900 per 6 months full-time, $18,900 per 9 months part-time. Program-related fees include: $650 for knife set, tool kit, books, and uniforms.

Financial Aid
In 1996, 1 scholarship was awarded (award was $5000); 47 loans were granted (average loan was $6847). Employment placement assistance is available. Employment opportunities within the program are available.

Application Information
Students are accepted for enrollment year-round. In 1996, 558 applied; 458 were accepted. Applicants must submit a formal application, letters of reference, and a portfolio.

Contact
Steven Chinni, Enrollment Manager, Classic Culinary Arts, 462 Broadway, New York, NY 10013; 888-FCI-CHEF; Fax: 212-431-3054; World Wide Web: http://www.frenchculinary.com

See color display following page 140.

Jefferson Community College

Hospitality and Tourism Department

Watertown, New York

General Information
Public, coeducational, two-year college. Urban campus. Founded in 1961. Accredited by Middle States Association of Colleges and Schools.

Program Information
Offered since 1973. Member of Council on Hotel, Restaurant, and Institutional Education. Program calendar is divided into semesters. 2-year Associate degree in Hospitality and Tourism.

Areas of Study
Buffet catering; computer applications; controlling costs in food service; culinary skill development; food preparation; food purchasing; food service math; garde-manger; international cuisine; introduction to food service; kitchen management; management and human resources; meat cutting; meat fabrication; menu and facilities design; nutrition; nutrition and food service; sanitation; saucier; seafood processing; soup, stock, sauce, and starch production; wines and spirits.

Facilities
Cafeteria; catering service; food production kitchen; gourmet dining room; learning resource center; library; student lounge; teaching kitchen.

Culinary Student Profile
85 total: 70 full-time; 15 part-time.

Faculty
3 full-time; 3 part-time. 3 are industry professionals; 1 is a culinary-accredited teacher.

Expenses
Application fee: $30. Tuition: $1303 per semester full-time, $88 per credit hour part-time.

Financial Aid
In 1996, 3 scholarships were awarded (average award was $750); 42 loans were granted (average loan was $2300). Employment placement assistance is available.

Application Information
Students are accepted for enrollment in January and August. Applicants must submit a formal application and academic transcripts.

Contact
Rosanne Weir, Director of Admissions, Hospitality and Tourism Department, Outer Coffeen Street, Watertown, NY 13601; 315-786-2277; Fax: 315-786-2459; E-mail: admissions@ccmgate.sunyjefferson.edu

Julie Sahni's School of Indian Cooking

Brooklyn, New York

General Information
Private, coeducational, culinary institute. Urban campus. Founded in 1974.

Program Information
Offered since 1974. Member of International Association of Culinary Professionals. Program calendar is divided into weekends. 3-day Diplomas in Spices and Herbs; Indian Cooking.

Areas of Study
International cuisine; Indian cooking; spices and herbs.

Culinary Student Profile
3 full-time.

Faculty
1 full-time. Prominent faculty: Julie Sahni.

Expenses
Application fee: $35. Tuition: $985 per diploma.

Application Information
Students are accepted for enrollment in January, February, March, April, May, June, September, October, November, and December. Applicants must submit a formal application.

Contact
Julie Sahni, President, 101 Clark Street, Brooklyn, NY 11201-2746; 718-625-3958; Fax: 718-625-3456; E-mail: jshani@worldnet.att.net

Mohawk Valley Community College

Hospitality Programs

Utica, New York

General Information
Public, coeducational, two-year college. Urban campus. Founded in 1946. Accredited by Middle States Association of Colleges and Schools.

Program Information
Offered since 1980. Member of American Culinary Federation; American Culinary Federation Educational Institute; Council on Hotel, Restaurant, and Institutional Education; Educational Foundation of the NRA; National Restaurant Association. Program calendar is divided into semesters. 1-year Certificate in Chef Training. 2-year Associate degrees in Food Service; Nutrition and Dietetics; Food Service Administration: Restaurant Management.

Areas of Study
Baking; beverage management; buffet catering; computer applications; controlling costs in food service; culinary skill development; food preparation; food purchasing; garde-manger; international cuisine; management and human resources; menu and facilities design; nutrition; patisserie; sanitation; saucier; soup, stock, sauce, and starch production.

Facilities
Bake shop; catering service; 6 classrooms; computer laboratory; demonstration laboratory; 2 food production kitchens; 2 gourmet dining rooms; learning resource center; library; public restaurant; snack shop; teaching kitchen.

Culinary Student Profile
120 total: 100 full-time; 20 part-time. 16 are under

Mohawk Valley Community College ***(continued)***

25 years old; 96 are between 25 and 44 years old; 8 are over 44 years old.

Faculty

3 full-time; 4 part-time. 2 are industry professionals; 1 is a culinary-accredited teacher. Prominent faculty: Dennis R. Baumeyer.

Special Programs

Annual participation in National Restaurant Association show.

Expenses

Application fee: $30. In-state tuition: $2600 per year full-time, $90 per credit hour part-time. Out-of-state tuition: $5200 per year full-time, $180 per credit hour part-time.

Financial Aid

Employment placement assistance is available. Employment opportunities within the program are available.

Housing

Average off-campus housing cost per month: $200.

Application Information

Students are accepted for enrollment in January, May, and August. Application deadline for fall is continuous with a recommended date of August 15. Application deadline for spring is continuous with a recommended date of January 15. Application deadline for summer is continuous with a recommended date of May 15. In 1996, 50 applied; 45 were accepted. Applicants must submit a formal application.

Contact

Dennis Kennelty, Assistant Dean for Enrollment Management, Hospitality Programs, 1101 Sherman Drive, Utica, NY 13501-5394; 315-792-5354; Fax: 315-792-5527; E-mail: dkennelty@mvcc.edu

Monroe Community College

Department of Food, Hotel, and Tourism Management

Rochester, New York

General Information
Public, coeducational, two-year college. Suburban campus. Founded in 1961. Accredited by Middle States Association of Colleges and Schools.

Program Information
Offered since 1967. Member of American Culinary Federation; American Culinary Federation Educational Institute; American Dietetic Association; Council on Hotel, Restaurant, and Institutional Education; Interntional Food Service Executives Association; National Restaurant Association; New York Restaurant Association. Program calendar is divided into semesters. 2-year Associate degrees in Travel and Tourism Manager; Hotel Technology; Food Service Administration.

Areas of Study
Beverage management; buffet catering; computer applications; controlling costs in food service; culinary French; culinary skill development; food preparation; food purchasing; introduction to food service; management and human resources; meal planning; menu and facilities design; nutrition; nutrition and food service; restaurant opportunities; sanitation; soup, stock, sauce, and starch production.

Facilities
2 cafeterias; 2 catering services; coffee shop; 12 computer laboratories; 4 demonstration laboratories; 3 food production kitchens; gourmet dining room; 10 learning resource centers; 2 libraries; public restaurant; snack shop; 2 student lounges; 3 teaching kitchens.

Culinary Student Profile
160 total: 128 full-time; 32 part-time.

Faculty
5 full-time; 10 part-time. 5 are industry professionals; 2 are culinary-accredited teachers. Prominent faculty: Eddy F. Callens.

Special Programs
Visits to various culinary conventions.

Expenses
Application fee: $20. Tuition: $250 per year full-time, $105 per credit hour part-time. Program-related fees include: $50 for lab fees.

Financial Aid
In 1996, 15 scholarships were awarded (average award was $1000). Employment placement assistance is available. Employment opportunities within the program are available.

Application Information
Students are accepted for enrollment in January and September. In 1996, 220 applied. Applicants must submit a formal application.

Contact
Eddy F. Callens, Chairperson, Department of Food, Hotel, and Tourism Management, 1000 East Henrietta Road, Rochester, NY 14623-5780; 716-292-2586; Fax: 716-427-2749.

New School for Social Research

New School Culinary Arts

New York, New York

General Information
Private, coeducational, university. Urban campus. Founded in 1919. Accredited by Middle States Association of Colleges and Schools.

Program Information
Offered since 1979. Member of American Institute of Wine & Food; James Beard Foundation, Inc.; National Restaurant Association. Program calendar is divided into trimesters. 1-week Certificate in Cake Decorating (Master Class). 2-week Certificates in Bread Baking (Master Class); Italian Cooking (Master Class); Catering (Master Class). 3-week Certificate in Baking (Master Class). 5-week Certificate in Cooking (Master Class).

New School for Social Research *(continued)*

Areas of Study

Baking; buffet catering; controlling costs in food service; convenience cookery; culinary skill development; food preparation; international cuisine; kitchen management; meal planning; meat cutting; menu and facilities design; nutrition; patisserie; restaurant opportunities; wines and spirits.

Facilities

Bake shop; 10 classrooms; garden; gourmet dining room; 3 lecture rooms; library; teaching kitchen.

Culinary Student Profile

330 total: 180 full-time; 150 part-time.

Faculty

60 part-time. 40 are industry professionals; 20 are culinary-accredited teachers. Prominent faculty: Karen Snyder Kadish and Richard Glavin.

Special Programs

One-session: Behind the Scenes at the Great Restaurants of New York.

Expenses

Tuition: $2275 per 5 weeks full-time, $450 per week part-time. Program-related fees include: $85 for materials per week.

Financial Aid

In 1996, 3 scholarships were awarded (average award was $2000). Employment placement assistance is available.

Application Information

Students are accepted for enrollment year-round. Applicants must interview.

Contact

Gary A. Goldberg, Executive Director, New School Culinary Arts, 100 Greenwich Avenue, New York, NY 10011-8603; 212-255-4141; Fax: 212-807-0406; E-mail: admissions@newschool.edu

New York Institute of Technology

Culinary Arts Center

Central Islip, New York

General Information

Private, coeducational, comprehensive institution. Rural campus. Founded in 1955. Accredited by Middle States Association of Colleges and Schools.

Program Information

Offered since 1989. Member of Council on Hotel, Restaurant, and Institutional Education; National Restaurant Association. Program calendar is divided into semesters. 3-week Certificates in Baking and Pastry Arts; Health Professionals in Culinary Arts. 4-semester Associate degree in Culinary Arts. 6-month Certificates in Culinary Arts; Pastry Arts.

Areas of Study

Baking; beverage management; buffet catering; computer applications; confectionery show pieces; controlling costs in food service; convenience cookery; culinary skill development; food preparation; food purchasing; food service math; garde-manger; international cuisine; kitchen management; menu and facilities design; nutrition; patisserie; wines and spirits.

Facilities

Bakery; cafeteria; catering service; 10 classrooms; coffee shop; 3 computer laboratories; food production kitchen; garden; learning resource center; library; public restaurant; snack shop; student lounge; 3 teaching kitchens.

Culinary Student Profile

179 total: 164 full-time; 15 part-time.

Faculty

9 full-time; 6 part-time. 1 is a culinary-accredited teacher.

Special Programs

Summer program in Switzerland.

Expenses
Application fee: $40. Tuition: $5090 per semester full-time, $325 per credit part-time. Program-related fees include: $575 for equipment fees; $60 for uniforms; $475 for comprehensive food fees.

Financial Aid
In 1996, 3 scholarships were awarded (average award was $1000). Program-specific awards include Whitson's scholarship ($500), Timothy Ryan scholarship ($5000–$6000), J. King scholarship ($10,000).

Housing
Coed housing available. Average off-campus housing cost per month: $600.

Application Information
Students are accepted for enrollment in January and September. In 1996, 198 applied; 100 were accepted. Applicants must submit a formal application and have a high school diploma.

Contact
Admissions Office, Culinary Arts Center, PO Box 8000, Old Westbury, NY 11568-8000; World Wide Web: http://www.nyit.edu/culinary

New York Restaurant School

Culinary Arts and Restaurant Management

New York, New York

General Information
Private, coeducational, culinary institute. Urban campus. Founded in 1980.

Program Information
Offered since 1980. Member of American Culinary Federation Educational Institute; American Institute of Wine & Food; Educational Foundation of the NRA; International Association of Culinary Professionals; James Beard Foundation, Inc.; National Restaurant Association; Women Chefs and Restaurateurs; New York State Restaurant Association. 12-month Certificates in Culinary Skills; Pastry Arts; Restaurant Management. 18-month Associate degree in Culinary Arts and Restaurant Management. 9-month Certificate in Culinary Arts.

Areas of Study
Baking; beverage management; controlling costs in food service; culinary skill development; food preparation; food purchasing; food service communication; food service math; garde-manger; international cuisine; introduction to food service; kitchen management; management and human resources; meal planning; meat fabrication; menu and facilities design; nutrition; nutrition and food service; patisserie; sanitation; saucier; soup, stock, sauce, and starch production; wines and spirits.

Facilities
Bake shop; 5 classrooms; computer laboratory; 7 food production kitchens; learning resource center; library; student lounge; 7 teaching kitchens.

Culinary Student Profile
1,170 total: 800 full-time; 370 part-time. 351 are under 25 years old; 702 are between 25 and 44 years old; 117 are over 44 years old.

Faculty
30 full-time.

Special Programs
360-hour externship in the area of the student's interest.

Expenses
Application fee: $40. Tuition: $10,000–$25,000, depending on program.

Financial Aid
In 1996, 1 scholarship was awarded (award was $14,000). Employment placement assistance is available. Employment opportunities within the program are available.

Housing
Average off-campus housing cost per month: $650.

Application Information
Students are accepted for enrollment year-round. In 1996, 1,300 applied; 1,170 were accepted.

New York Restaurant School *(continued)*

Applicants must submit a formal application and letters of reference and take an entrance exam.

Contact

Director of Admissions, Culinary Arts and Restaurant Management, 75 Varick Street, 16th Floor, New York, NY 10013; 212-226-5500; Fax: 212-226-5644.

See affiliated programs: Colorado Institute of Art; The Art Institute of Atlanta; The Art Institute of Fort Lauderdale; The Art Institute of Houston; The Art Institute of Philadelphia; The Art Institute of Phoenix; The Art Institute of Seattle.

See display on page 48.

New York University

Department of Nutrition and Food Studies

New York, New York

General Information

Private, coeducational, university. Urban campus. Founded in 1831. Accredited by Middle States Association of Colleges and Schools.

Program Information

Member of National Restaurant Association. Program calendar is divided into semesters. 1-year Master's degrees in Food Studies; Food Management. 3-year Doctoral degree in Food and Food Management. 4-year Bachelor's degrees in Food Studies; Food and Restaurant Management.

Areas of Study
Beverage management; computer applications; controlling costs in food service; food preparation; food purchasing; food service communication; international cuisine; introduction to food service; management and human resources; menu and facilities design; nutrition; nutrition and food service; restaurant opportunities; sanitation; food history and culture; food writing.

Facilities
Classroom; computer laboratory; demonstration laboratory; lecture room; library; teaching kitchen.

Culinary Student Profile
350 total: 150 full-time; 200 part-time.

Faculty
8 full-time; 50 part-time. 20 are industry professionals; 5 are culinary-accredited teachers. Prominent faculty: Marion Nestle and Amy Bentley.

Expenses
Tuition: $10,865 per semester full-time, $632 per credit part-time.

Financial Aid
Employment placement assistance is available. Employment opportunities within the program are available.

Housing
Coed, apartment-style, and single-sex housing available. Average on-campus housing cost per month: $800.

Application Information
Students are accepted for enrollment in January, May, June, July, and September. Application deadline for fall is January 15. Applicants must submit a formal application.

Contact
Amy Bentley, Associate Professor, Department of Nutrition and Food Studies, 35 West 4th Street, 10th Floor, New York, NY 10012-1172; 212-998-5591; E-mail: ab51@155.nyu.edu; World Wide Web: http://www.nyu.edu/education/

Onondaga Community College

Culinary Arts Department

Syracuse, New York

General Information
Public, coeducational, two-year college. Suburban campus. Founded in 1962. Accredited by Middle States Association of Colleges and Schools.

Program Information
Program calendar is divided into semesters. 1-year Certificate in Professional Cooking. 2-year Associate degrees in Hotel Technology; Food Service Administration/Restaurant Management.

Areas of Study
Buffet catering; controlling costs in food service; food preparation; food purchasing; management and human resources; meal planning; nutrition; nutrition and food service; sanitation.

Facilities
Catering service; 2 food production kitchens; learning resource center; library; 2 teaching kitchens.

Culinary Student Profile
60 total: 50 full-time; 10 part-time. 46 are under 25 years old; 12 are between 25 and 44 years old; 2 are over 44 years old.

Faculty
3 full-time; 2 part-time. 2 are industry professionals. Prominent faculty: James P. Drake and Jillann Neely.

Special Programs
2-day trip to New York City; 5-day tour of restaurants, hotels, and casinos in Las Vegas; Walt Disney World College Program (internship).

Expenses
Application fee: $30. Tuition: $2500 per year full-time, $98 per credit hour part-time. Program-related fees include: $12 for lab fees.

Financial Aid
Employment placement assistance is available.

Onondaga Community College *(continued)*

Application Information
Students are accepted for enrollment in January and August. Application deadline for fall is August 8. Application deadline for spring is January 8. In 1996, 30 applied; 30 were accepted. Applicants must submit a formal application.

Contact
Jim Drake, Professer, Culinary Arts Department, 4941 Onondago Road, Syracuse, NY 13215; 315-469-2231; Fax: 315-464-6775; E-mail: drakej@goliath.sunyocc.edu

Paul Smith's College of Arts and Sciences

Culinary Arts and Culinary Baking Concentrations

Paul Smiths, New York

General Information
Private, coeducational, two-year college. Rural campus. Founded in 1937. Accredited by Middle States Association of Colleges and Schools.

Program Information
Offered since 1980. Accredited by American Culinary Federation Education Institute. Member of American Culinary Federation; American Culinary Federation Educational Institute; Council on Hotel, Restaurant, and Institutional Education; Educational Foundation of the NRA; National Restaurant Association; Retail Bakers Association of America. Program calendar is divided into semesters. 1-year Certificate in Baking and Pastry Arts. 2-year Associate degrees in Hotel and Restaurant Management; Culinary Arts-Baking Track; Culinary Arts.

Areas of Study
Baking; buffet catering; computer applications; controlling costs in food service; culinary skill development; food preparation; food purchasing; garde-manger; international cuisine; kitchen management; management and human resources; meal planning; meat fabrication; menu and facilities design; nutrition; nutrition and food service; patisserie; restaurant opportunities; sanitation; soup, stock, sauce, and starch production.

Facilities
Bake shop; 2 bakeries; cafeteria; catering service; 15 classrooms; 3 computer laboratories; demonstration laboratory; 3 food production kitchens; gourmet dining room; 5 laboratories; learning resource center; lecture room; library; 2 public restaurants; snack shop; student lounge; 5 teaching kitchens.

Culinary Student Profile
165 full-time. 153 are under 25 years old; 12 are between 25 and 44 years old.

Faculty
16 full-time. 9 are culinary-accredited teachers; 1 is a master baker. Prominent faculty: Paul Sorgule and Robert Brown.

Special Programs
10-week internship in Burgundy region of France, 1 week at Le Cordon Bleu in Paris, over 300 externship opportunities.

Expenses
Application fee: $25. Tuition: $12,000 per year full-time, $385 per credit hour part-time. Program-related fees include: $500 for uniforms, clothing, and equipment.

Financial Aid
In 1996, 12 scholarships were awarded (average award was $2000). Program-specific awards include Cooking for Scholarships competition. Employment placement assistance is available.

Housing
Coed and single-sex housing available. Average on-campus housing cost per month: $160.

Application Information
Students are accepted for enrollment in January and September. Applicants must submit a formal application and SAT/ACT scores.

Contact
Jennifer Grillo, Director of Admissions, Culinary Arts and Culinary Baking Concentrations, Routes

86 and 30, PO Box 265, Paul Smiths, NY 12970; 518-327-6227; Fax: 518-327-6161; E-mail: grilloj@paulsmiths.edu; World Wide Web: http://www.paulsmiths.edu/

See color display following page 172.

Peter Kump's New York Cooking School

New York, New York

General Information
Private, coeducational, culinary institute. Urban campus. Founded in 1974.

Program Information
Offered since 1974. Member of International Association of Culinary Professionals; James Beard Foundation, Inc.; Women Chefs and Restaurateurs. 20-week Diplomas in Full-time Career Pastry and Baking Arts; Full-time Career Culinary Arts. 26-week Diplomas in Part-time Career Pastry and Baking Arts; Part-time Career Culinary Arts.

Areas of Study
Baking; controlling costs in food service; culinary skill development; food preparation; garde-manger; international cuisine; kitchen management; meal planning; meat cutting; nutrition; patisserie; sanitation; seafood processing; soup, stock, sauce, and starch production; wines and spirits.

Facilities
2 demonstration laboratories; library; 8 teaching kitchens.

Culinary Student Profile
239 total: 128 full-time; 111 part-time.

Faculty
5 full-time; 16 part-time. 21 are industry professionals. Prominent faculty: Nick Malgieri.

Peter Kump's New York Cooking School *(continued)*

Special Programs
210-hour externship in restaurant, pastry shop, or caterer in the United States or France.

Expenses
Application fee: $100. Tuition: $9950 per diploma full-time, $7600 per diploma part-time.

Financial Aid
In 1996, 2 scholarships were awarded (average award was $2000); 60 loans were granted (average loan was $5000). Program-specific awards include work-study program. Employment placement assistance is available.

Application Information
Students are accepted for enrollment year-round. In 1996, 272 applied; 239 were accepted. Applicants must submit a formal application and interview.

Contact
Bill Grant, Director, 307 East 92 Street, New York, NY 10128; 800-522-4610; Fax: 212-348-6360; World Wide Web: http://www.pkcookschool.com

Schenectady County Community College

Hotel, Culinary Arts, and Tourism

Schenectady, New York

General Information
Public, coeducational, two-year college. Urban campus. Founded in 1968. Accredited by Middle States Association of Colleges and Schools.

Program Information
Offered since 1969. Accredited by American Culinary Federation Education Institute. Member of American Culinary Federation Educational Institute. Program calendar is divided into semesters. 1-year Certificate in Assistant Chef. 2-year Associate degree in Culinary Arts.

Areas of Study
Baking; beverage management; buffet catering; computer applications; confectionery show pieces; controlling costs in food service; culinary French; culinary skill development; food preparation; food purchasing; food service math; garde-manger; international cuisine; introduction to food service; kitchen management; management and human resources; meal planning; meat cutting; menu and facilities design; nutrition; patisserie; restaurant opportunities; sanitation; saucier; seafood processing; soup, stock, sauce, and starch production; wines and spirits.

Facilities
2 bake shops; cafeteria; 6 demonstration laboratories; gourmet dining room; library; public restaurant; 6 teaching kitchens.

Culinary Student Profile
360 total: 264 full-time; 96 part-time. 178 are under 25 years old; 164 are between 25 and 44 years old; 18 are over 44 years old.

Faculty
12 full-time; 17 part-time. 2 are culinary-accredited teachers.

Special Programs
4-10 days' work experience at Kentucky Derby, semester internships at Disney World.

Expenses
Application fee: $25. Tuition: $1170 per semester full-time, $97 per credit hour part-time.

Financial Aid
Employment placement assistance is available. Employment opportunities within the program are available.

Application Information
Students are accepted for enrollment in January, June, and August. In 1996, 288 applied; 95 were accepted. Applicants must submit a formal application.

Contact
Robert Dinello, Director of Admissions, Hotel, Culinary Arts, and Tourism, 78 Washington Avenue, Schenectady, NY 12305-2294; 518-381-1370; Fax: 518-346-0379; E-mail: dinellre@gw.sunysccc.edu

State University of New York College of Agriculture and Technology at Cobleskill

Culinary Arts, Hospitality, and Tourism

Cobleskill, New York

General Information
Public, coeducational, two-year coilege. Rural campus. Founded in 1916. Accredited by Middle States Association of Colleges and Schools.

Program Information
Offered since 1971. Accredited by American Culinary Federation Education Institute. Member of American Culinary Federation; American Culinary Federation Educational Institute; Confrerie de la Chaine des Rotisseurs; Council on Hotel, Restaurant, and Institutional Education; National Restaurant Association. Program calendar is divided into semesters. 1-year Certificate in Culinary Arts. 2-year Associate degree in Culinary Arts.

Areas of Study
Baking; beverage management; buffet catering; computer applications; confectionery show pieces; controlling costs in food service; culinary French; culinary skill development; food preparation; food purchasing; food service math; garde-manger; international cuisine; introduction to food service; kitchen management; management and human resources; meat cutting; meat fabrication; menu and facilities design; nutrition; patisserie; restaurant opportunities; sanitation; wines and spirits.

Facilities
Bake shop; bakery; cafeteria; catering service; coffee shop; 4 computer laboratories; gourmet dining room; learning resource center; public restaurant.

Culinary Student Profile
100 full-time.

Faculty
8 full-time; 2 part-time. 6 are industry professionals; 2 are culinary-accredited teachers.

Special Programs
Exchange program with Birmingham College, Birmingham, England.

Expenses
Application fee: $25. In-state tuition: $3300 per semester. Out-of-state tuition: $5000 per semester. Program-related fees include: $120 for knives; $100 for uniforms.

Financial Aid
In 1996, 10 scholarships were awarded (average award was $1500). Employment placement assistance is available.

Housing
Coed and single-sex housing available.

Application Information
Students are accepted for enrollment in January and August. Application deadline for fall is continuous with a recommended date of April 30. Application deadline for spring is continuous with a recommended date of November 30. In 1996, 250 applied; 100 were accepted. Applicants must submit a formal application.

Contact
Alan Roer, Department Chair, Culinary Arts, Hospitality, and Tourism, Cobleskill, NY 12043; 518-234-5425; Fax: 518-234-5333; E-mail: roerah@cobleskill.edu

State University of New York College of Technology at Alfred

Culinary Arts Program

Wellsville, New York

General Information
Public, coeducational, two-year college. Rural campus. Founded in 1908. Accredited by Middle States Association of Colleges and Schools.

State University of New York College of Technology at Alfred *(continued)*

PROGRAM INFORMATION
Offered since 1966. Member of American Culinary Federation; Council on Hotel, Restaurant, and Institutional Education; National Restaurant Association. Program calendar is divided into semesters. 2-year Associate degrees in Food Production; Baking Production and Management.

AREAS OF STUDY
Baking; beverage management; buffet catering; controlling costs in food service; culinary skill development; food preparation; food purchasing; food service math; introduction to food service; kitchen management; management and human resources; meal planning; menu and facilities design; nutrition; sanitation.

FACILITIES
Bake shop; bakery; cafeteria; catering service; 2 classrooms; coffee shop; computer laboratory; demonstration laboratory; food production kitchen; gourmet dining room; laboratory; learning resource center; 2 lecture rooms; 2 libraries; public restaurant; snack shop; teaching kitchen.

CULINARY STUDENT PROFILE
60 full-time. 40 are under 25 years old; 18 are between 25 and 44 years old; 2 are over 44 years old.

FACULTY
3 full-time; 2 part-time. 5 are industry professionals.

SPECIAL PROGRAMS
Annual Culinary Arts Food Show, restaurant and trade show field trips.

EXPENSES
Application fee: $50. Tuition: $10,000 per year.

FINANCIAL AID
In 1996, 30 scholarships were awarded (average award was $250); 15 loans were granted (average loan was $500). Program-specific awards include work-study program. Employment placement assistance is available. Employment opportunities within the program are available.

HOUSING
50 culinary students housed on campus. Coed housing available. Average on-campus housing cost per month: $350. Average off-campus housing cost per month: $250.

APPLICATION INFORMATION
Students are accepted for enrollment in January and August. Application deadline for fall is continuous with a recommended date of April 30. In 1996, 85 applied; 73 were accepted. Applicants must submit a formal application.

CONTACT
Deb Goodvich, Admissions Director, Culinary Arts Program, Huntington Building, Alfred, NY 14802; 607587-4215; E-mail: good~idj@alfredtech.edu

STATE UNIVERSITY OF NEW YORK COLLEGE OF TECHNOLOGY AT DELHI

Hospitality Department

Delhi, New York

GENERAL INFORMATION
Public, coeducational, two-year college. Rural campus. Founded in 1913. Accredited by Middle States Association of Colleges and Schools.

PROGRAM INFORMATION
Offered since 1994. Member of Council on Hotel, Restaurant, and Institutional Education; National Restaurant Association. Program calendar is divided into semesters. 2-year Associate degree in Culinary Arts.

AREAS OF STUDY
Baking; beverage management; buffet catering; controlling costs in food service; food preparation; food purchasing; garde-manger; international cuisine; management and human resources; meal planning; nutrition; sanitation; wines and spirits; restaurant management.

Facilities

Catering service; 2 classrooms; computer laboratory; demonstration laboratory; 2 food production kitchens; gourmet dining room; learning resource center; library; public restaurant; 2 teaching kitchens; beverage lounge.

Culinary Student Profile

47 total: 45 full-time; 2 part-time. 43 are under 25 years old; 2 are between 25 and 44 years old.

Faculty

2 full-time. Prominent faculty: Thomas Emerick and Lucie Costa.

Expenses

Application fee: $30. Tuition: $1600 per semester full-time, $145 per credit hour part-time. Program-related fees include: $90 for lab fees.

Financial Aid

In 1996, 1 scholarship was awarded (award was $250). Employment placement assistance is available.

Housing

Coed and single-sex housing available. Average on-campus housing cost per month: $300.

Application Information

Students are accepted for enrollment in January and September. In 1996, 84 applied. Applicants must submit a formal application.

Contact

Gary Cole, Director of Enrollment Services, Hospitality Department, Main Street, Delhi, NY 13753; 607-746-4550; Fax: 607-746-4104.

Sullivan County Community College

Hospitality Division

Loch Sheldrake, New York

General Information

Public, coeducational, two-year college. Rural campus. Founded in 1962. Accredited by Middle States Association of Colleges and Schools.

Program Information

Offered since 1965. Accredited by American Culinary Federation Education Institute. Member of American Culinary Federation; American Culinary Federation Educational Institute; Council on Hotel, Restaurant, and Institutional Education; Educational Foundation of the NRA; National Restaurant Association. Program calendar is divided into semesters. 1-year Certificate in Food Service. 2-year Associate degrees in Hotel Technology; Professional Chef.

Areas of Study

Baking; beverage management; buffet catering; computer applications; confectionery show pieces; controlling costs in food service; convenience cookery; culinary skill development; food preparation; food purchasing; food service math; garde-manger; international cuisine; kitchen management; management and human resources; meat fabrication; nutrition; nutrition and food service; patisserie; restaurant opportunities; sanitation; saucier; soup, stock, sauce, and starch production; wines and spirits.

Facilities

2 bake shops; classroom; 3 demonstration laboratories; 2 food production kitchens; gourmet dining room.

Culinary Student Profile

127 total: 102 full-time; 25 part-time. 99 are under 25 years old; 24 are between 25 and 44 years old; 4 are over 44 years old.

Faculty

6 full-time; 2 part-time. 2 are industry professionals; 4 are culinary-accredited teachers. Prominent faculty: Edmund Nadeau and Mark Sanok.

Expenses

Application fee: $30. In-state tuition: $2500 per year full-time, $90 per credit hour part-time. Out-of-state tuition: $3700 per year full-time, $133 per credit hour part-time. Program-related fees include: $155 for knife set; $90 for uniforms.

Financial Aid

In 1996, 12 scholarships were awarded (average award was $1000). Employment placement

Sullivan County Community College *(continued)*

assistance is available. Employment opportunities within the program are available.

Application Information
Students are accepted for enrollment in January and September. In 1996, 219 applied; 70 were accepted. Applicants must submit a formal application.

Contact
Admissions Department, Hospitality Division, Box 4002, Loch Sheldrake, NY 12759-4002; 800-577-5243; Fax: 914-434-4806.

Syracuse University

Restaurant and Food Service Management

Syracuse, New York

General Information
Private, coeducational, university. Urban campus. Founded in 1870. Accredited by Middle States Association of Colleges and Schools.

Program Information
Offered since 1986. Member of American Dietetic Association; Council on Hotel, Restaurant, and Institutional Education; Educational Foundation of the NRA; National Restaurant Association. Program calendar is divided into semesters. 4-year Bachelor's degrees in Coordinated Dietetics; Nutrition Science; Nutrition; Restaurant Food Service Management.

Areas of Study
Baking; beverage management; buffet catering; computer applications; culinary skill development; food preparation; food purchasing; kitchen management; management and human resources; menu and facilities design; nutrition; nutrition and food service; sanitation; wines and spirits; restaurant development.

Facilities
Bake shop; bakery; 8 cafeterias; catering service; 25 classrooms; coffee shop; 5 computer laboratories; 2 demonstration laboratories; 2 food production kitchens; 2 gardens; gourmet dining room; 5 learning resource centers; 12 lecture rooms; 8 libraries; 2 public restaurants; 6 snack shops; 9 student lounges; 2 teaching kitchens.

Culinary Student Profile
45 full-time. 45 are under 25 years old.

Faculty
6 full-time; 8 part-time. 3 are industry professionals; 1 is a master chef; 2 are culinary-accredited teachers.

Expenses
Application fee: $40. Tuition: $25,340 per year.

Financial Aid
In 1996, 2 scholarships were awarded (average award was $500). Employment placement assistance is available.

Housing
Coed, apartment-style, and single-sex housing available. Average on-campus housing cost per month: $436. Average off-campus housing cost per month: $300.

Application Information
Students are accepted for enrollment in January and August. Application deadline for fall is January 15. Applicants must submit a formal application and letters of reference.

Contact
Bradley Beran, Director, Restaurant and Food Service Management, 034 Slocum Hall, Syracuse, NY 13244; 315-443-2386; Fax: 315-443-2562; E-mail: bcberan@mailbox.syr.edu

Westchester Community College

Restaurant Management

Valhalla, New York

General Information
Public, coeducational, two-year college. Suburban

campus. Founded in 1946. Accredited by Middle States Association of Colleges and Schools.

Program Information
Offered since 1946. Member of American Dietetic Association; Council on Hotel, Restaurant, and Institutional Education; Educational Foundation of the NRA. Program calendar is divided into semesters. 2-year Associate degree in Food Service Administration: Restaurant Management.

Areas of Study
Baking; beverage management; buffet catering; computer applications; controlling costs in food service; food preparation; food purchasing; food service math; garde-manger; international cuisine; introduction to food service; kitchen management; management and human resources; meal planning; menu and facilities design; nutrition; nutrition and food service; restaurant opportunities; sanitation; soup, stock, sauce, and starch production; wines and spirits.

Facilities
Catering service; classroom; computer laboratory; demonstration laboratory; 2 food production kitchens; gourmet dining room; 2 laboratories; learning resource center; lecture room; library; public restaurant; student lounge; 2 teaching kitchens.

Culinary Student Profile
120 total: 70 full-time; 50 part-time.

Faculty
2 full-time; 2 part-time. 1 is an industry professional; 1 is a culinary-accredited teacher. Prominent faculty: Daryl Nosek.

Expenses
Application fee: $25. Tuition: $1175 per semester full-time, $98 per credit part-time. Program-related fees include: $10 for lab fees.

Financial Aid
In 1996, 7 scholarships were awarded (average award was $850). Employment placement assistance is available. Employment opportunities within the program are available.

Application Information
Students are accepted for enrollment in January, May, and September. Applicants must submit a formal application.

Contact
Daryl Nosek, Curriculum Chair, Restaurant Management, 75 Grasslands Road, Valhalla, NY 10595-1698; 914-785-6551; Fax: 914-785-6423.

Asheville-Buncombe Technical Community College

Culinary Technology

Asheville, North Carolina

General Information
Public, coeducational, two-year college. Urban campus. Founded in 1959. Accredited by Southern Association of Colleges and Schools.

Program Information
Offered since 1967. Member of American Culinary Federation; Council on Hotel, Restaurant, and Institutional Education; Educational Foundation of the NRA; National Restaurant Association. Program calendar is divided into semesters. 2-year Associate degrees in Hotel and Restaurant Management; Culinary Technology.

Areas of Study
Baking; beverage management; computer applications; confectionery show pieces; controlling costs in food service; culinary French; culinary skill development; food preparation; food purchasing; food service math; garde-manger; international cuisine; introduction to food service; management and human resources; meal planning; meat cutting; meat fabrication; menu and facilities design; nutrition; nutrition and food service; patisserie; sanitation; saucier; seafood processing; soup, stock, sauce, and starch production; wines and spirits.

Facilities
Bake shop; 4 classrooms; computer laboratory; demonstration laboratory; food production kitchen; gourmet dining room; laboratory; learning resource center; public restaurant; snack shop; student lounge; teaching kitchen; hotel.

Asheville-Buncombe Technical Community College *(continued)*

Culinary Student Profile
90 total: 70 full-time; 20 part-time.

Faculty
5 full-time; 5 part-time. 2 are industry professionals; 3 are culinary-accredited teachers. Prominent faculty: Scott Gerken and Brian McDonald.

Expenses
In-state tuition: $280 per semester full-time, $20 per credit part-time. Out-of-state tuition: $2282 per semester full-time, $163 per credit part-time. Program-related fees include: $200 for knives; $100 for uniforms.

Financial Aid
In 1996, 5 scholarships were awarded (average award was $500). Employment placement assistance is available. Employment opportunities within the program are available.

Application Information
Students are accepted for enrollment in January and August. In 1996, 65 applied; 35 were accepted. Applicants must submit a formal application.

Contact
Sheila Tillman, Chairperson, Culinary Technology, 340 Victoria Road, Asheville, NC 28801-4897; 704-254-1921 Ext. 232; Fax: 704-251-6355.

Central Piedmont Community College

Hospitality Education Division

Charlotte, North Carolina

General Information
Public, coeducational, two-year college. Urban campus. Founded in 1963. Accredited by Southern Association of Colleges and Schools.

Program Information
Offered since 1977. Program calendar is divided into semesters. 1-year Certificate in Culinary Technology. 2-year Associate degrees in Hotel and Restaurant Management; Culinary Technology.

Areas of Study
Baking; beverage management; computer applications; confectionery show pieces; controlling costs in food service; culinary skill development; food preparation; food purchasing; food service communication; garde-manger; international cuisine; introduction to food service; kitchen management; management and human resources; meal planning; meat cutting; nutrition; patisserie; restaurant opportunities; sanitation; soup, stock, sauce, and starch production; wines and spirits.

Culinary Student Profile
600 total: 300 full-time; 300 part-time.

Faculty
4 full-time; 10 part-time.

Expenses
In-state tuition: $180 per semester full-time, $20 per credit part-time. Out-of-state tuition: $1467 per semester full-time, $163 per credit part-time.

Application Information
Students are accepted for enrollment in January, June, and August. Applicants must submit a formal application and have a high school diploma or GED.

Contact
Robert G. Boll, Division Director, Hospitality Education Division, PO Box 35009, Charlotte, NC 28235-5009; 704-330-6721; Fax: 704-330-6581.

Guilford Technical Community College

Culinary Technology

Jamestown, North Carolina

General Information
Public, coeducational, two-year college. Suburban campus. Founded in 1958. Accredited by Southern Association of Colleges and Schools.

Program Information
Offered since 1976. Accredited by American

Culinary Federation Education Institute. Program calendar is divided into semesters. 1-year Certificate in Culinary Technology. 1-year Diploma in Culinary Technology. 2-year Associate degree in Culinary Technology.

Areas of Study
Baking; computer applications; controlling costs in food service; culinary skill development; food preparation; food purchasing; food service communication; food service math; garde-manger; international cuisine; introduction to food service; kitchen management; management and human resources; meal planning; meat cutting; meat fabrication; menu and facilities design; nutrition; patisserie; restaurant opportunities; sanitation; saucier; seafood processing; soup, stock, sauce, and starch production; wines and spirits; food and beverage service.

Culinary Student Profile
130 total: 90 full-time; 40 part-time.

Faculty
2 full-time; 6 part-time.

Expenses
In-state tuition: $280 per semester full-time, $20 per credit hour part-time. Out-of-state tuition: $2282 per semester full-time, $163 per credit hour part-time.

Application Information
Students are accepted for enrollment in January and August. Applicants must submit a formal application.

Contact
Keith Gardiner, Program Chair, Culinary Technology, PO Box 309, Jamestown, NC 27282-0309; 910-454-1126 Ext. 2302; Fax: 910-454-2510.

Sandhills Community College

Culinary Technology

Pinehurst, North Carolina

General Information
Public, coeducational, two-year college. Rural campus. Founded in 1963. Accredited by Southern Association of Colleges and Schools.

Program Information
Offered since 1993. Member of American Hotel and Motel Association. Program calendar is divided into semesters. 1-year Diploma in Basic Culinary Technology. 2-year Certificate in Culinary Technology.

Areas of Study
Baking; beverage management; computer applications; controlling costs in food service; culinary skill development; food preparation; food purchasing; introduction to food service; management and human resources; nutrition; sanitation.

Facilities
Cafeteria; 5 classrooms; 3 computer laboratories; food production kitchen; garden; learning resource center; lecture room; library; student lounge; teaching kitchen.

Culinary Student Profile
8 total: 4 full-time; 4 part-time. 8 are between 25 and 44 years old.

Faculty
1 full-time; 1 part-time. 1 is an industry professional; 1 is a culinary-accredited teacher. Prominent faculty: A. M. Stratta.

Special Programs
5-month internship at leading resorts, 2-3 year apprenticeship at a leading resort.

Expenses
Tuition: $20 per credit hour. Program-related fees include: $10 for lab fees.

Financial Aid
Employment placement assistance is available.

Application Information
Students are accepted for enrollment in January, May, and August. Application deadline for fall is August 21. Application deadline for spring is January 7. Application deadline for summer is May 18. In 1996, 3 applied; 3 were accepted. Applicants must submit a formal application.

Sandhills Community College *(continued)*

Contact
A. M. Stratta, Coordinator, Culinary Technology, 2200 Airport Road, Pinehurst, NC 28374-8299; 910-695-3756; Fax: 910-695-1823.

Wake Technical Community College

Culinary Technology

Raleigh, North Carolina

General Information
Public, coeducational, two-year college. Rural campus. Founded in 1958. Accredited by Southern Association of Colleges and Schools.

Program Information
Member of American Culinary Federation; American Institute of Wine & Food; Council on Hotel, Restaurant, and Institutional Education; Educational Foundation of the NRA; National Restaurant Association; North Carolina Restaurant and Hotel Management Association. Program calendar is divided into semesters. 2-year Associate degree in Culinary Technology.

Areas of Study
Baking; beverage management; buffet catering; computer applications; confectionery show pieces; controlling costs in food service; culinary French; culinary skill development; food preparation; food purchasing; food service communication; garde-manger; international cuisine; kitchen management; management and human resources; menu and facilities design; nutrition; patisserie; sanitation; seafood processing; soup, stock, sauce, and starch production; wines and spirits; table service.

Facilities
Bake shop; demonstration laboratory; 4 food production kitchens; gourmet dining room; learning resource center; 2 lecture rooms; library; public restaurant; student lounge; 7 teaching kitchens.

Culinary Student Profile
80 total: 70 full-time; 10 part-time. 20 are under 25 years old; 50 are between 25 and 44 years old; 10 are over 44 years old.

Faculty
5 full-time. 5 are culinary-accredited teachers. Prominent faculty: Fredi Mort and Caralyn House.

Special Programs
3-week work-study in France.

Expenses
In-state tuition: $280 per semester full-time, $20 per semester hour part-time. Out-of-state tuition: $2282 per semester full-time, $163 per semester hour part-time. Program-related fees include: $15 for lab fees for specific courses.

Financial Aid
Employment placement assistance is available. Employment opportunities within the program are available.

Application Information
Students are accepted for enrollment in January, May, and August. In 1996, 30 were accepted. Applicants must submit a formal application and SAT scores or equivalent.

Contact
Office of the Registrar, Culinary Technology, 9101 Fayetteville Road, Raleigh, NC 27603-5696; 919-662-3400; Fax: 919-779-3360.

North Dakota State College of Science

Culinary Arts

Wahpeton, North Dakota

General Information
Public, coeducational, two-year college. Rural campus. Founded in 1903. Accredited by North Central Association of Colleges and Schools.

Program Information
Offered since 1971. Member of American Culinary Federation; National Restaurant Association;

North Dakota Hospitality Association. Program calendar is divided into semesters. 2-year Associate degree in Chef Training and Management Technology. 2-year Diploma in Chef Training and Management Technology.

Areas of Study
Baking; buffet catering; computer applications; controlling costs in food service; culinary skill development; food preparation; food purchasing; food service math; garde-manger; introduction to food service; kitchen management; management and human resources; meal planning; meat cutting; meat fabrication; menu and facilities design; nutrition and food service; patisserie; sanitation; saucier; seafood processing; soup, stock, sauce, and starch production.

Facilities
Bake shop; bakery; cafeteria; catering service; classroom; coffee shop; 7 computer laboratories; demonstration laboratory; food production kitchen; learning resource center; lecture room; 2 libraries; public restaurant; snack shop; 3 student lounges; teaching kitchen.

Culinary Student Profile
30 full-time. 28 are under 25 years old; 2 are between 25 and 44 years old.

Faculty
2 full-time. 2 are industry professionals. Prominent faculty: Neil Rittenour and Mary Uhren.

Special Programs
Cooperative Education Program (paid internship).

Expenses
Application fee: $25. In-state tuition: $1751 per year. Out-of-state tuition: $2139 per year. Program-related fees include: $95 for uniforms; $115 for cutlery.

Financial Aid
Employment placement assistance is available.

Housing
Coed, apartment-style, and single-sex housing available. Average on-campus housing cost per month: $96. Average off-campus housing cost per month: $125.

Application Information
Students are accepted for enrollment in January and August. Application deadline for fall is August 25. Application deadline for spring is January 5. In 1996, 30 applied; 30 were accepted. Applicants must submit a formal application and ACT scores.

Contact
Neil Rittenour, Director of Culinary Arts, Culinary Arts, 800 North Sixth Street, Wahpeton, ND 58076; 701-671-2264; Fax: 701-671-2126; E-mail: rittenou@plains.nodak.edu; World Wide Web: http://www.ndscs.edu

Ashland County-West Holmes Career Center

Culinary Chef Institute

Ashland, Ohio

General Information
Public, coeducational, adult vocational school. Rural campus. Founded in 1995.

Program Information
9-month Diploma in Culinary Arts.

Areas of Study
Baking; buffet catering; controlling costs in food service; convenience cookery; culinary French; culinary skill development; food preparation; food purchasing; food service math; garde-manger; international cuisine; introduction to food service; meal planning; meat fabrication; nutrition; patisserie; restaurant opportunities; sanitation; saucier; soup, stock, sauce, and starch production.

Culinary Student Profile
15 full-time.

Faculty
1 full-time; 2 part-time.

Expenses
Tuition: $3550 per diploma.

Ashland County-West Holmes Career Center *(continued)*

Application Information
Students are accepted for enrollment in January, April, and September.

Contact
Adult Education, Culinary Chef Institute, 1783 State Route 60, Ashland, OH 44805-9377; 419-289-3313; Fax: 419-289-3729.

Cincinnati State Technical and Community College

Business Division

Cincinnati, Ohio

General Information
Public, coeducational, two-year college. Urban campus. Founded in 1966. Accredited by North Central Association of Colleges and Schools.

Program Information
Offered since 1978. Accredited by American Culinary Federation Education Institute. Member of American Culinary Federation; American Culinary Federation Educational Institute; Council on Hotel, Restaurant, and Institutional Education; Educational Foundation of the NRA; National Restaurant Association. Program calendar is divided into quarters. 1-year Certificate in Culinary Arts. 2-year Associate degree in Culinary Arts.

Areas of Study
Baking; beverage management; buffet catering; computer applications; controlling costs in food service; food preparation; food purchasing; garde-manger; international cuisine; management and human resources; meat cutting; menu and facilities design; nutrition; nutrition and food service; restaurant opportunities; sanitation; seafood processing; soup, stock, sauce, and starch production; wines and spirits.

Facilities
Demonstration laboratory; 2 food production kitchens; laboratory; lecture room; public restaurant; 2 teaching kitchens.

Culinary Student Profile
170 total: 145 full-time; 25 part-time.

Faculty
4 full-time; 3 part-time. 3 are industry professionals; 1 is a master chef; 3 are culinary-accredited teachers.

Special Programs
Cooperative education experience.

Expenses
Tuition: $61.85 per credit hour. Program-related fees include: $400 for knives; $300 for lab fees.

Financial Aid
In 1996, 8 scholarships were awarded (average award was $800). Employment placement assistance is available. Employment opportunities within the program are available.

Application Information
Students are accepted for enrollment in February, April, June, September, and November. In 1996, 90 applied; 90 were accepted. Applicants must submit a formal application.

Contact
Richard Hendrix, Program Chair, Business Division, 3520 Central Parkway, Cincinnati, OH 45223-2690; 513-569-1662; Fax: 513-569-1467; E-mail: hendrixr@cinstate.cc.oh.us

Columbus State Community College

Hospitality Management Department

Columbus, Ohio

General Information
Public, coeducational, two-year college. Urban campus. Founded in 1963. Accredited by North Central Association of Colleges and Schools.

PROGRAM INFORMATION
Offered since 1966. Accredited by American Culinary Federation Education Institute. Member of American Culinary Federation; American Dietetic Association; American Institute of Baking; Council on Hotel, Restaurant, and Institutional Education; Educational Foundation of the NRA; National Restaurant Association. Program calendar is divided into quarters. 2-year Associate degree in Food Service/Restaurant Management. 3-year Associate degree in Chef Apprenticeship.

AREAS OF STUDY
Baking; beverage management; computer applications; controlling costs in food service; food preparation; food purchasing; garde-manger; introduction to food service; kitchen management; management and human resources; meal planning; menu and facilities design; nutrition; sanitation.

FACILITIES
3 classrooms; computer laboratory; demonstration laboratory; 2 food production kitchens; 2 laboratories; learning resource center; 3 lecture rooms; library; 2 teaching kitchens.

CULINARY STUDENT PROFILE
200 total: 75 full-time; 125 part-time.

FACULTY
6 full-time; 8 part-time. 2 are industry professionals; 3 are culinary-accredited teachers.

SPECIAL PROGRAMS
Opportunity for matriculated students to assist with noncredit courses.

EXPENSES
Application fee: $10. Tuition: $708 per semester full-time, $59 per credit hour part-time. Program-related fees include: $100 for lab fees.

FINANCIAL AID
In 1996, 20 scholarships were awarded (average award was $1500). Employment placement assistance is available.

APPLICATION INFORMATION
Students are accepted for enrollment in January, March, June, and September. Applicants must submit a formal application.

CONTACT
Carol Kizer, Chairperson, Hospitality Management Department, Box 1609, Columbus, OH 43216-1609; 614-227-2579; Fax: 614-227-5146; E-mail: ckizer@cscc.edu

HOCKING COLLEGE

Culinary Arts Department

Nelsonville, Ohio

GENERAL INFORMATION
Public, coeducational, two-year college. Rural campus. Founded in 1968. Accredited by North Central Association of Colleges and Schools.

PROGRAM INFORMATION
Offered since 1978. Accredited by American Culinary Federation Education Institute. Member of American Culinary Federation; American Culinary Federation Educational Institute; Educational Foundation of the NRA; National Restaurant Association; American Hotel/Motel Association. Program calendar is divided into quarters. 2-year Associate degree in Culinary Arts.

AREAS OF STUDY
Baking; beverage management; buffet catering; computer applications; controlling costs in food service; culinary skill development; food preparation; food purchasing; garde-manger; international cuisine; introduction to food service; kitchen management; management and human resources; meal planning; meat cutting; meat fabrication; menu and facilities design; nutrition; patisserie; sanitation; seafood processing; soup, stock, sauce, and starch production; wines and spirits.

FACILITIES
Bake shop; bakery; 2 catering services; 5 classrooms; coffee shop; computer laboratory; demonstration laboratory; food production kitchen; garden; laboratory; learning resource center; 3 lecture rooms; library; public restaurant; student lounge; teaching kitchen.

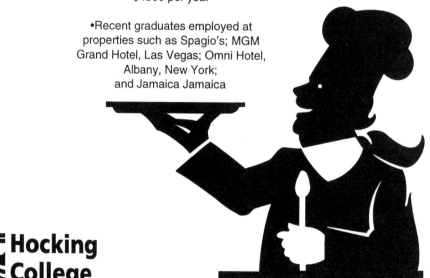

Hocking College *(continued)*

Culinary Student Profile
120 total: 110 full-time; 10 part-time.

Faculty
3 full-time; 1 part-time. 4 are industry professionals. Prominent faculty: Thomas J. Landusky.

Special Programs
Internships in the Caribbean and South America, hot food competition, trip to Chicago Food Show.

Expenses
Application fee: $50. Tuition: $720 per quarter full-time, $60 per credit part-time. Program-related fees include: $145 for knives; $200 for books; $80 for uniforms.

Financial Aid
In 1996, 10 scholarships were awarded (average award was $1000). Employment placement assistance is available. Employment opportunities within the program are available.

Housing
Coed and apartment-style housing available. Average on-campus housing cost per month: $200. Average off-campus housing cost per month: $250.

Application Information
Students are accepted for enrollment year-round. In 1996, 55 applied; 55 were accepted. Applicants must submit a formal application.

Contact
Thomas J. Landusky, Chef/Instructor, Culinary Arts Department, 3301 Hocking Parkway, Nelsonville, OH 45764-9588; 614-753-3531 Ext. 304; Fax: 614-753-9158; World Wide Web: http://www.hocking.cc.oh.us

See display on page 192.

The Loretta Paganini School of Cooking

Chesterland, Ohio

General Information
Private, coeducational, culinary institute. Suburban campus. Founded in 1981.

Program Information
Offered since 1989. Accredited by American Culinary Federation Education Institute. Member of American Culinary Federation; American Institute of Wine & Food; International Association of Culinary Professionals. Program calendar is divided into trimesters. 6-month Diplomas in Pastry Arts; Culinary Arts.

Areas of Study
Baking; confectionery show pieces; culinary French; culinary skill development; food preparation; garde-manger; international cuisine; meal planning; meat cutting; meat fabrication; patisserie; saucier; seafood processing; soup, stock, sauce, and starch production; wines and spirits.

Facilities
Classroom; food production kitchen; garden; library; teaching kitchen.

Culinary Student Profile
220 total: 110 full-time; 110 part-time.

Faculty
20 full-time; 50 part-time.

Special Programs
Two 10-day gastronomic tours of Italy, 9-day Mediterranean cruise, 7-day cooking program in Italy.

Expenses
Tuition: $2240 per diploma full-time, $195 per 4 classes part-time.

Application Information
Students are accepted for enrollment in January, March, and August. Applicants must submit a formal application and interview.

Contact
Loretta Paganini, Director, 8613 Mayfield Road, Chesterland, OH 44026; 440-729-1110; Fax: 440-729-6459.

Owens Community College

Hospitality Management Technology

Toledo, Ohio

General Information
Public, coeducational, two-year college. Suburban campus. Founded in 1965. Accredited by North Central Association of Colleges and Schools.

Program Information
Offered since 1968. Member of American Culinary Federation; Council on Hotel, Restaurant, and Institutional Education; Educational Foundation of the NRA. Program calendar is divided into semesters. 1-year Certificate in Food Service. 2-year Associate degree in Food Service Management.

Areas of Study
Baking; beverage management; buffet catering; controlling costs in food service; food preparation; food purchasing; introduction to food service; management and human resources; nutrition; sanitation; advanced food production.

Facilities
Food production kitchen.

Culinary Student Profile
29 total: 15 full-time; 14 part-time.

Faculty
1 full-time; 4 part-time. 4 are industry professionals; 1 is a registered dietitian.

Special Programs
Cooperative work experience.

Expenses
Tuition: $948 per semester full-time, $79 per credit hour part-time. Program-related fees include: $230 for lab fees.

Financial Aid
In 1996, 5 scholarships were awarded (average award was $500). Employment placement assistance is available.

Application Information
Students are accepted for enrollment in January, June, and August. In 1996, 22 applied; 22 were accepted. Applicants must submit a formal application.

Contact
Dolores Dobelbower, Facilitator, Hospitality Management Technology, PO Box 10000, Oregon Road, Toledo, OH 43699-1947; 419-661-7359; Fax: 419-661-7665.

Sinclair Community College

Hospitality Management/Culinary Arts Option

Dayton, Ohio

General Information
Public, coeducational, two-year college. Urban campus. Founded in 1887. Accredited by North Central Association of Colleges and Schools.

Program Information
Offered since 1993. Accredited by American Culinary Federation Education Institute. Member of American Culinary Federation; American Culinary Federation Educational Institute; American Institute of Baking; Council on Hotel, Restaurant, and Institutional Education; Educational Foundation of the NRA; National Restaurant Association; The Bread Bakers Guild of America. Program calendar is divided into quarters. 1-year Certificate in Food Service Management. 2-year Associate degrees in Culinary Arts Option; Hospitality Management.

Areas of Study
Baking; beverage management; controlling costs in food service; culinary skill development; food preparation; food purchasing; garde-manger; management and human resources; meat cutting; meat fabrication; menu and facilities design; nutrition; patisserie; sanitation; seafood processing; soup, stock, sauce, and starch production.

Facilities
Bake shop; cafeteria; catering service; 5 computer laboratories; demonstration laboratory; food production kitchen; gourmet dining room; 3 laboratories; learning resource center; library; 3 snack shops; 3 teaching kitchens.

Culinary Student Profile
225 total: 25 full-time; 200 part-time. 50 are under 25 years old; 150 are between 25 and 44 years old; 25 are over 44 years old.

Faculty
3 full-time; 8 part-time. 8 are industry professionals; 2 are culinary-accredited teachers. Prominent faculty: Steven K. Cornelius and Frank Leibold.

Expenses
Application fee: $10. Tuition: $31 per credit hour. Program-related fees include: $35 for lab fees; $125 for uniforms and knives.

Financial Aid
In 1996, 7 scholarships were awarded (average award was $400). Employment placement assistance is available. Employment opportunities within the program are available.

Application Information
Students are accepted for enrollment in January, June, and September. In 1996, 225 applied; 225 were accepted. Applicants must submit a formal application.

Contact
Steven Cornelius, Department Chair, Hospitality Management/Culinary Arts Option, 444 West Third Street, Dayton, OH 45402-1460; 937-449-5197; Fax: 937-449-4530; E-mail: scorneli@sinclair.edu

University of Akron

Akron, Ohio

General Information
Public, coeducational, university. Urban campus. Founded in 1870. Accredited by North Central Association of Colleges and Schools.

Program Information
Member of American Culinary Federation; National Restaurant Association; Ohio Hotel/Motel Association. Program calendar is divided into semesters. 1-year Certificates in Hotel/Motel Management; Restaurant Management; Culinary Arts. 2-year Associate degrees in Hospitality Marketing and Sales; Hotel/Motel Management; Culinary Arts; Restaurant Management.

Areas of Study
Baking; beverage management; computer applications; controlling costs in food service; culinary skill development; food preparation; food purchasing; food service communication; garde-manger; international cuisine; introduction to food service; management and human resources; meal planning; menu and facilities design; nutrition; sanitation; soup, stock, sauce, and starch production; wines and spirits.

Culinary Student Profile
125 total: 50 full-time; 75 part-time.

Faculty
5 full-time; 4 part-time.

Special Programs
Internships, field trips to local and national professional shows.

Expenses
In-state tuition: $1576.50 per semester full-time, $122 per credit part-time. Out-of-state tuition: $4171.50 per semester full-time, $295 per credit part-time. Program-related fees include: $30 for lab fees; $50 for uniforms; $75 for knife set.

Financial Aid
Employment placement assistance is available. Employment opportunities within the program are available.

Housing
Coed and single-sex housing available.

Application Information
Students are accepted for enrollment in January and August. Applicants must submit a formal application and have a high school diploma or GED.

University of Akron *(continued)*

Contact
Lawrence Gilpatric, Assistant Professor, Business Technology, 102 Gallucci Hall, Akron, OH 44325-7907; 330-972-5393; Fax: 330-972-5525; E-mail: gilpatr@uakron.edu

Great Plains Area Vocational Technical Center

Commercial Foods

Lawton, Oklahoma

General Information
Public, coeducational, technical institute. Urban campus. Founded in 1971. Accredited by North Central Association of Colleges and Schools.

Program Information
Offered since 1971. Program calendar is divided into semesters. 18-month Certificates in Baker studies; Hot Food Cook studies; Cold Food Cook studies.

Areas of Study
Baking; beverage management; buffet catering; controlling costs in food service; food preparation; food purchasing; food service math; garde-manger; introduction to food service; kitchen management; management and human resources; meal planning; nutrition; sanitation; soup, stock, sauce, and starch production.

Faculty
6 full-time.

Expenses
Tuition: $1600 per certificate.

Application Information
Students are accepted for enrollment in January and August. Applicants must submit a formal application.

Contact
Brenda Ronio, Head, Commercial Foods, 4500 West Lee Boulevard, Lawton, OK 73505; 405-250-5622; Fax: 405-250-5677.

Meridian Technology Center

Commercial Food Production

Stillwater, Oklahoma

General Information
Public, coeducational, two-year college. Rural campus. Founded in 1975.

Program Information
Offered since 1975. Member of National Restaurant Association. Program calendar is divided into semesters. 2-year Certificate in Commercial Food Production.

Areas of Study
Baking; buffet catering; controlling costs in food service; culinary skill development; food preparation; food purchasing; food service math; garde-manger; introduction to food service; kitchen management; management and human resources; meal planning; meat cutting; menu and facilities design; nutrition; restaurant opportunities; sanitation; soup, stock, sauce, and starch production.

Facilities
Bake shop; cafeteria; catering service; classroom; coffee shop; 2 computer laboratories; food production kitchen; 2 gourmet dining rooms; laboratory; learning resource center; lecture room; 2 public restaurants; snack shop; student lounge; teaching kitchen.

Culinary Student Profile
36 full-time. 20 are under 25 years old; 16 are between 25 and 44 years old.

Faculty
6 full-time. 4 are industry professionals; 2 are culinary-accredited teachers. Prominent faculty: Scott Galloway and Nancy Mitchell.

Expenses
Tuition: $400 per semester.

Financial Aid
Employment placement assistance is available. Employment opportunities within the program are available.

Application Information
Students are accepted for enrollment year-round. In 1996, 36 applied; 36 were accepted. Applicants must submit a formal application.

Contact
Student Services, Commercial Food Production, 1312 South Sangre Road, Stillwater, OK 74074; 405-377-3333; World Wide Web: http://www.meridian-technology.com

Metro Area Vocational Technical School District 22

Culinary Arts

Oklahoma City, Oklahoma

General Information
Public, coeducational, technical institute. Urban campus. Founded in 1980. Accredited by North Central Association of Colleges and Schools.

Program Information
Offered since 1980. Program calendar is divided into quarters. 9-month Certificate in Culinary Arts.

Areas of Study
Baking; buffet catering; computer applications; controlling costs in food service; food preparation; food purchasing; food service communication; food service math; introduction to food service; kitchen management; meal planning; nutrition; nutrition and food service; sanitation.

Facilities
Bake shop; cafeteria; catering service; classroom; demonstration laboratory; food production kitchen; learning resource center; library.

Culinary Student Profile
40 total: 20 full-time; 20 part-time.

Faculty
2 full-time.

Expenses
Tuition: $1050 per program.

Financial Aid
Employment placement assistance is available. Employment opportunities within the program are available.

Application Information
Students are accepted for enrollment in January, February, March, August, September, October, November, and December. In 1996, 50 applied; 40 were accepted. Applicants must submit a formal application.

Contact
Lois T. Jones, Counselor, Culinary Arts, 1720 Springlake Drive, Oklahoma City, OK 73111; 405-424-8324 Ext. 642; Fax: 405-424-9403.

Oklahoma State University, Okmulgee

Hospitality Services Management

Okmulgee, Oklahoma

General Information
Public, coeducational, two-year college. Rural campus. Founded in 1946. Accredited by North Central Association of Colleges and Schools.

Program Information
Offered since 1946. Member of American Dietetic Association; Educational Foundation of the NRA; National Restaurant Association; Oklahoma Restaurant Association. Program calendar is divided into trimesters. 30-month Associate degrees in Dietetic Technology; Culinary Arts; Baking.

Areas of Study
Baking; buffet catering; computer applications; controlling costs in food service; culinary French;

Oklahoma State University, Okmulgee
(continued)

culinary skill development; food preparation; food purchasing; food service math; garde-manger; introduction to food service; management and human resources; meal planning; meat cutting; nutrition; patisserie; sanitation; saucier; seafood processing; soup, stock, sauce, and starch production.

Facilities
Bake shop; bakery; cafeteria; catering service; 8 classrooms; coffee shop; computer laboratory; 3 food production kitchens; gourmet dining room; learning resource center; 2 lecture rooms; library; 2 public restaurants; snack shop; student lounge; 3 teaching kitchens.

Culinary Student Profile
200 total: 170 full-time; 30 part-time. 60 are under 25 years old; 120 are between 25 and 44 years old; 20 are over 44 years old.

Faculty
7 full-time; 2 part-time. 3 are industry professionals; 2 are culinary-accredited teachers; 2 are certified executive chefs.

Special Programs
Trip to the National Restaurant Association convention.

Expenses
Tuition: $50 per credit hour. Program-related fees include: $120 for beginning knife set; $90 for 3 uniforms.

Financial Aid
In 1996, 27 scholarships were awarded (average award was $500). Employment placement assistance is available.

Housing
Coed, apartment-style, and single-sex housing available. Average on-campus housing cost per month: $300. Average off-campus housing cost per month: $300.

Application Information
Students are accepted for enrollment in January, April, and September. Application deadline for spring is December 10. Application deadline for summer is April 25. Application deadline for fall is September 1. In 1996, 200 applied; 200 were accepted. Applicants must have a high school diploma or GED and submit SAT s.

Contact
Michael Turner, Director of Admissions, Hospitality Services Management, 1801 East Fourth Street, Okmulgee, OK 74447-3901; 800-772-4471.

Oklahoma State University–Okmulgee is the only college in Oklahoma offering an Associate in Applied Science degree in baking, culinary arts, and dietetic technology. The program's baking director (34 years of experience) has been listed twice in the Guiness Book of World Records. Students maintain a full-service bakery. Culinary Arts has 2 CEC's (from Switzerland and France) and 2 Executive Chefs, both graduates of CIA. Students operate 2 restaurants from the newly remodeled electric and gas kitchens. The dietetic technology degree (supervised by a PhD) is approved by the American Dietetic Association to include 450 supervised hours in clinical nutrition and food service management at various health-care and food service facilities.

Central Oregon Community College

Cascade Culinary Institute

Bend, Oregon

General Information
Public, coeducational, two-year college. Rural campus. Founded in 1949. Accredited by Northwest Association of Schools and Colleges.

Program Information
Offered since 1993. Program calendar is divided into quarters. 4-term Certificate in Culinary Arts.

Areas of Study
Baking; controlling costs in food service; culinary skill development; food preparation; food service communication; food service math; garde-manger; introduction to food service; management and human resources; meal planning; nutrition and food service; sanitation; soup, stock, sauce, and starch production.

Facilities
Bake shop; cafeteria; catering service; computer laboratory; food production kitchen; 2 lecture rooms; library; public restaurant; snack shop; teaching kitchen.

Culinary Student Profile
24 full-time. 12 are under 25 years old; 8 are between 25 and 44 years old; 4 are over 44 years old.

Faculty
2 full-time; 2 part-time. 1 is a certified executive chef. Prominent faculty: Julian Darwin and Timothy H. Hill.

Expenses
In-state tuition: $37 per credit hour. Out-of-state tuition: $133 per credit hour.

Financial Aid
In 1996, 3 scholarships were awarded (average award was $1000). Employment placement assistance is available. Employment opportunities within the program are available.

Housing
Coed housing available. Average on-campus housing cost per month: $250. Average off-campus housing cost per month: $300.

Application Information
Students are accepted for enrollment in January, March, and September. In 1996, 30 applied; 30 were accepted. Applicants must submit a formal application.

Contact
Timothy H. Hill, Associate Professor, Cascade Culinary Institute, 2600 Northwest College Way, Bend, OR 97701-5998; 541-383-7713; Fax: 541-383-7708.

International School of Baking

Bend, Oregon

General Information
Private, coeducational, culinary institute. Urban campus. Founded in 1986.

Program Information
Offered since 1986. Member of American Culinary Federation; International Association of Culinary Professionals; James Beard Foundation, Inc.; The Bread Bakers Guild of America; Retail Bakers Association. Program calendar is customized to meet students needs. Certificate in Individualized Specialization. 1-month Certificate in Start Up Bakery Course. 2-week Certificates in European Pastries; Artisan Breads.

Areas of Study
Baking; patisserie; bakery start-up; custom designed.

Facilities
Bakery; classroom; demonstration laboratory; food production kitchen; learning resource center; teaching kitchen.

Culinary Student Profile
4 total: 2 full-time; 2 part-time.

Faculty
1 full-time; 2 part-time. 1 is an industry professional. Prominent faculty: Marda E. Stoliar.

Expenses
Tuition: $5950 per month full-time, $350 per day part-time.

Financial Aid
Employment placement assistance is available.

Application Information
Students are accepted for enrollment in January, February, March, May, June, July, August, September, and October. Applicants must submit a formal application.

Contact
Marda Stoliar, Director, 1971 N. W. Juniper Avenue, Bend, OR 97701; 541-389-8553; Fax: 541-389-3736; E-mail: domocorp@empnet.com

Under the International School of Baking's unique approach to teaching, a maximum of 2 students work, totally hands-on, side by side with the instructor. Classes can be customized to meet specific date requirements and interests. The School teaches methods for control of the product by temperature and weight of ingredients to guarantee consistent results. In addition to teaching at the International School of Baking for the past 12 years, Director Marda Stoliar has owned and operated a French bakery, "Breads of France," in Bend, Oregon, and been a consultant for U.S. Wheat Associates, Washington, DC, working in Hong Kong, Macao, and China.

Linn-Benton Community College

Culinary Arts/Restaurant Management

Albany, Oregon

General Information
Public, coeducational, two-year college. Rural campus. Founded in 1966. Accredited by Northwest Association of Schools and Colleges.

Program Information
Offered since 1969. Member of American Culinary Federation; Council on Hotel, Restaurant, and Institutional Education; National Restaurant Association; Women Chefs and Restaurateurs. Program calendar is divided into quarters. 2-year Associate degrees in Restaurant and Catering Management; Chef Training.

Areas of Study
Baking; beverage management; buffet catering; computer applications; controlling costs in food service; culinary French; culinary skill development; food preparation; food purchasing; garde-manger; international cuisine; introduction to food service; kitchen management; management and human resources; meal planning; meat

cutting; meat fabrication; menu and facilities design; nutrition; patisserie; sanitation; saucier; seafood processing; soup, stock, sauce, and starch production; wines and spirits.

Facilities

Bake shop; cafeteria; catering service; classroom; coffee shop; computer laboratory; food production kitchen; garden; gourmet dining room; laboratory; learning resource center; lecture room; library; public restaurant; snack shop.

Culinary Student Profile

40 full-time. 10 are under 25 years old; 25 are between 25 and 44 years old; 5 are over 44 years old.

Faculty

2 full-time; 4 part-time. 2 are culinary-accredited teachers. Prominent faculty: Scott Anselm and Mark Whitehead.

Expenses

Tuition: $525 per term full-time, $35 per credit part-time.

Financial Aid

In 1996, 7 scholarships were awarded (average award was $500). Employment placement assistance is available.

Housing

Average off-campus housing cost per month: $225.

Application Information

Students are accepted for enrollment in January, March, and September. Application deadline for fall is continuous with a recommended date of September 15. Application deadline for winter is continuous with a recommended date of December 1. Application deadline for spring is continuous with a recommended date of April 15. In 1996, 46 applied; 40 were accepted. Applicants must submit a formal application.

Contact

Scott Anselm, Program Coordinator, Culinary Arts/Restaurant Management, 6500 Southwest Pacific Boulevard, Albany, OR 97321; 541-917-4388; Fax: 541-917-4395; E-mail: anselms@gw.lbcc.cc.or.us

---

Students get a great head start on their careers in the hospitality industry with a degree from Linn-Benton Community College in Albany, Oregon. The professional chefs provide an extensive hands-on and theory-based program to prepare students for professional chef, restaurant, and catering management positions. The demanding 2-year curriculum covers all aspects of food preparation, including pantry, bakery, garde-manger, grill, sandwich making, la carte, quantity food, production, soups, sauces, and meat preparation as well as the fundamentals of money, personnel, and facilities management, all in an outstanding full-service kitchen with a wide variety of modern equipment.

Western Culinary Institute

Portland, Oregon

General Information

Private, coeducational, culinary institute. Urban campus. Founded in 1983.

Program Information

Offered since 1983. Accredited by American Culinary Federation Education Institute. Member of American Culinary Federation; American Culinary Federation Educational Institute. Program calendar is divided into six-week cycles. 12-month Diploma in Culinary Arts.

Areas of Study

Baking; beverage management; confectionery show pieces; controlling costs in food service; convenience cookery; culinary skill development; food preparation; food purchasing; food service math; garde-manger; international cuisine; introduction to food service; kitchen management; management and human resources; meal planning; meat cutting; meat fabrication; menu and facilities design; nutrition; nutrition and food service; patisserie; sanitation; saucier; soup, stock, sauce, and starch production; wines and spirits; internship.

Western Culinary Institute *(continued)*

Facilities
Bake shop; bakery; catering service; coffee shop; 6 food production kitchens; garden; gourmet dining room; learning resource center; 4 lecture rooms; library; 3 public restaurants; student lounge; bakery/delicatessen.

Culinary Student Profile
450 full-time.

Faculty
18 full-time; 5 part-time. 4 are industry professionals; 14 are culinary-accredited teachers.

Special Programs
6-week internship required at US or international location.

Expenses
Application fee: $125. Tuition: $13,511 per year. Program-related fees include: $300 for books and handouts; $195 for cutlery; $264 for uniforms.

Financial Aid
In 1996, 7 scholarships were awarded (average award was $500). Employment placement assistance is available. Employment opportunities within the program are available.

Housing
Average off-campus housing cost per month: $300.

Application Information
Students are accepted for enrollment in January, February, April, May, July, August, October, and November. Applicants must submit a formal application and have a high school diploma.

Contact
Mary Harris, Director of Admissions, 1316 Southwest 13th Avenue, Portland, OR 97201-3355; 800-666-0312; Fax: 503-223-0126; World Wide Web: http://www.westernculinary.com

---

The core of the culinary curriculum at Western Culinary Institute is the hands-on teaching of cooking and baking skills as well as the theoretical knowledge that must underlie competency in both fields. It endeavors to expose students to the different styles and experiences of the school's chefs and instructors, to acquaint students with a wide variety of equipment, and to prepare them for whatever area of the food service/hospitality industry they choose to enter. WCI provides placement assistance throughout the career of its graduates. The Institute's outstanding reputation attracts prestigious employers from all over the world.

See color display following page 172.

The Art Institute of Philadelphia

School of Culinary Arts

Philadelphia, Pennsylvania

General Information
Private, coeducational, two-year college. Urban campus. Founded in 1966.

Program Information
Offered since 1977. Member of American Culinary Federation; Council on Hotel, Restaurant, and Institutional Education; National Restaurant Association. Program calendar is divided into quarters. 18-month Associate degree in Culinary Arts.

Areas of Study
Baking; beverage management; computer applications; controlling costs in food service; culinary French; culinary skill development; food preparation; food purchasing; food service communication; food service math; garde-manger; international cuisine; introduction to food service; kitchen management; management and human resources; meat cutting; menu and facilities design; nutrition; patisserie; restaurant opportunities; sanitation; saucier; seafood processing; soup, stock, sauce, and starch production; wines and spirits; a la carte; safety.

Facilities
Bake shop; 5 classrooms; 3 computer laboratories; 2 food production kitchens; gourmet dining room;

learning resource center; library; public restaurant; 2 student lounges; 3 teaching kitchens.

Faculty
3 full-time; 2 part-time. 5 are industry professionals.

Special Programs
Individualized instruction.

Expenses
Application fee: $50. Tuition: $19,500 per 18 months full-time, $217 per credit part-time. Program-related fees include: $250 for lab fees per quarter.

Financial Aid
Program-specific awards include Pennsylvania Restaurant Association scholarships. Employment placement assistance is available. Employment opportunities within the program are available.

Housing
Average off-campus housing cost per month: $425.

Application Information
Students are accepted for enrollment in January, April, July, and October. Applicants must submit a formal application and letters of reference.

Contact
Christine Spotts, Associate Director of Admissions, School of Culinary Arts, 2300 Market Street, Philadelphia, PA 19103; 215-567-7080 Ext. 423; Fax: 215-564-0241.

See affiliated programs: Colorado Institute of Art; New York Restaurant School; The Art Institute of Atlanta; The Art Institute of Fort Lauderdale; The Art Institute of Houston; The Art Institute of Phoenix; The Art Institute of Seattle.

Few occupations offer the creativity, excitement, and growth found in culinary arts. The curriculum for The Art Institute of Philadelphia's School of Culinary Arts is based on classical principles that emphasize modern techniques and trends. The faculty is nationally known in the culinary industry. Faculty members offer students the expertise and knowledge needed to become true professionals in this field. Classes are taught in new state-of-the-art kitchen facilities. Courses in basic skills and advanced techniques include international cooking, a la carte, garde-manger, baking and pastry, safety, sanitation, and nutrition. For more information, students should contact the Admissions Office at 800-275-2474.

See display on page 48.

Bucks County Community College

Business Department

Newtown, Pennsylvania

General Information
Public, coeducational, two-year college. Suburban campus. Founded in 1964. Accredited by Middle States Association of Colleges and Schools.

Program Information
Offered since 1967. Member of American Culinary Federation; Confrerie de la Chaine des Rotisseurs; Council on Hotel, Restaurant, and Institutional Education; Educational Foundation of the NRA; Interntional Food Service Executives Association; National Restaurant Association. Program calendar is divided into semesters. 1-year Certificate in Hospitality/Restaurant/Institutional Supervision. 2-year Associate degree in Hospitality/Restaurant/ Institutional Management. 3-year Associate degrees in Chef Apprenticeship-Pastry Emphasis; Chef Apprenticeship-Foods Emphasis.

Areas of Study
Baking; buffet catering; computer applications; confectionery show pieces; controlling costs in food service; culinary skill development; food preparation; food purchasing; food service communication; food service math; garde-manger; introduction to food service; kitchen management; management and human resources; meal planning; meat cutting; meat fabrication; menu and facilities design; nutrition; nutrition and food service; patisserie; restaurant opportunities; sanitation; saucier; seafood processing; soup, stock, sauce, and starch production.

Bucks County Community College *(continued)*

FACILITIES
Cafeteria; 3 classrooms; computer laboratory; demonstration laboratory; food production kitchen; gourmet dining room; laboratory; learning resource center; library; snack shop; student lounge; teaching kitchen.

CULINARY STUDENT PROFILE
49 total: 5 full-time; 44 part-time.

FACULTY
3 full-time; 3 part-time. 3 are industry professionals; 3 are culinary-accredited teachers.

SPECIAL PROGRAMS
Required paid cooperative education and paid summer internship in management.

EXPENSES
Application fee: $30. In-state tuition: $66 per credit. Out-of-state tuition: $198 per credit.

FINANCIAL AID
In 1996, 3 scholarships were awarded (average award was $200). Employment placement assistance is available. Employment opportunities within the program are available.

APPLICATION INFORMATION
Students are accepted for enrollment in January and August. Applicants must submit a formal application.

CONTACT
Earl R. Arrowood, Coordinator, Business Department, Swamp Road, Newtown, PA 18940-1525; 215-968-8241; World Wide Web: http://www.bucks.edu

See affiliated apprenticeship program.

COMMUNITY COLLEGE OF ALLEGHENY COUNTY ALLEGHENY CAMPUS

Culinary Arts Program

Pittsburgh, Pennsylvania

GENERAL INFORMATION
Public, coeducational, two-year college. Urban campus. Founded in 1966. Accredited by Middle States Association of Colleges and Schools.

PROGRAM INFORMATION
Offered since 1974. Member of American Culinary Federation; American Culinary Federation Educational Institute; The Bread Bakers Guild of America. Program calendar is divided into semesters. 2-year Associate degree in Culinary Arts.

AREAS OF STUDY
Baking; beverage management; buffet catering; controlling costs in food service; culinary French; culinary skill development; food preparation; food purchasing; food service communication; food service math; garde-manger; international cuisine; introduction to food service; kitchen management; meal planning; menu and facilities design; nutrition; sanitation; saucier; soup, stock, sauce, and starch production.

FACILITIES
Bake shop; bakery; cafeteria; catering service; 6 classrooms; coffee shop; computer laboratory; demonstration laboratory; food production kitchen; laboratory; learning resource center; lecture room; library; student lounge; teaching kitchen.

CULINARY STUDENT PROFILE
25 full-time. 15 are under 25 years old; 5 are between 25 and 44 years old; 5 are over 44 years old.

FACULTY
6 part-time. 6 are culinary-accredited teachers.

EXPENSES
In-state tuition: $1020 per semester full-time, $68 per credit part-time. Out-of-state tuition: $3060 per semester full-time, $204 per credit part-time.

FINANCIAL AID
Employment placement assistance is available. Employment opportunities within the program are available.

APPLICATION INFORMATION
Students are accepted for enrollment in January and August. In 1996, 150 applied; 18 were accepted. Applicants must submit a formal application.

Contact
Willie F. Stinson, Coordinator, Culinary Arts Program, 808 Ridge Avenue, Pittsburgh, PA 15212-6003; 412-237-2698; Fax: 412-237-4678.

Community College of Philadelphia

Hospitality Technologies Programs

Philadelphia, Pennsylvania

General Information
Public, coeducational, two-year college. Urban campus. Founded in 1964. Accredited by Middle States Association of Colleges and Schools.

Program Information
Offered since 1964. Program calendar is divided into semesters. 2-year Associate degrees in Chef Option; Restaurant Management; Hotel Management. 3-year Associate degree in Chef Apprenticeship Option.

Areas of Study
Baking; beverage management; computer applications; controlling costs in food service; culinary skill development; food preparation; food purchasing; food service math; garde-manger; international cuisine; kitchen management; menu and facilities design; nutrition; patisserie; restaurant opportunities; sanitation; soup, stock, sauce, and starch production; wines and spirits; introduction to hospitality; catering.

Culinary Student Profile
250 total: 188 full-time; 62 part-time.

Faculty
2 full-time; 4 part-time.

Special Programs
Required 100-6,000 hour internships.

Expenses
In-state tuition: $3000 per degree full-time, $69 per credit part-time. Out-of-state tuition: $9000 per degree full-time, $207 per credit part-time.

Application Information
Students are accepted for enrollment in January and September. Applicants must submit a formal application.

Contact
Mark Kushner, Curriculum Coordinator, Hospitality Technologies Programs, 1700 Spring Garden Street, Philadelphia, PA 19130-3991; 215-751-8797; Fax: 215-972-6388.

Drexel University

Hotel Management and Culinary Arts

Philadelphia, Pennsylvania

General Information
Private, coeducational, university. Urban campus. Founded in 1891. Accredited by Middle States Association of Colleges and Schools.

Program Information
Member of American Culinary Federation; American Culinary Federation Educational Institute; Confrerie de la Chaine des Rotisseurs; Council on Hotel, Restaurant, and Institutional Education; Educational Foundation of the NRA; International Association of Culinary Professionals; James Beard Foundation, Inc.; National Restaurant Association. Program calendar is divided into quarters. 4-year Bachelor's degrees in Culinary Arts; Hotel and Restaurant Management.

Areas of Study
Baking; beverage management; computer applications; confectionery show pieces; culinary French; culinary skill development; food preparation; food purchasing; food service communication; garde-manger; international cuisine; introduction to food service; kitchen management; management and human resources; meal planning; meat cutting; meat fabrication; menu and facilities design; nutrition and food service; patisserie; restaurant opportunities; sanitation; saucier; seafood processing; soup, stock, sauce, and starch production; wines and spirits.

Drexel University *(continued)*

Facilities

Bake shop; bakery; 10 classrooms; computer laboratory; demonstration laboratory; 3 food production kitchens; 2 gourmet dining rooms; 2 laboratories; learning resource center; 10 lecture rooms; 2 libraries; public restaurant; student lounge; 3 teaching kitchens.

Culinary Student Profile

195 total: 120 full-time; 75 part-time.

Faculty

5 full-time; 15 part-time. Prominent faculty: Francis McFadden and Lynn Hoffman.

Expenses

Application fee: $35. Tuition: $14,000 per year full-time, $180 per credit part-time.

Financial Aid

In 1996, 20 scholarships were awarded (average award was $6000); 80 loans were granted (average loan was $5000). Employment placement assistance is available. Employment opportunities within the program are available.

Housing

Coed, apartment-style, and single-sex housing available. Average on-campus housing cost per month: $400. Average off-campus housing cost per month: $500.

Application Information

Students are accepted for enrollment in January, March, June, and September. Applicants must submit a formal application.

Contact

Francis McFadden, Chef, Hotel Management and Culinary Arts, Nesbitt College, 33rd and Market Streets, Philadelphia, PA 19104-2875; 215-895-4919; Fax: 215-895-4917; E-mail: mcfaddfm@duvm.ocs.drexel.edu; World Wide Web: http://www.drexel.edu

Harrisburg Area Community College

Hospitality, Restaurant, and Institutional Management Department

Harrisburg, Pennsylvania

General Information

Public, coeducational, two-year college. Suburban campus. Founded in 1964. Accredited by Middle States Association of Colleges and Schools.

Program Information

Offered since 1989. Member of National Restaurant Association; College Restaurant Hospitality Institute Educators. Program calendar is divided into semesters. 1-year Diploma in Culinary Arts/Catering. 21-month Associate degree in Culinary Arts. 21-month Certificate in Culinary Arts.

Areas of Study

Baking; computer applications; controlling costs in food service; culinary skill development; food preparation; food purchasing; kitchen management; management and human resources; meal planning; menu and facilities design; nutrition; sanitation; introduction to hospitality.

Facilities

Classroom; computer laboratory; demonstration laboratory; food production kitchen; gourmet dining room; learning resource center; lecture room; library; teaching kitchen; herb garden.

Culinary Student Profile

113 total: 66 full-time; 47 part-time. 57 are under 25 years old; 48 are between 25 and 44 years old; 8 are over 44 years old.

Faculty

4 full-time; 3 part-time. 2 are industry professionals; 1 is a culinary-accredited teacher. Prominent faculty: Marcia W. Shore and Ruth Anne McGinley.

Expenses

Application fee: $25. Tuition: $1542 per semester full-time, $128.50 per credit hour part-time.

Program-related fees include: $250 for knives, garnishing tools, and uniforms; $12 per hour for lab fees.

Financial Aid
Employment placement assistance is available.

Application Information
Students are accepted for enrollment in January, May, and August. Application deadline for fall is continuous with a recommended date of May 1. In 1996, 117 applied; 48 were accepted. Applicants must submit a formal application and letters of reference and have a health certificate.

Contact
Marcia W. Shore, Program Coordinator, Hospitality, Restaurant, and Institutional Management Department, M250A One HACC Drive, Harrisburg, PA 17110; 717-780-2674; Fax: 717-231-7670.

Hiram G. Andrews Center

Culinary Arts Program

Johnstown, Pennsylvania

General Information
Private, coeducational, technical institute. Suburban campus. Founded in 1959.

Program Information
Offered since 1974. Member of Council on Hotel, Restaurant, and Institutional Education; National Restaurant Association. Program calendar is divided into trimesters. 20-month Associate degree in Culinary Arts. 8-month Diploma in Kitchen Helper studies.

Areas of Study
Baking; computer applications; controlling costs in food service; food preparation; food purchasing; food service math; introduction to food service; management and human resources; meal planning; menu and facilities design; nutrition; sanitation; soup, stock, sauce, and starch production.

Facilities
Bake shop; cafeteria; 3 classrooms; 2 computer laboratories; food production kitchen; laboratory; learning resource center; library; teaching kitchen.

Culinary Student Profile
45 full-time. 10 are under 25 years old; 25 are between 25 and 44 years old; 10 are over 44 years old.

Faculty
3 full-time. 1 is an industry professional; 2 are culinary-accredited teachers. Prominent faculty: Noel B. Graham and Robert H. Lawson.

Expenses
Tuition: $2940 per 16 weeks.

Financial Aid
Employment placement assistance is available.

Housing
Single-sex housing available.

Application Information
Students are accepted for enrollment in January, May, and September. In 1996, 45 applied; 45 were accepted. Applicants must submit a formal application.

Contact
Albert Hromulak, Assistant Director, Student Services, Culinary Arts Program, 727 Goucher Street, Johnstown, PA 15905; 814-255-8372; Fax: 814-255-3406.

Indiana University of Pennsylvania

Academy of Culinary Arts

Punxsutawney, Pennsylvania

General Information
Public, coeducational, university. Rural campus. Founded in 1875. Accredited by Middle States Association of Colleges and Schools.

Program Information
Offered since 1989. Accredited by American Culinary Federation Education Institute. Member

Indiana University of Pennsylvania *(continued)*

of American Culinary Federation Educational Institute; Council on Hotel, Restaurant, and Institutional Education; International Association of Culinary Professionals. Program calendar is divided into semesters. 16-month Certificate in Culinary Arts. 4-year Bachelor's degree in Hotel, Restaurant, and Institutional Management.

Areas of Study
Baking; beverage management; buffet catering; computer applications; controlling costs in food service; culinary skill development; food preparation; food purchasing; food service math; garde-manger; international cuisine; kitchen management; meal planning; meat fabrication; menu and facilities design; nutrition; nutrition and food service; sanitation; soup, stock, sauce, and starch production; wines and spirits.

Facilities
Bake shop; 3 classrooms; computer laboratory; 2 demonstration laboratories; 5 food production kitchens; gourmet dining room; learning resource center; library; student lounge.

Culinary Student Profile
100 full-time.

Faculty
8 full-time; 1 part-time. Prominent faculty: Albert Wutsch and Hilary DeMane.

Special Programs
18-month international studies option.

Expenses
Application fee: $30. Tuition: $4600 per semester. Program-related fees include: $1200 for supply package.

Financial Aid
In 1996, 18 scholarships were awarded (average award was $1500). Program-specific awards include private scholarship support for program. Employment placement assistance is available. Employment opportunities within the program are available.

Housing
Coed and single-sex housing available. Average on-campus housing cost per month: $225. Average off-campus housing cost per month: $250.

Application Information
Students are accepted for enrollment in September. In 1996, 250 applied; 188 were accepted. Applicants must submit a formal application and letters of reference.

Contact
Kelly Barry, Admissions Coordinator, Academy of Culinary Arts, Reschini Building, Indiana, PA 15705; 800-727-0997; Fax: 412-357-6200; E-mail: culinary-arts@grove.iup.edu; World Wide Web: http://www.iup.edu/cularts

JNA Institute of Culinary Arts

Philadelphia, Pennsylvania

General Information
Private, coeducational, culinary institute. Urban campus. Founded in 1988.

Program Information
Offered since 1988. Member of Interntional Food Service Executives Association; National Restaurant Association. Program calendar is divided into ten-week cycles. 2-year Associate degree in Culinary Arts/Restaurant Management. 30-week Diploma in Food Service Training.

Areas of Study
Baking; beverage management; buffet catering; computer applications; controlling costs in food service; culinary French; culinary skill development; food preparation; food purchasing; food service communication; food service math; garde-manger; international cuisine; introduction to food service; kitchen management; management and human resources; menu and facilities design; nutrition; nutrition and food service; patisserie; sanitation; saucier; seafood processing; soup, stock, sauce, and starch production.

Facilities
2 classrooms; food production kitchen; lecture room; library; student lounge; teaching kitchen.

Culinary Student Profile
75 total: 60 full-time; 15 part-time. 30 are under 25 years old; 30 are between 25 and 44 years old; 15 are over 44 years old.

Faculty
6 full-time; 3 part-time. 9 are industry professionals.

Special Programs
Paid externships.

Expenses
Tuition: $5000 per year full-time, $133 per credit part-time. Program-related fees include: $75 for registration.

Financial Aid
Employment placement assistance is available. Employment opportunities within the program are available.

Housing
Average off-campus housing cost per month: $450.

Application Information
Students are accepted for enrollment in January, March, April, June, July, September, October, and December. In 1996, 98 applied; 80 were accepted. Applicants must submit a formal application and give an interview or letters of reference.

Contact
Diane Goldstein, Director of Admissions, 1212 South Broad Street, Philadelphia, PA 19146; 215-468-8800; Fax: 215-468-8838; E-mail: dsg@culinaryarts.com; World Wide Web: http://www.culinaryarts.com

---

JNA Institute of Culinary Arts, located in Philadelphia, PA, offers an associate degree program in culinary arts/restaurant management. A shorter diploma program is also offered. Both programs include a paid externship. Both programs offer high-quality training by current and former chefs with numerous years in the food service industry. The curriculum is diverse and affordable and provides the necessary mixture of core and specialized classes to fully prepare the student for a career in the culinary arts. Extracurricular activities include ice carving and pastry decorating. Students are encouraged to participate in Coquina Academia, an association of students, alumni, and instructors working as peers to further their culinary education.

Keystone College

Culinary Arts

La Plume, Pennsylvania

General Information
Private, coeducational, two-year college. Rural campus. Founded in 1868. Accredited by Middle States Association of Colleges and Schools.

Program Information
Offered since 1995. Member of American Culinary Federation; Council on Hotel, Restaurant, and Institutional Education. Program calendar is divided into semesters. 2-year Associate degrees in Restaurant/Food Service Management; Culinary Arts.

Areas of Study
Baking; beverage management; computer applications; controlling costs in food service; food preparation; food purchasing; garde-manger; introduction to food service; kitchen management; management and human resources; meal planning; meat cutting; nutrition; nutrition and food service; sanitation; saucier; soup, stock, sauce, and starch production; wines and spirits.

Facilities
Bakery; cafeteria; catering service; 3 classrooms; 3 computer laboratories; 7 demonstration laboratories; food production kitchen; 3 gardens; gourmet dining room; 5 laboratories; 2 learning resource centers; 38 lecture rooms; library; snack shop; 6 student lounges; teaching kitchen.

Culinary Student Profile
41 total: 24 full-time; 17 part-time. 18 are under 25 years old; 15 are between 25 and 44 years old; 8 are over 44 years old.

Keystone College ***(continued)***

Faculty
2 full-time; 3 part-time. 5 are industry professionals.

Expenses
Application fee: $25. Tuition: $4660 per semester full-time, $230 per credit part-time. Program-related fees include: $175 for cooking courses.

Financial Aid
In 1996, 3 scholarships were awarded (average award was $2000). Employment placement assistance is available. Employment opportunities within the program are available.

Housing
Coed, apartment-style, and single-sex housing available.

Application Information
Students are accepted for enrollment in January and August. Application deadline for fall is August 1. Application deadline for spring is January 2. In 1996, 37 applied; 23 were accepted. Applicants must submit a formal application and letters of reference and interview if possible.

Contact
Patricia Davis, Division Chair, Culinary Arts, Harris Hall 204, LaPlume, PA 18440; 717-945-5141 Ext. 3860; Fax: 717-945-6960.

Mercyhurst College

The Culinary and Wine Institute

Erie, Pennsylvania

General Information
Private, coeducational, comprehensive institution. Urban campus. Founded in 1926. Accredited by Middle States Association of Colleges and Schools.

Program Information
Offered since 1995. Member of American Culinary Federation; American Culinary Federation Educational Institute; Council on Hotel, Restaurant, and Institutional Education; National Restaurant Association. Program calendar is divided into terms. 2-year Associate degree in Hotel, Restaurant, and Institutional Management.

Areas of Study
Baking; beverage management; computer applications; culinary skill development; food preparation; food purchasing; food service communication; food service math; garde-manger; international cuisine; introduction to food service; kitchen management; management and human resources; meal planning; meat cutting; meat fabrication; menu and facilities design; nutrition; nutrition and food service; patisserie; sanitation; saucier; seafood processing; soup, stock, sauce, and starch production; wines and spirits; wine making.

Facilities
Bake shop; cafeteria; 4 classrooms; 2 computer laboratories; 4 demonstration laboratories; 4 food production kitchens; gourmet dining room; learning resource center; 2 lecture rooms; library; 2 student lounges; teaching kitchen; vineyard.

Culinary Student Profile
45 full-time. 15 are under 25 years old; 25 are between 25 and 44 years old; 5 are over 44 years old.

Faculty
3 full-time; 7 part-time. 1 is an industry professional; 2 are culinary-accredited teachers. Prominent faculty: Stephen C. Fernald and John Harrison.

Expenses
Application fee: $10. Tuition: $7264 per year full-time, $681 per course part-time. Program-related fees include: $333 for culinary fee (knives, uniforms, and baking kit).

Financial Aid
Program-specific awards include institution grants and service grants ($1275). Employment placement assistance is available.

Housing
Coed housing available. Average on-campus housing cost per month: $463.

Application Information
Students are accepted for enrollment in March, September, and November. In 1996, 62 applied;

35 were accepted. Applicants must submit a formal application and have a high school diploma or GED.

Contact

James Theeuwes, Director of Admissions, The Culinary and Wine Institute, 501 East 38th Street, Erie, PA 16546; 800-825-1926 Ext. 2238; Fax: 814-824-2179; E-mail: jtheeuwe@paradise.mercy.edu

See color display following page 204.

Northampton County Area Community College

Culinary Arts

Bethlehem, Pennsylvania

General Information

Public, coeducational, two-year college. Suburban campus. Founded in 1967. Accredited by Middle States Association of Colleges and Schools.

Program Information

Offered since 1993. Program calendar is divided into semesters. 2-year Associate degree in Restaurant/Hotel Management. 45-week Diploma in Culinary Arts.

Areas of Study

Baking; beverage management; controlling costs in food service; culinary skill development; food preparation; food purchasing; garde-manger; introduction to food service; meat cutting; meat fabrication; restaurant opportunities; sanitation; seafood processing; soup, stock, sauce, and starch production; wines and spirits.

Facilities

Bakery; food production kitchen; gourmet dining room; teaching kitchen.

Culinary Student Profile

40 full-time. 10 are under 25 years old; 25 are between 25 and 44 years old; 5 are over 44 years old.

Faculty

2 full-time; 1 part-time.

Expenses

Application fee: $20. Tuition: $73 per credit.

Financial Aid

In 1996, 1 scholarship was awarded (award was $250); 10 loans were granted (average loan was $2000). Employment placement assistance is available.

Housing

Coed and apartment-style housing available. Average on-campus housing cost per month: $600. Average off-campus housing cost per month: $800.

Application Information

Students are accepted for enrollment in March and September. Application deadline for fall is continuous with a recommended date of May 30. Application deadline for spring is continuous with a recommended date of November 20. In 1996, 100 applied; 46 were accepted. Applicants must submit a formal application.

Contact

Admissions, Culinary Arts, 3835 Green Pond Road, Bethlehem, PA 18017-7599; 610-861-5593.

Pennsylvania College of Technology

Hospitality Division

Williamsport, Pennsylvania

General Information

Public, coeducational, two-year college. Urban campus. Founded in 1989. Accredited by Middle States Association of Colleges and Schools.

Program Information

Offered since 1989. Accredited by American Culinary Federation Education Institute. Member of American Culinary Federation; American Culinary Federation Educational Institute; American Institute of Baking; Council on Hotel, Restaurant, and Institutional Education; Educational Foundation of the NRA; National Restaurant Association; Retail Bakers Association.

Pennsylvania College of Technology ***(continued)***

Program calendar is divided into semesters. 2-year Associate degrees in Food/Hospitality Management; Baking/Pastry Arts; Culinary Arts. 8-month Certificates in Dining Room Service; Professional Baking; Professional Cooking.

Areas of Study
Baking; beverage management; buffet catering; computer applications; confectionery show pieces; controlling costs in food service; culinary French; culinary skill development; food preparation; food purchasing; garde-manger; international cuisine; introduction to food service; management and human resources; meal planning; meat cutting; meat fabrication; nutrition; nutrition and food service; restaurant opportunities; sanitation; soup, stock, sauce, and starch production; wines and spirits.

Facilities
Bake shop; 2 bakeries; catering service; 3 classrooms; computer laboratory; demonstration laboratory; 5 food production kitchens; garden; 2 gourmet dining rooms; learning resource center; lecture room; library; public restaurant.

Culinary Student Profile
114 total: 94 full-time; 20 part-time.

Faculty
8 full-time; 5 part-time. 7 are industry professionals; 5 are culinary-accredited teachers.

Special Programs
Distinguished Visiting Chefs/Pastry Chefs Series.

Expenses
Application fee: $40. Tuition: $6540 per year full-time, $218 per credit part-time.

Financial Aid
In 1996, 4 scholarships were awarded (average award was $300). Employment placement assistance is available. Employment opportunities within the program are available.

Housing
Coed housing available.

Application Information
Students are accepted for enrollment in January and August. In 1996, 156 applied; 143 were accepted. Applicants must submit a formal application.

Contact
William Butler, Dean of Hospitality, Hospitality Division, One College Avenue, Williamsport, PA 17701-5778; 717-326-3761 Ext. 7630; E-mail: wbutler@pct.edu; World Wide Web: http://www.pct.edu

See color display following page 204.

Pennsylvania Institute of Culinary Arts

Pennsylvania Culinary

Pittsburgh, Pennsylvania

General Information
Private, coeducational, culinary institute. Urban campus. Founded in 1986.

Program Information
Offered since 1986. Accredited by American Culinary Federation Education Institute. Member of American Culinary Federation; American Culinary Federation Educational Institute; Confrerie de la Chaine des Rotisseurs; Council on Hotel, Restaurant, and Institutional Education; Educational Foundation of the NRA; National Restaurant Association; Women Chefs and Restaurateurs. Program calendar is divided into semesters. 16-month Associate degrees in Hotel-Restaurant Management; Culinary Arts.

Areas of Study
Baking; beverage management; computer applications; controlling costs in food service; convenience cookery; culinary skill development; food preparation; food purchasing; food service communication; food service math; garde-manger; international cuisine; introduction to food service; kitchen management; management and human resources; meal planning; meat cutting; meat fabrication; menu and facilities design; nutrition;

nutrition and food service; patisserie; restaurant opportunities; sanitation; saucier; seafood processing; soup, stock, sauce, and starch production; wines and spirits.

Facilities
Bake shop; 5 classrooms; computer laboratory; 9 food production kitchens; gourmet dining room; 5 lecture rooms; library; public restaurant; 3 student lounges; 9 teaching kitchens.

Culinary Student Profile
1,000 full-time.

Faculty
20 full-time; 1 part-time. 1 is a master chef; 20 are culinary-accredited teachers. Prominent faculty: Dieter Kiessling.

Special Programs
4-month paid externship.

Expenses
Application fee: $50. Tuition: $5000 per semester.

Financial Aid
In 1996, 2 scholarships were awarded (average award was $4000). Employment placement assistance is available. Employment opportunities within the program are available.

Housing
500 culinary students housed on campus. Coed and apartment-style housing available.

Application Information
Students are accepted for enrollment in January, March, May, June, September, and October. Applicants must submit a formal application and letters of reference.

Contact
Debra Maurizio, Director of Marketing, Pennsylvania Culinary, 717 Liberty Avenue, Pittsburgh, PA 15222; 800-432-2433; Fax: 412-566-2434; World Wide Web: http://www.pacul.com

The Restaurant School

School of Culinary Arts

Philadelphia, Pennsylvania

General Information
Private, coeducational, culinary institute. Urban campus. Founded in 1974.

Program Information
Offered since 1982. Member of American Culinary Federation; American Institute of Wine & Food; Confrerie de la Chaine des Rotisseurs; Council on Hotel, Restaurant, and Institutional Education; Educational Foundation of the NRA; International Association of Culinary Professionals; National Restaurant Association. 15-month Associate degree in Specialized Technology.

Areas of Study
Baking; beverage management; buffet catering; computer applications; confectionery show pieces; controlling costs in food service; culinary French; culinary skill development; food preparation; food purchasing; food service communication; food service math; garde-manger; international cuisine; introduction to food service; kitchen management; management and human resources; meal planning; meat cutting; meat fabrication; menu and facilities design; nutrition; nutrition and food service; patisserie; restaurant opportunities; sanitation; saucier; seafood processing; soup, stock, sauce, and starch production; wines and spirits.

Facilities
Bakery; 2 classrooms; computer laboratory; 2 demonstration laboratories; 4 food production kitchens; 2 gourmet dining rooms; learning resource center; 2 lecture rooms; library; 2 public restaurants; student lounge; wine cellar.

Culinary Student Profile
610 total: 450 full-time; 160 part-time.

Faculty
12 full-time; 6 part-time. 13 are industry professionals; 1 is a master chef; 4 are certified executive chefs.

Special Programs
8-day gastronomic tour of France (culinary students), 8-day Florida and Bahamas cruise and resort tour (management students).

Expenses
Application fee: $50. Tuition: $17,900 per 15 months full-time, $17,900 per 24 months part-time. Program-related fees include: $1500 for tools, books, equipment, and uniforms; $150 for registration fees.

Financial Aid
In 1996, 20 scholarships were awarded (average award was $1000). Employment placement assistance is available.

Housing
25 culinary students housed on campus. Coed and apartment-style housing available. Average on-campus housing cost per month: $350. Average off-campus housing cost per month: $350.

Application Information
Students are accepted for enrollment in April, May, October, and November. Applicants must submit a formal application and letters of reference.

Contact
Susan Salmon Fad, Director of Admissions, School of Culinary Arts, 4207 Walnut Street, Philadelphia, PA 19104-3518; 215-222-4200 Ext. 3011; Fax: 215-222-4219.

The Restaurant School, founded in 1974, is one of the nation's first schools to offer programs that specialize in the art of fine dining. It is the only school in the country to include an 8-day gastronomic tour of France as part of the school curriculum for the chef and pastry chef students and an 8-day tour of Florida and the Bahamas for the restaurant and hotel management students. The Restaurant School believes in giving students every opportunity and venue for learning. The inspiration students receive and the marks these tours make on a student's resume are invaluable.

Westmoreland County Community College

Hospitality Department

Youngwood, Pennsylvania

General Information
Public, coeducational, two-year college. Rural campus. Founded in 1970. Accredited by Middle States Association of Colleges and Schools.

Program Information
Offered since 1980. Accredited by American Culinary Federation Education Institute. Member of American Culinary Federation; American Culinary Federation Educational Institute; American Dietetic Association; Council on Hotel, Restaurant, and Institutional Education; National Restaurant Association. Program calendar is divided into semesters. 2-year Associate degrees in Travel and Tourism; Hotel/Motel Management; Dietetic Technician; Food Service Management; Baking and Pastry; Culinary Arts–Nonapprenticeship. 3-year Associate degree in Culinary Arts–Apprenticeship. 5-month Certificate in Baking and Pastry.

Areas of Study
Baking; beverage management; buffet catering; computer applications; confectionery show pieces; controlling costs in food service; culinary French; culinary skill development; food preparation; food purchasing; food service communication; food service math; garde-manger; international cuisine; introduction to food service; kitchen management; management and human resources; meal planning; menu and facilities design; nutrition; nutrition and food service; patisserie; restaurant opportunities; sanitation; saucier; seafood processing; soup, stock, sauce, and starch production; wines and spirits.

Facilities
2 bake shops; cafeteria; 10 classrooms; 20 computer laboratories; demonstration laboratory; food production kitchen; gourmet dining room; laboratory; learning resource center; 10 lecture rooms; library; 2 student lounges; teaching kitchen.

Culinary Student Profile
273 total: 106 full-time; 167 part-time. 152 are under 25 years old; 94 are between 25 and 44 years old; 27 are over 44 years old.

Faculty
4 full-time; 18 part-time. 17 are industry professionals; 5 are culinary-accredited teachers. Prominent faculty: Mary Zappone and Carl Dunkel.

Special Programs
10-day hospitality study tour of Italy.

Expenses
Application fee: $10. In-state tuition: $552 per semester full-time, $46 per credit part-time. Out-of-state tuition: $1656 per semester full-time, $138 per credit part-time. Program-related fees include: $150 for uniforms; $125 for knives; $200 for lab fees.

Financial Aid
In 1996, 3 scholarships were awarded (average award was $600); 20 loans were granted (average loan was $1300). Employment placement assistance is available. Employment opportunities within the program are available.

Application Information
Students are accepted for enrollment in January, May, and August. In 1996, 290 applied; 290 were accepted. Applicants must submit a formal application and take a physical exam.

Contact
Susan Kuhn, Admissions Coordinator, Hospitality Department, Armbrust Road, Youngwood, PA 15697; 412-925-4064; Fax: 412-925-1150; E-mail: kuhnsl@wccc.westmoreland.cc.pa.us; World Wide Web: http://www.westmoreland.cc.pa.us

Johnson & Wales University

College of Culinary Arts

Providence, Rhode Island

General Information
Private, coeducational, comprehensive institution.

JOHNSON & WALES UNIVERSITY

PROVIDENCE, RHODE ISLAND U.S.A.

America's Career University®

Johnson & Wales University, founded in 1914, is a private, coeducational institution of more than 6,500 full-time students offering programs in hospitality, business, technology and foodservice. Often referred to as "America's Career University," Johnson & Wales is known for its practical, hands-on approach to career education.

Some of the unique features that make a Johnson & Wales education very much in demand are:

- flexible, specialized programs
- four-day school week
- three-term calendar schedule
- upside-down curriculum where students take courses in their major field of study immediately
- very high employment rate

A University in the City

Johnson & Wales University's main campus is located in Providence, Rhode Island. Providence is the capital city of Rhode Island, which is nicknamed the "Ocean State." Only an hour from Boston and three hours from New York, Providence is a city of historic charm and culture and is within easy reach of several well-known vacation destinations like Newport, Rhode Island and Cape Cod, Massachusetts.

World's Largest Hospitality Educator

Johnson & Wales University has been recognized by CHRIE (Council of Hotel and Restaurant Institutional Educators) for enrolling the most hospitality (hotel-restaurant and travel-tourism) students in the world. The school also holds the distinction of being the largest culinary arts and foodservice educator in the world.

English Language Institute

Johnson & Wales is committed to the success of each and every enrolled student. The English Language Institute offers year-round intensive English language instruction with course credit at the intermediate and advanced levels.

Graduate School

In addition to associate and bachelor's degrees, Johnson & Wales offers master's degrees (M.S., M.A., M.Ed. and M.B.A.) through its Graduate School. One-year accelerated programs in International Business, Hospitality Administration and Management are offered, in addition to two-year evening programs.

Career-focused Education

Johnson & Wales students graduate with hands-on experience in their majors, thanks to the University's commitment to providing quality education and professional development. For one term, students experience various industries through "practicum assignments" at the University's public training facilities. These facilities include the University-owned and -operated Radisson Airport Hotel, the Johnson & Wales Inn, an American Express Travel Service, a women's retail store, and the Culinary Archives & Museum. All provide real-life professional experience for Johnson & Wales students.

Branch Campuses

In addition to Providence, Johnson & Wales calls four other U.S. cities home. Campuses are located in Charleston, South Carolina; Norfolk, Virginia; and North Miami, Florida. In addition, the University offers programs in Gothenburg, Sweden and Lucerne, Switzerland.

Diverse Student Body

Among J&W's student body are more than 800 students representing every U.S. state and 80 countries.

Easy Enrollment

Applications for admissions are accepted for terms beginning in September, December, March and June. Johnson & Wales' International Admissions Office is available to assist students with the entire admissions process.

Accreditation

Johnson & Wales University is fully accredited by the New England Association of Schools and Colleges and the Accrediting Council for Independent Colleges and Schools.

For more information:
International Admissions Office
Johnson & Wales University
8 Abbott Park Place
Providence, Rhode Island 02903
U.S.A.

telephone: (401) 598-1074
fax: (401) 598-4773

Johnson & Wales University *(continued)*

Urban campus. Founded in 1914. Accredited by New England Association of Schools and Colleges.

Program Information
Offered since 1973. Member of American Culinary Federation; American Dietetic Association; American Institute of Baking; American Institute of Wine & Food; Confrerie de la Chaine des Rotisseurs; Council on Hotel, Restaurant, and Institutional Education; Educational Foundation of the NRA; Institute of Food Technologists; International Association of Culinary Professionals; International Foodservice Editorial Council; Interntional Food Service Executives Association; James Beard Foundation, Inc.; National Restaurant Association; Oldways Preservation and Exchange Trust; The Bread Bakers Guild of America. Program calendar is divided into quarters. 1-year Master's degree in Hospitality Administration. 2-year Associate degrees in Culinary Arts; Baking and Pastry Arts. 4-year Bachelor's degrees in Baking and Pastry Arts; Culinary Arts.

Areas of Study
Baking; beverage management; buffet catering; computer applications; confectionery show pieces; controlling costs in food service; convenience cookery; culinary French; culinary skill development; food preparation; food purchasing; food service communication; food service math; garde-manger; international cuisine; introduction to food service; kitchen management; management and human resources; meal planning; meat cutting; meat fabrication; menu and facilities design; nutrition; nutrition and food service; patisserie; sanitation; saucier; seafood processing; soup, stock, sauce, and starch production; wines and spirits.

Facilities
11 bake shops; bakery; cafeteria; catering service; 20 classrooms; coffee shop; 4 computer laboratories; 20 demonstration laboratories; 30 food production kitchens; 2 gardens; 4 gourmet dining rooms; 10 laboratories; 2 learning resource centers; 20 lecture rooms; 2 libraries; 5 public restaurants; snack shop; 5 student lounges.

Culinary Student Profile
2,295 total: 2,102 full-time; 193 part-time.

Faculty
68 full-time; 2 part-time.

Special Programs
Customized corporate and commercial training programs.

Expenses
Tuition: $14,376 per year.

Financial Aid
Employment placement assistance is available. Employment opportunities within the program are available.

Housing
Coed housing available.

Application Information
Students are accepted for enrollment in March, June, July, September, and December. In 1996, 3,137 applied; 2,644 were accepted. Applicants must submit a formal application.

Contact
Licia Dwyer, Director, Culinary Admissions, College of Culinary Arts, 8 Abbott Park Place, Providence, RI 02903-3703; 800-342-5598; Fax: 401-598-1835; E-mail: admissions@jwu.edu; World Wide Web: http://www.jwu.edu

See affiliated programs at Charleston, South Carolina; North Miami, Florida; Vail, Colorado; Norfolk, Virginia.

See display on page 216.

Greenville Technical College

Hospitality Education

Greenville, South Carolina

General Information
Public, coeducational, two-year college. Suburban campus. Founded in 1962. Accredited by Southern Association of Colleges and Schools.

Program Information
Accredited by American Culinary Federation Education Institute. Member of American

Greenville Technical College *(continued)*

Culinary Federation; Interntional Food Service Executives Association; National Restaurant Association. Program calendar is divided into semesters. 1-year Certificates in Dietary Manager; Culinary Education. 2-year Associate degrees in Food Service Management-Culinary Option; Food Service Management.

Areas of Study

Buffet catering; computer applications; controlling costs in food service; food preparation; food purchasing; introduction to food service; kitchen management; management and human resources; meal planning; menu and facilities design; nutrition; restaurant opportunities; sanitation.

Culinary Student Profile

100 total: 40 full-time; 60 part-time.

Faculty

2 full-time; 2 part-time.

Expenses

Application fee: $25. In-state tuition: $525 per semester full-time, $44 per credit part-time. Out-of-state tuition: $1525 per semester full-time, $127 per credit part-time. Program-related fees include: $110 for knife set; $50 for uniforms.

Application Information

Students are accepted for enrollment in January, May, and August. Applicants must submit a formal application.

Contact

Rita Richards, Lab Specialist, Hospitality Education, PO Box 5616, Greenville, SC 29606-5616; 864-250-8030; Fax: 864-250-8455.

Horry-Georgetown Technical College

Culinary Arts Department

Conway, South Carolina

General Information

Public, coeducational, two-year college. Suburban campus. Founded in 1965. Accredited by Southern Association of Colleges and Schools.

Program Information

Offered since 1987. Accredited by American Culinary Federation Education Institute. Member of American Culinary Federation; American Culinary Federation Educational Institute; Council on Hotel, Restaurant, and Institutional Education; Educational Foundation of the NRA. Program calendar is divided into semesters. 5-semester Associate degree in Culinary Arts Technology-Business Major.

Areas of Study

Baking; beverage management; buffet catering; computer applications; controlling costs in food service; culinary French; food preparation; food purchasing; food service communication; food service math; garde-manger; international cuisine; introduction to food service; management and human resources; meat fabrication; menu and facilities design; nutrition; sanitation; saucier; seafood processing; soup, stock, sauce, and starch production.

Facilities

Bake shop; cafeteria; 2 classrooms; computer laboratory; demonstration laboratory; 3 food production kitchens; 2 gourmet dining rooms; learning resource center; lecture room; library; student lounge.

Culinary Student Profile

100 total: 85 full-time; 15 part-time.

Faculty

4 full-time; 3 part-time. 1 is an industry professional; 2 are culinary-accredited teachers; 1 is a registered dietitian.

Special Programs

Student exchange program with Bahamas Hotel Training College, Nassau, Bahamas.

Expenses

Application fee: $20. In-state tuition: $550 per semester full-time, $45 per credit hour part-time. Out-of-state tuition: $1194 per semester full-time, $90 per credit hour part-time. Program-related fees include: $165 for knives; $60 for uniforms; $200 for books per semester.

Financial Aid
In 1996, 50 scholarships were awarded (average award was $680); 3 loans were granted (average loan was $300). Employment placement assistance is available.

Application Information
Students are accepted for enrollment in January, May, and August. Application deadline for fall is continuous with a recommended date of August 1. Application deadline for spring is continuous with a recommended date of December 15. Application deadline for summer is continuous with a recommended date of May 1. In 1996, 45 applied; 45 were accepted. Applicants must submit a formal application and SAT, CPT, or ACT scores.

Contact
Carmen Catino, Department Head, Culinary Arts Department, 2050 Highway 501, PO Box 1966, Conway, SC 29526; 803-347-3186; Fax: 803-347-4207.

Johnson & Wales University

College of Culinary Arts

Charleston, South Carolina

General Information
Private, coeducational, four-year college. Urban campus. Founded in 1984. Accredited by New England Association of Schools and Colleges.

Program Information
Member of American Culinary Federation; Council on Hotel, Restaurant, and Institutional Education; Interntional Food Service Executives Association; National Restaurant Association; International Food Service Executives Association. Program calendar is divided into trimesters. 2-year Associate degrees in Baking and Pastry Arts; Culinary Arts. 4-year Bachelor's degree in Food Service Management.

Areas of Study
Baking; beverage management; confectionery show pieces; controlling costs in food service; culinary French; culinary skill development; food preparation; food purchasing; food service communication; food service math; garde-manger; international cuisine; management and human resources; meal planning; meat cutting; meat fabrication; menu and facilities design; nutrition; nutrition and food service; patisserie; sanitation; soup, stock, sauce, and starch production; wines and spirits.

Facilities
2 bake shops; 4 classrooms; coffee shop; 2 computer laboratories; demonstration laboratory; 2 food production kitchens; garden; 2 gourmet dining rooms; 10 laboratories; learning resource center; 4 lecture rooms; library; student lounge; storeroom.

Culinary Student Profile
1,274 total: 1,246 full-time; 28 part-time.

Faculty
20 full-time; 4 part-time. 1 is a master chef. Prominent faculty: Karl Guggenmos.

Special Programs
Required externships for 11 weeks during sophomore year, cooperative education experience, career days.

Expenses
Tuition: $12,348 per year full-time, $217.78 per credit hour part-time.

Financial Aid
In 1996, 7 scholarships were awarded (average award was $875). Program-specific awards include National Student Organization scholarships, Recipe Contest Winners - Finalists' scholarships, Career Exploration scholarships. Employment placement assistance is available. Employment opportunities within the program are available.

Housing
Apartment-style housing available.

Application Information
Students are accepted for enrollment in March, June, September, and December. In 1996, 1,050 applied; 928 were accepted. Applicants must submit a formal application and academic transcripts.

Johnson & Wales University *(continued)*

Contact

Mary Hovis, Director of Admissions, College of Culinary Arts, PCC Box 1409, 701 East Bay Street, Charleston, SC 29403; 803-727-3000; Fax: 803-763-0318; E-mail: admissions@jwu.edu-sc

See affiliated programs at North Miami, Florida; Providence, Rhode Island; Vail, Colorado; Norfolk, Virginia.

See display on page 216.

Trident Technical College

Culinary Arts Technology

Charleston, South Carolina

General Information

Public, coeducational, two-year college. Urban campus. Founded in 1964. Accredited by Southern Association of Colleges and Schools.

Program Information

Offered since 1988. Accredited by American Culinary Federation Education Institute. Member of American Culinary Federation; Council on Hotel, Restaurant, and Institutional Education; Educational Foundation of the NRA; National Restaurant Association. Program calendar is divided into semesters. 1-year Diploma in Culinary Arts. 2-year Associate degrees in Hospitality Tourism Management; Culinary Arts Technology.

Areas of Study

Baking; beverage management; buffet catering; computer applications; controlling costs in food service; culinary skill development; food preparation; food purchasing; garde-manger; introduction to food service; kitchen management; management and human resources; nutrition; sanitation; saucier; soup, stock, sauce, and starch production.

Facilities

Bake shop; 10 classrooms; 2 computer laboratories; 2 demonstration laboratories; 2 food production kitchens; gourmet dining room; learning resource center; library; 2 teaching kitchens.

Culinary Student Profile

135 total: 95 full-time; 40 part-time.

Faculty

3 full-time; 7 part-time. 8 are industry professionals; 2 are culinary-accredited teachers.

Expenses

Application fee: $50. Tuition: $512 per semester full-time, $44 per credit hour part-time. Program-related fees include: $12 for student activities and labs.

Financial Aid

In 1996, 4 scholarships were awarded (average award was $500). Employment placement assistance is available.

Application Information

Students are accepted for enrollment in January, May, and September. Applicants must submit a formal application.

Contact

Scott Roark, Program Coordinator, Culinary Arts Technology, PO Box 118067, Charleston, SC 29423-8067; 803-722-5571; Fax: 803-720-5614; E-mail: zproark@trident.tec.sc.us

Mitchell Technical Institute

Culinary Arts Program

Mitchell, South Dakota

General Information

Public, coeducational, two-year college. Rural campus. Founded in 1968. Accredited by North Central Association of Colleges and Schools.

Program Information

Member of Educational Foundation of the NRA. Program calendar is divided into semesters. 52-week Diploma in Culinary Arts.

Areas of Study

Baking; computer applications; controlling costs in food service; culinary skill development; food preparation; food purchasing; food service math; introduction to food service; management and human resources; meal planning; meat cutting; meat fabrication; nutrition; patisserie; sanitation; saucier; seafood processing; soup, stock, sauce, and starch production.

Facilities

Bake shop; cafeteria; 2 classrooms; computer laboratory; demonstration laboratory; 2 food production kitchens; gourmet dining room; learning resource center; public restaurant.

Culinary Student Profile

28 full-time.

Faculty

2 full-time. 2 are industry professionals. Prominent faculty: John H. Weber and Randy Doescher.

Expenses

Application fee: $25. Tuition: $2500 per 52 weeks. Program-related fees include: $680 for department fees; $300 for knife set; $375 for books.

Financial Aid

Employment placement assistance is available.

Application Information

Students are accepted for enrollment in January, June, and September. In 1996, 21 applied; 17 were accepted. Applicants must submit a formal application.

Contact

Lance Carter, Director of Student Services, Culinary Arts Program, 821 North Capital, Mitchell, SD 57301; 605-995-3024; Fax: 605-996-3299.

Opryland Hotel Culinary Institute

Nashville, Tennessee

General Information

Private, coeducational, culinary institute. Urban campus. Founded in 1987.

Program Information

Offered since 1987. Accredited by American Culinary Federation Education Institute. Member of American Culinary Federation; American Culinary Federation Educational Institute; Educational Foundation of the NRA; National Restaurant Association. 3-year Associate degree in Culinary Arts.

Areas of Study

Baking; beverage management; computer applications; controlling costs in food service; culinary French; culinary skill development; food preparation; food purchasing; food service communication; food service math; garde-manger; international cuisine; introduction to food service; kitchen management; management and human resources; meal planning; meat cutting; meat fabrication; menu and facilities design; nutrition; nutrition and food service; patisserie; restaurant opportunities; sanitation; saucier; seafood processing; soup, stock, sauce, and starch production; wines and spirits.

Facilities

Bake shop; cafeteria; catering service; classroom; 2 coffee shops; computer laboratory; demonstration laboratory; food production kitchen; 4 gourmet dining rooms; laboratory; learning resource center; lecture room; library; 5 public restaurants; 5 snack shops; teaching kitchen.

Culinary Student Profile

25 full-time.

Faculty

1 full-time; 15 part-time. 10 are industry professionals; 6 are chefs.

Expenses

Tuition is covered by the Opryland Hotel. Program-related fees include: $500 for registration.

Financial Aid

Employment placement assistance is available. Employment opportunities within the program are available.

Housing

Average off-campus housing cost per month: $400.

Opryland Hotel Culinary Institute *(continued)*

Application Information
Students are accepted for enrollment in August. Application deadline for fall is March 1. In 1996, 95 applied; 25 were accepted. Applicants must submit a formal application, letters of reference, and academic transcripts.

Contact
Dina Starks, Culinary Apprenticeship Coordinator, 2800 Opryland Drive, Nashville, TN 37214; 615-871-7765; Fax: 615-871-7872.

See affiliated apprenticeship program.

The Art Institute of Houston

The School of Culinary Arts

Houston, Texas

General Information
Private, coeducational, two-year college. Urban campus. Founded in 1978.

Program Information
Offered since 1992. Accredited by American Culinary Federation Education Institute. Member of American Culinary Federation; Texas Restaurant Association. Program calendar is divided into quarters. 18-month Associate degree in Culinary Arts.

Areas of Study
Baking; computer applications; culinary skill development; food preparation; food purchasing; garde-manger; introduction to food service; kitchen management; meal planning; meat cutting; menu and facilities design; nutrition; sanitation; soup, stock, sauce, and starch production.

Facilities
Bakery; catering service; classroom; computer laboratory; demonstration laboratory; food production kitchen; gourmet dining room; learning resource center; lecture room; library; public restaurant; snack shop; student lounge; teaching kitchen.

Culinary Student Profile
220 full-time.

Faculty
12 full-time. 7 are industry professionals; 1 is a master chef; 4 are culinary-accredited teachers.

Special Programs
6-day restaurant tour and meeting with chefs in New York City.

Expenses
Application fee: $50. Tuition: $3285 per quarter.

Financial Aid
Employment placement assistance is available. Employment opportunities within the program are available.

Housing
150 culinary students housed on campus. Apartment-style housing available. Average on-campus housing cost per month: $377.

Application Information
Students are accepted for enrollment in January, April, July, and October. Applicants must submit a formal application.

Contact
Rick Simmons, Director of Admissions, The School of Culinary Arts, 1900 Yorktown, Houston, TX 77056-4115; 800-275-4244; Fax: 713-966-2797; E-mail: simmonsr@aii.edu

See affiliated programs: Colorado Institute of Art; New York Restaurant School; The Art Institute of Atlanta; The Art Institute of Fort Lauderdale; The Art Institute of Philadelphia; The Art Institute of Phoenix; The Art Institute of Seattle.

See display on page 48.

Del Mar College

Department of Hospitality Management

Corpus Christi, Texas

General Information
Public, coeducational, two-year college. Urban

campus. Founded in 1935. Accredited by Southern Association of Colleges and Schools.

Program Information
Member of American Culinary Federation; Council on Hotel, Restaurant, and Institutional Education; National Restaurant Association; Texas Restaurant Association. Program calendar is divided into semesters. 1-year Certificates in Restaurant Supervisor; Kitchen Manager. 2-year Associate degrees in Restaurant Management; Culinary Arts. 9-month Certificate in Cook/Baker.

Areas of Study
Baking; beverage management; buffet catering; computer applications; confectionery show pieces; controlling costs in food service; culinary skill development; food preparation; food purchasing; food service math; garde-manger; introduction to food service; kitchen management; management and human resources; menu and facilities design; nutrition; nutrition and food service; patisserie; sanitation; saucier; soup, stock, sauce, and starch production.

Facilities
Bake shop; cafeteria; catering service; classroom; computer laboratory; demonstration laboratory; food production kitchen; learning resource center; lecture room; library; public restaurant.

Culinary Student Profile
95 total: 60 full-time; 35 part-time. 18 are under 25 years old; 63 are between 25 and 44 years old; 14 are over 44 years old.

Faculty
3 full-time; 12 part-time. 15 are industry professionals.

Special Programs
Paid internships in local restaurants, hotels, and clubs.

Expenses
Tuition: $400 per semester full-time, $100 per course part-time.

Financial Aid
In 1996, 12 scholarships were awarded (average award was $750). Employment placement assistance is available. Employment opportunities within the program are available.

Housing
Average off-campus housing cost per month: $500.

Application Information
Students are accepted for enrollment in January, June, and September. Applications are accepted continuously for spring, summer, and fall. In 1996, 54 applied; 54 were accepted. Applicants must submit a formal application, academic transcripts, and SAT, ACT, or ASSET scores.

Contact
D. W. Haven, Chair, Department of Hospitality Management, 101 Baldwin Boulevard, Corpus Christi, TX 78404-3897; 512-886-1734; Fax: 512-886-1829.

El Centro College

Food and Hospitality Services Institute

Dallas, Texas

General Information
Public, coeducational, two-year college. Urban campus. Founded in 1966. Accredited by Southern Association of Colleges and Schools.

Program Information
Offered since 1968. Member of American Culinary Federation; Council on Hotel, Restaurant, and Institutional Education; Educational Foundation of the NRA; National Restaurant Association; Educational Institute-American Hotel and Motel Association. Program calendar is divided into semesters. 1-year Certificate in Food and Hospitality Services. 24-month Certificate in Baking and Pastry. 3-year Certificate in Chef Apprenticeship. 30-month Associate degree in Food and Hospitality Services.

Areas of Study
Baking; beverage management; buffet catering; computer applications; controlling costs in food service; culinary skill development; food preparation; food purchasing; food service communication; food service math; garde-manger; introduction to food service; kitchen management;

El Centro College ***(continued)***

management and human resources; meal planning; menu and facilities design; nutrition; nutrition and food service; restaurant opportunities; sanitation; saucier; soup, stock, sauce, and starch production; wines and spirits.

Facilities
2 bake shops; cafeteria; 3 classrooms; computer laboratory; demonstration laboratory; 2 food production kitchens; gourmet dining room; learning resource center; 3 lecture rooms; library; student lounge.

Culinary Student Profile
410 total: 130 full-time; 280 part-time.

Faculty
3 full-time; 15 part-time. 17 are industry professionals; 1 is a culinary-accredited teacher.

Expenses
Tuition: $300 per semester full-time, $89 per course part-time.

Financial Aid
In 1996, 10 scholarships were awarded (average award was $600). Employment placement assistance is available.

Application Information
Students are accepted for enrollment in January, June, and August. Applicants must submit a formal application.

Contact
Gus Katsigris, Director, Food and Hospitality Services Institute, Main and Lamar Streets, Dallas, TX 75202-3604; 214-860-2202; Fax: 214-860-2335.

El Paso Community College

Hospitality and Travel Services

El Paso, Texas

General Information
Public, coeducational, two-year college. Urban campus. Founded in 1969. Accredited by Southern Association of Colleges and Schools.

Program Information
Offered since 1989. Member of Council on Hotel, Restaurant, and Institutional Education; Educational Foundation of the NRA. Program calendar is divided into semesters. 1-year Certificates in Advanced Cuisine; Culinary Arts. 2-year Associate degrees in Restaurant Management; Culinary Arts.

Areas of Study
Baking; computer applications; controlling costs in food service; food preparation; food purchasing; garde-manger; international cuisine; management and human resources; meal planning; meat cutting; menu and facilities design; nutrition; sanitation; food and beverage management; confectionery.

Facilities
Cafeteria; food production kitchen; 6 lecture rooms; library; teaching kitchen.

Culinary Student Profile
180 total: 90 full-time; 90 part-time.

Faculty
1 full-time; 3 part-time. 4 are industry professionals.

Expenses
In-state tuition: $75 per credit hour. Out-of-state tuition: $200 per credit hour.

Financial Aid
Employment placement assistance is available. Employment opportunities within the program are available.

Housing
Average off-campus housing cost per month: $700.

Application Information
Students are accepted for enrollment in January, May, and August. Applicants must submit a formal application and have a high school diploma or GED.

Contact
M. J. Linney, Instructional Coordinator, Hospitality and Travel Services, PO Box 20500, El Paso, TX 79998-0500; 915-594-2217; Fax: 915-594-2155.

Galveston College

Culinary Arts Department

Galveston, Texas

General Information
Public, coeducational, two-year college. Urban campus. Founded in 1967. Accredited by Southern Association of Colleges and Schools.

Program Information
Offered since 1987. Member of American Culinary Federation; Educational Foundation of the NRA; National Restaurant Association; Texas Chef's Association. Program calendar is divided into semesters. 2-semester Certificates in Hospitality Management; Food Preparation. 2-year Associate degree in Culinary Arts/Hospitality.

Areas of Study
Beverage management; computer applications; controlling costs in food service; food preparation; food purchasing; menu and facilities design; nutrition; sanitation.

Facilities
Bake shop; 3 classrooms; 2 computer laboratories; food production kitchen; gourmet dining room; learning resource center; library; snack shop; student lounge.

Culinary Student Profile
60 total: 40 full-time; 20 part-time. 10 are under 25 years old; 25 are between 25 and 44 years old; 25 are over 44 years old.

Faculty
3 full-time. 3 are certified Chefs de Cuisine. Prominent faculty: Leslie Bartosh and Charles Collins.

Special Programs
Southern and Northern Italian cooking (3-hour courses), Baking and Desserts (3-hour courses).

Expenses
Tuition: $550 per semester full-time, $300 per semester part-time. Program-related fees include: $150 for knives; $70 for books; $100 for uniforms.

Financial Aid
Employment placement assistance is available.

Application Information
Students are accepted for enrollment in January, May, July, and September. Application deadline for fall is continuous with a recommended date of August 30. Application deadline for spring is continuous with a recommended date of January 5. In 1996, 60 applied; 60 were accepted. Applicants must submit a formal application.

Contact
Bruce Ozga, Culinary Arts Director, Culinary Arts Department, 4015 Avenue Q, Galveston, TX 77550-7496; 409-763-6551 Ext. 304; Fax: 409-762-9367.

Houston Community College System

Culinary Services

Houston, Texas

General Information
Public, coeducational, two-year college. Urban campus. Founded in 1971. Accredited by Southern Association of Colleges and Schools.

Program Information
Offered since 1980. Program calendar is divided into semesters. 45-week Certificates in Baking and Pastry; Culinary Arts.

Areas of Study
Baking; confectionery show pieces; controlling costs in food service; culinary skill development; food preparation; food purchasing; food service math; garde-manger; international cuisine; introduction to food service; kitchen management; management and human resources; nutrition; patisserie; sanitation; soup, stock, sauce, and starch production; breads; croissant, Danish, and puff pastry.

Facilities
Bake shop; cafeteria; computer laboratory; demonstration laboratory; 2 food production kitchens; gourmet dining room; learning resource center; library; public restaurant; student lounge.

Houston Community College System *(continued)*

Culinary Student Profile
200 total: 100 full-time; 100 part-time. 50 are under 25 years old; 100 are between 25 and 44 years old; 50 are over 44 years old.

Faculty
5 full-time; 5 part-time. 4 are industry professionals; 3 are culinary-accredited teachers. Prominent faculty: Eddy Van Damme.

Expenses
In-state tuition: $420 per semester full-time, $101 per semester part-time. Out-of-state tuition: $1530 per semester full-time, $360 per semester part-time. Program-related fees include: $6 for lab fees.

Financial Aid
In 1996, 5 scholarships were awarded (average award was $125). Employment placement assistance is available. Employment opportunities within the program are available.

Application Information
Students are accepted for enrollment in January, May, and September. Application deadline for fall is continuous with a recommended date of August 14. Application deadline for spring is continuous with a recommended date of November 14. Application deadline for summer is continuous with a recommended date of April 14. Applicants must submit a formal application.

Contact
Culinary Services, Culinary Services, 1300 Holman, Houston, TX 77004; 713-718-6046; Fax: 713-718-6054.

Le Chef College of Hospitality Careers

Culinary Arts and Food and Beverage Management Program

Austin, Texas

General Information
Private, coeducational, two-year college. Urban campus. Founded in 1981.

Program Information
Offered since 1992. Member of American Culinary Federation. Program calendar is divided into semesters. 17-month Diploma in Culinary Arts. 2-year Associate degree in Culinary Arts/Food and Beverage Management.

Areas of Study
Baking; beverage management; buffet catering; computer applications; confectionery show pieces; controlling costs in food service; convenience cookery; culinary French; culinary skill development; food preparation; food purchasing; food service communication; food service math; garde-manger; international cuisine; introduction to food service; kitchen management; management and human resources; meal planning; meat cutting; meat fabrication; menu and facilities design; nutrition; nutrition and food service; patisserie; restaurant opportunities; sanitation; saucier; seafood processing; soup, stock, sauce, and starch production; wines and spirits.

Facilities
3 classrooms; computer laboratory; food production kitchen; learning resource center; library; 2 student lounges.

Culinary Student Profile
180 full-time.

Faculty
6 full-time. 2 are master chefs; 4 are culinary-accredited teachers.

Expenses
Tuition: $21,168 per degree full-time, $294 per credit part-time.

Financial Aid
Employment placement assistance is available. Employment opportunities within the program are available.

Housing
Average off-campus housing cost per month: $600.

Application Information
Students are accepted for enrollment year-round. Applicants must submit a formal application.

Contact
Shawn Fortner, Administration Assistant, Culinary Arts and Food and Beverage Management Program, 6020 Dillard Circle, Austin, TX 78752; 888-5Le-CHEF; Fax: 512-323-2126; E-mail: lechef@onr.com; World Wide Web: http://www.lechef.org

Odessa College

Culinary Arts

Odessa, Texas

General Information
Public, coeducational, two-year college. Urban campus. Founded in 1946. Accredited by Southern Association of Colleges and Schools.

Program Information
Offered since 1990. Member of Council on Hotel, Restaurant, and Institutional Education. Program calendar is divided into semesters. 1-semester Certificate in Food Preparation Cook. 16-month Associate degree in Culinary Arts. 2-semester Certificate in Food Production Cook.

Areas of Study
Baking; beverage management; buffet catering; computer applications; confectionery show pieces; controlling costs in food service; convenience cookery; culinary skill development; food preparation; food purchasing; food service communication; food service math; garde-manger; international cuisine; introduction to food service; kitchen management; management and human resources; meal planning; meat cutting; menu and facilities design; nutrition; nutrition and food service; patisserie; restaurant opportunities; sanitation; saucier; seafood processing; soup, stock, sauce, and starch production; wines and spirits.

Facilities
Bake shop; cafeteria; classroom; food production kitchen; gourmet dining room; laboratory; learning resource center; library; public restaurant.

Culinary Student Profile
38 total: 30 full-time; 8 part-time.

Faculty
2 full-time. 1 is an industry professional. Prominent faculty: Peter Lewis.

Special Programs
4 day practicum in Las Vegas, Nevada.

Expenses
Tuition: $424 per semester full-time, $100 per semester part-time.

Financial Aid
Employment placement assistance is available. Employment opportunities within the program are available.

Application Information
Students are accepted for enrollment in January and August. Applicants must submit a formal application.

Contact
Peter Lewis, Department Chair, Culinary Arts, 201 West University, Odessa, TX 79764-7127; 915-335-6320; Fax: 915-335-6860; E-mail: plewis@odessa.edu

St. Philip's College

Hospitality Operations

San Antonio, Texas

General Information
Public, coeducational, two-year college. Urban campus. Founded in 1898. Accredited by Southern Association of Colleges and Schools.

Program Information
Accredited by American Culinary Federation Education Institute. Member of American Culinary Federation; American Culinary Federation Educational Institute; American Dietetic Association; Educational Foundation of the NRA; National Restaurant Association. Program calendar is divided into semesters. 2-year Associate degrees in Hotel Management; Dietetic Technology; Restaurant Management; Culinary Arts.

Areas of Study
Baking; beverage management; buffet catering; controlling costs in food service; culinary skill

St. Philip's College ***(continued)***

development; food preparation; food purchasing; garde-manger; international cuisine; introduction to food service; kitchen management; management and human resources; menu and facilities design; nutrition and food service; sanitation.

Facilities
Bake shop; bakery; 4 classrooms; food production kitchen; garden; gourmet dining room; learning resource center; library; public restaurant; student lounge.

Culinary Student Profile
250 total: 200 full-time; 50 part-time.

Faculty
5 full-time; 6 part-time. 3 are industry professionals; 1 is a culinary-accredited teacher.

Expenses
Tuition: $400 per semester full-time, $120 per credit hour part-time.

inancial Aid
Employment placement assistance is available. Employment opportunities within the program are available.

Application Information
Students are accepted for enrollment in January and August. Application deadline for fall is August 21. Application deadline for spring is January 7. Applicants must submit a formal application.

Contact
Mary Kunz, Chair, Hospitality Operations, 1801 Martin Luther King Drive; 210-531-3315; Fax: 210-531-3351; E-mail: m.kunz@accd.edu

Texas State Technical College-Waco/Marshall Campus

Culinary Arts/Food Service

Waco, Texas

General Information
Public, coeducational, two-year college. Suburban campus. Founded in 1965. Accredited by Southern Association of Colleges and Schools.

Program Information
Offered since 1968. Member of Texas Restaurant Association. Program calendar is divided into quarters. 12-month Certificate in Food Service/ Culinary Arts. 6-month Certificate in General Food Service Worker.

Areas of Study
Baking; buffet catering; computer applications; controlling costs in food service; culinary skill development; food preparation; food purchasing; food service communication; food service math; garde-manger; introduction to food service; kitchen management; management and human resources; meal planning; meat cutting; meat fabrication; menu and facilities design; nutrition; nutrition and food service; sanitation; saucier; seafood processing; soup, stock, sauce, and starch production.

Facilities
Bake shop; bakery; cafeteria; catering service; 2 classrooms; computer laboratory; demonstration laboratory; food production kitchen; laboratory; learning resource center; lecture room; library; public restaurant; student lounge; teaching kitchen.

Culinary Student Profile
90 total: 75 full-time; 15 part-time. 50 are under 25 years old; 25 are between 25 and 44 years old; 15 are over 44 years old.

Faculty
3 full-time; 2 part-time. 5 are industry professionals.

Expenses
Tuition: $800 per quarter full-time, $27.50 per credit hour part-time. Program-related fees include: $250 for tools, equipment, and uniforms.

Financial Aid
In 1996, 10 scholarships were awarded (average award was $250); 60 loans were granted (average loan was $1250). Employment placement assistance is available. Employment opportunities within the program are available.

Housing
Apartment-style and single-sex housing available.

Application Information
Students are accepted for enrollment in March and September. In 1996, 60 applied; 60 were accepted. Applicants must submit a formal application.

Contact
Homer Jones, Department Chair, Culinary Arts/Food Service, 3801 Campus Drive, Waco, TX 76705-1695; 254-867-4868; Fax: 254-867-3663.

Salt Lake Community College

Apprenticeship Division

Salt Lake City, Utah

General Information
Public, coeducational, two-year college. Suburban campus. Founded in 1948. Accredited by Northwest Association of Schools and Colleges.

Program Information
Offered since 1985. Accredited by American Culinary Federation Education Institute. Member of American Culinary Federation; American Culinary Federation Educational Institute; Educational Foundation of the NRA. Program calendar is divided into quarters. 2-year Associate degree in Apprentice Chef. 2-year Certificate in Professional Management Development Program.

Areas of Study
Baking; beverage management; buffet catering; computer applications; controlling costs in food service; culinary French; culinary skill development; food preparation; food purchasing; food service math; garde-manger; international cuisine; introduction to food service; kitchen management; meal planning; meat cutting; meat fabrication; nutrition; restaurant opportunities; sanitation; saucier; seafood processing; soup, stock, sauce, and starch production; wines and spirits; marketing for restaurants.

Facilities
Catering service; 10 classrooms; garden; laboratory; learning resource center; library; student lounge.

Culinary Student Profile
90 total: 60 full-time; 30 part-time.

Faculty
2 full-time; 9 part-time. 3 are industry professionals; 6 are culinary-accredited teachers.

Expenses
Application fee: $20. Tuition: $240 per class full-time, $82 per class part-time.

Financial Aid
Program-specific awards include 5 Gertrude Marshall Scholarships, 1 NRA Scholarship ($500-$2000). Employment placement assistance is available.

Application Information
Students are accepted for enrollment in January, March, June, and September. In 1996, 35 applied; 30 were accepted. Applicants must submit a formal application.

Contact
Joe Mulvey, Apprentice Director, Apprenticeship Division, PO Box 30808, Salt Lake City, UT 84130-0808; 801-957-4066; Fax: 801-957-4895.

Utah Valley State College

Hospitality Management

Orem, Utah

General Information
Public, coeducational, two-year college. Suburban campus. Founded in 1941. Accredited by Northwest Association of Schools and Colleges.

Program Information
Offered since 1990. Member of American Culinary Federation; Educational Foundation of the NRA; National Restaurant Association. Program calendar is divided into semesters. 1-year Certificates in Hospitality Management; Culinary Arts. 24-month

Utah Valley State College *(continued)*

Associate degrees in Hospitality Management; Culinary Arts. 4-year Bachelor's degree in Hospitality Management.

Areas of Study
Baking; beverage management; buffet catering; computer applications; controlling costs in food service; culinary French; culinary skill development; food preparation; food purchasing; food service communication; food service math; garde-manger; international cuisine; introduction to food service; kitchen management; management and human resources; meal planning; menu and facilities design; nutrition; nutrition and food service; patisserie; restaurant opportunities; sanitation; saucier; seafood processing; soup, stock, sauce, and starch production; wines and spirits.

Facilities
Bake shop; 3 classrooms; computer laboratory; food production kitchen; gourmet dining room; learning resource center; lecture room; library; public restaurant; teaching kitchen.

Culinary Student Profile
30 total: 25 full-time; 5 part-time.

Faculty
3 full-time; 2 part-time. 3 are industry professionals; 2 are culinary-accredited teachers. Prominent faculty: Gref Forte and Hans Zulliger.

Special Programs
ACF Conference and competitions, field trips to food service operations in Las Vegas, NV, co-op experience in Europe.

Expenses
Application fee: $20. Tuition: $900 per semester full-time, $400 per semester part-time. Program-related fees include: $120 for uniforms; $200 for knives; $100 for kitchen fees.

Financial Aid
In 1996, 12 scholarships were awarded (average award was $500); 10 loans were granted (average loan was $1500). Program-specific awards include work-study program in college banquet halls. Employment placement assistance is available. Employment opportunities within the program are available.

Housing
Average off-campus housing cost per month: $200.

Application Information
Students are accepted for enrollment in January, May, and August. Applicants must submit a formal application and letters of reference.

Contact
Greg Forte, Chair, Hospitality Management, 800 West 1200 South Street, Orem, UT 84058-0001; 801-222-8087; Fax: 801-222-8769; E-mail: fortegr@uvsc.edu

New England Culinary Institute

Occupational Studies in Culinary Arts

Montpelier, Vermont

General Information
Private, coeducational, culinary institute. Rural campus. Founded in 1980.

Program Information
Offered since 1980. Member of American Culinary Federation; American Institute of Wine & Food; Council on Hotel, Restaurant, and Institutional Education; Educational Foundation of the NRA; International Association of Culinary Professionals; James Beard Foundation, Inc.; National Restaurant Association. Program calendar is divided into quarters. 10-month Certificate in Basic Cooking. 2-year Associate degree in Culinary Arts. 3-year Bachelor's degree in Food and Beverage Management.

Areas of Study
Baking; beverage management; buffet catering; computer applications; controlling costs in food service; culinary French; culinary skill development; food preparation; food purchasing; food service communication; food service math;

garde-manger; introduction to food service; kitchen management; management and human resources; meal planning; meat cutting; meat fabrication; menu and facilities design; nutrition; nutrition and food service; patisserie; restaurant opportunities; sanitation; saucier; soup, stock, sauce, and starch production; wines and spirits; non-commercial preparation.

Facilities

2 bake shops; bakery; 3 cafeterias; 2 catering services; 20 classrooms; coffee shop; 2 computer laboratories; 12 food production kitchens; 2 gardens; 2 gourmet dining rooms; 2 laboratories; 2 learning resource centers; 2 libraries; 7 public restaurants; non-commercial food kitchen.

Culinary Student Profile

650 full-time.

Faculty

58 full-time. Prominent faculty: Michel LeBorgne and Robert Barral.

Special Programs

6-month paid internships following 6-month residency each year.

Expenses

Application fee: $25. Tuition: $16,750 per year. Program-related fees include: $230 for books; $320 for knives.

Financial Aid

In 1996, 280 scholarships were awarded (average award was $500). Employment placement assistance is available. Employment opportunities within the program are available.

Housing

275 culinary students housed on campus. Coed, apartment-style, and single-sex housing available. Average on-campus housing cost per month: $285. Average off-campus housing cost per month: $450.

Application Information

Students are accepted for enrollment in March, June, September, and December. In 1996, 950

New England Culinary Institute ***(continued)***

applied; 760 were accepted. Applicants must submit a formal application, letters of reference, and an essay and have a high school diploma or a GED.

Contact
Shari L. McLaughlin, Associate Director, Admissions, Occupational Studies in Culinary Arts, 250 Main Street, Montpelier, VT 05602-9720; 802-223-6324; Fax: 802-223-0634; World Wide Web: http://www.neculinary.com/

ATI-Career Institute-School of Culinary Arts

Falls Church, Virginia

General Information
Private, coeducational, culinary institute. Urban campus. Founded in 1989.

Program Information
Offered since 1989. Accredited by American Culinary Federation Education Institute. Member of American Culinary Federation; American Culinary Federation Educational Institute; Council on Hotel, Restaurant, and Institutional Education; Educational Foundation of the NRA; International Association of Culinary Professionals; Interntional Food Service Executives Association; National Restaurant Association; The Bread Bakers Guild of America. Program calendar is divided into six-week cycles. 1-year Certificates in Hotel/ Restaurant Management; Culinary Arts.

Areas of Study
Baking; beverage management; buffet catering; controlling costs in food service; convenience cookery; culinary skill development; food preparation; food purchasing; food service math; garde-manger; international cuisine; introduction to food service; kitchen management; management and human resources; meal planning; meat fabrication; menu and facilities design; nutrition; nutrition and food service; patisserie; sanitation; saucier; seafood processing; soup, stock, sauce, and starch production; wines and spirits.

Facilities
Bake shop; bakery; catering service; 6 classrooms; demonstration laboratory; 3 food production kitchens; gourmet dining room; learning resource center; 6 lecture rooms; library; student lounge; 3 teaching kitchens.

Culinary Student Profile
175 full-time. 65 are under 25 years old; 85 are between 25 and 44 years old; 25 are over 44 years old.

Faculty
12 full-time; 4 part-time. 4 are industry professionals; 12 are culinary-accredited teachers. Prominent faculty: John W. Martin and Glenn Walden.

Expenses
Application fee: $60. Tuition: $13,650 per certificate.

Financial Aid
In 1996, 10 scholarships were awarded (average award was $2000); 125 loans were granted (average loan was $6000). Employment placement assistance is available. Employment opportunities within the program are available.

Housing
Average off-campus housing cost per month: $600.

Application Information
Students are accepted for enrollment year-round. Applicants must submit a formal application.

Contact
John W. Martin, Director, 7777 Leesburg Pike, Falls Church, VA 22043; 703-821-8570; Fax: 703-821-9289; E-mail: jmarti7031@aol.com

J. Sargeant Reynolds Community College

Culinary Arts

Richmond, Virginia

General Information
Public, coeducational, two-year college. Urban

campus. Founded in 1972. Accredited by Southern Association of Colleges and Schools.

Program Information
Offered since 1973. Accredited by American Dietetic Association. Member of American Culinary Federation; American Culinary Federation Educational Institute; American Dietetic Association; Council on Hotel, Restaurant, and Institutional Education; Educational Foundation of the NRA; National Restaurant Association; Society of Wine Educators; Dietary Managers Association. Program calendar is divided into semesters. 2-year Associate degrees in Lodging Operations; Entrepreneurship; Food Service Management; Dietetic Technician studies; Culinary Arts.

Areas of Study
Baking; buffet catering; computer applications; controlling costs in food service; culinary skill development; food preparation; garde-manger; introduction to food service; kitchen management; management and human resources; meat cutting; meat fabrication; menu and facilities design; nutrition; nutrition and food service; patisserie; restaurant opportunities; sanitation; saucier; seafood processing; soup, stock, sauce, and starch production; wines and spirits; total quality management for hospitality; food and beverage service management.

Facilities
Bake shop; bakery; cafeteria; catering service; 5 classrooms; 4 computer laboratories; demonstration laboratory; food production kitchen; laboratory; 3 learning resource centers; 3 lecture rooms; 3 libraries; snack shop; student lounge; 2 teaching kitchens.

Culinary Student Profile
120 total: 70 full-time; 50 part-time. 60 are under 25 years old; 40 are between 25 and 44 years old; 20 are over 44 years old.

Faculty
2 full-time; 10 part-time. 9 are industry professionals; 1 is a culinary-accredited teacher; 1 is a registered dietitian. Prominent faculty: David J. Barrish and D. Bruce Clarke.

Special Programs
Semi-annual "President's Dinner" (multi-course, white glove gastronomic event), mentorship with Virginia Chefs Association Professional Chefs, advanced placement for previous experience.

Expenses
Tuition: $49.05 per credit hour. Program-related fees include: $600 for textbooks.

Financial Aid
In 1996, 6 scholarships were awarded (average award was $500); 60 loans were granted (average loan was $2000). Program-specific awards include Virginia Hospitality and Travel Association scholarship, Virginia Restaurant Association scholarship, American Hotel Foundation scholarship. Employment placement assistance is available.

Housing
Average off-campus housing cost per month: $400.

Application Information
Students are accepted for enrollment in January, May, and August. In 1996, 35 applied; 35 were accepted. Applicants must submit a formal application and academic transcripts.

Contact
David J. Barrish, Program Head, Culinary Arts, 701 East Jackson Street, Richmond, VA 23219; 804-786-2069; Fax: 804-786-5465; E-mail: srbarrd@jsr.cc.va.us; World Wide Web: http://www.jsrcc.va.us/dtcbusdiv/hospitality

Johnson & Wales University

College of Culinary Arts

Norfolk, Virginia

General Information
Private, coeducational, two-year college. Urban campus. Founded in 1986. Accredited by New England Association of Schools and Colleges.

Johnson & Wales University *(continued)*

Program Information
Offered since 1986. Member of American Culinary Federation; American Dietetic Association; American Institute of Baking; American Institute of Wine & Food; Confrerie de la Chaine des Rotisseurs; Council on Hotel, Restaurant, and Institutional Education; Educational Foundation of the NRA; Institute of Food Technologists; International Association of Culinary Professionals; International Foodservice Editorial Council; Interntional Food Service Executives Association; James Beard Foundation, Inc.; National Restaurant Association; Oldways Preservation and Exchange Trust; Tasters Guild International; The Bread Bakers Guild of America. Program calendar is divided into quarters. 1-year Certificate in Culinary Arts. 2-year Associate degree in Culinary Arts.

Areas of Study
Baking; beverage management; controlling costs in food service; culinary skill development; food preparation; garde-manger; international cuisine; meat cutting; nutrition; patisserie; sanitation; soup, stock, sauce, and starch production.

Facilities
Bake shop; 5 classrooms; computer laboratory; 3 food production kitchens; learning resource center; library; student lounge.

Culinary Student Profile
545 total: 441 full-time; 104 part-time.

Faculty
12 full-time; 12 part-time.

Expenses
Tuition: $12,024 per year.

Financial Aid
Employment placement assistance is available. Employment opportunities within the program are available.

Housing
134 culinary students housed on campus. Coed housing available.

Application Information
Students are accepted for enrollment in March, June, July, September, and December. In 1996, 623 applied; 513 were accepted. Applicants must submit a formal application.

Contact
Torri Butler, Director, Student Affairs, College of Culinary Arts, 2428 Almeda Avenue, Suite 316-318, Norfolk, VA 23513; 757-853-1906; Fax: 757-855-8271; World Wide Web: http://www.jwu.edu

See affiliated programs at Charleston, South Carolina; North Miami, Florida; Providence, Rhode Island; Vail, Colorado.

See display on page 216.

The Art Institute of Seattle

School of Culinary Arts

Seattle, Washington

General Information
Private, coeducational, two-year college. Urban campus. Founded in 1982. Accredited by Northwest Association of Schools and Colleges.

Program Information
Offered since 1996. Member of American Culinary Federation; National Restaurant Association. Program calendar is divided into quarters. 2-year Associate degrees in Northwest Food and Wine; Edible Visual Arts; Baking/Pastry.

Areas of Study
Baking; beverage management; computer applications; controlling costs in food service; food preparation; food purchasing; food service communication; food service math; garde-manger; kitchen management; meal planning; meat cutting; meat fabrication; menu and facilities design; nutrition; patisserie; sanitation; saucier; seafood processing; soup, stock, sauce, and starch production; wines and spirits.

Facilities
3 food production kitchens; gourmet dining room; library; 2 student lounges.

Culinary Student Profile
250 full-time.

Faculty
12 full-time. 9 are industry professionals; 1 is a master chef; 2 are culinary-accredited teachers.

Special Programs
Day tours to Pike Place Market.

Expenses
Application fee: $50. Tuition: $3450 per quarter. Program-related fees include: $620 for knives, uniforms, and books.

Financial Aid
In 1996, 38 scholarships were awarded (average award was $1000); 135 loans were granted (average loan was $6000). Employment placement assistance is available. Employment opportunities within the program are available.

Application Information
Students are accepted for enrollment in January, April, July, and October. In 1996, 225 applied; 156 were accepted. Applicants must submit a formal application.

Contact
Doug Worsley, Vice President, Director of Admissions, School of Culinary Arts, 2323 Elliott Avenue, Seattle, WA 98121-1642; 206-448-6600; Fax: 206-448-2501.

See affiliated programs: Colorado Institute of Art; New York Restaurant School; The Art Institute of Atlanta; The Art Institute of Fort Lauderdale; The Art Institute of Houston; The Art Institute of Philadelphia; The Art Institute of Phoenix.

See display on page 48.

Bates Technical College

Culinary Arts Program

Tacoma, Washington

General Information
Public, coeducational, two-year college. Urban campus. Founded in 1944. Accredited by Northwest Association of Schools and Colleges.

Program Information
Offered since 1950. Member of American Culinary Federation; American Culinary Federation Educational Institute; Washington State Chefs Association. Program calendar is divided into quarters. 2-year Associate degree in Culinary Arts.

Areas of Study
Baking; buffet catering; controlling costs in food service; convenience cookery; culinary French; culinary skill development; food preparation; food purchasing; food service communication; food service math; garde-manger; international cuisine; introduction to food service; kitchen management; management and human resources; meal planning; meat cutting; meat fabrication; menu and facilities design; nutrition; nutrition and food service; patisserie; restaurant opportunities; sanitation; saucier; seafood processing; soup, stock, sauce, and starch production; health and safety.

Facilities
2 bake shops; 2 cafeterias; 2 catering services; 2 classrooms; 2 coffee shops; 2 computer laboratories; 2 demonstration laboratories; 2 food production kitchens; 2 learning resource centers; 2 lecture rooms; 2 libraries; 2 teaching kitchens.

Culinary Student Profile
27 total: 22 full-time; 5 part-time.

Faculty
2 full-time; 2 part-time. 2 are industry professionals; 2 are culinary-accredited teachers. Prominent faculty: Ricardo Saenz and Roger Knapp.

Expenses
Application fee: $52. Tuition: $672.40 per quarter. Program-related fees include: $600 for books, tools, and supplies.

Financial Aid
Employment placement assistance is available.

Housing
Average off-campus housing cost per month: $400.

Application Information
Students are accepted for enrollment in January, February, March, April, May, June, July, September, October, November, and December. In

Bates Technical College *(continued)*

1996, 25 applied; 22 were accepted. Applicants must submit a formal application.

Contact

Ricardo Saenz, Chef Instructor, Culinary Arts Program, 1101 South Yakima Avenue, Tacoma, WA 98405; 253-596-1566; Fax: 253-596-1643.

Application Information

Students are accepted for enrollment in January, March, July, and September. Applicants must submit a formal application, be at least 16 years of age, and complete an entrance exam.

Contact

Michael Baldwin, Instructor, Culinary Arts, 3028 Lindbergh Avenue, Bellingham, WA 98225-1599; 360-715-8350; Fax: 360-676-2798.

Bellingham Technical College

Culinary Arts

Bellingham, Washington

General Information

Public, coeducational, two-year college. Urban campus. Founded in 1957.

Program Information

Accredited by American Culinary Federation Education Institute. Program calendar is divided into quarters. 1-year Certificates in Baking, Pastry, and Confection; Culinary Arts.

Areas of Study

Baking; beverage management; buffet catering; controlling costs in food service; culinary skill development; food preparation; food purchasing; food service communication; food service math; garde-manger; introduction to food service; kitchen management; meal planning; meat cutting; meat fabrication; nutrition; patisserie; restaurant opportunities; sanitation; saucier; seafood processing; soup, stock, sauce, and starch production.

Culinary Student Profile

28 total: 20 full-time; 8 part-time.

Faculty

2 full-time.

Expenses

In-state tuition: $2067 per year. Out-of-state tuition: $8068 per year.

Clark College

Culinary Arts

Vancouver, Washington

General Information

Public, coeducational, two-year college. Urban campus. Founded in 1933. Accredited by Northwest Association of Schools and Colleges.

Program Information

Offered since 1958. Member of American Culinary Federation; American Culinary Federation Educational Institute; Educational Foundation of the NRA; National Restaurant Association; The Bread Bakers Guild of America; Retail Bakers of America. Program calendar is divided into quarters. 18-month Certificates in Baking Management; Restaurant Management. 21-month Associate degrees in Restaurant Management; Baking. 9-month Certificate of Completions in Baking; Cooking.

Areas of Study

Baking; beverage management; buffet catering; computer applications; confectionery show pieces; controlling costs in food service; convenience cookery; culinary French; culinary skill development; food preparation; food purchasing; food service communication; food service math; garde-manger; introduction to food service; kitchen management; meal planning; meat cutting; meat fabrication; menu and facilities design; nutrition and food service; patisserie; restaurant opportunities; sanitation; saucier; seafood processing; soup, stock, sauce, and starch production; wines and spirits; specialty cake decorating; candy making.

Facilities
2 bake shops; bakery; cafeteria; catering service; 6 classrooms; 4 computer laboratories; 2 demonstration laboratories; 2 food production kitchens; garden; gourmet dining room; 2 learning resource centers; 3 lecture rooms; library; 2 public restaurants; 2 snack shops; student lounge; 5 teaching kitchens; bakery retail store.

Culinary Student Profile
75 full-time. 35 are under 25 years old; 30 are between 25 and 44 years old; 10 are over 44 years old.

Faculty
6 full-time; 12 part-time. 15 are industry professionals; 3 are culinary-accredited teachers. Prominent faculty: George Akav, Jr. and Per Zeeberg.

Special Programs
Winery and restaurant tours in Portland-Seattle area.

Expenses
Tuition: $665 per quarter. Program-related fees include: $77 for lunches and fees.

Financial Aid
In 1996, 10 scholarships were awarded (average award was $500); 30 loans were granted (average loan was $500). Program-specific awards include tuition reduction for summer quarter students. Employment placement assistance is available. Employment opportunities within the program are available.

Housing
Average off-campus housing cost per month: $450.

Application Information
Students are accepted for enrollment in January, April, June, and September. In 1996, 80 applied; 75 were accepted. Applicants must submit a formal application.

Contact
Larry Mains, Director, Culinary Arts, 1800 East McLoughlin Boulevard, Vancouver, WA 98663-3598; 360-992-2143; Fax: 360-992-2861.

Edmonds Community College

Culinary Arts

Lynnwood, Washington

General Information
Public, coeducational, two-year college. Suburban campus. Founded in 1967. Accredited by Northwest Association of Schools and Colleges.

Program Information
Offered since 1989. Member of American Culinary Federation; American Institute of Wine & Food; American Wine Society; Educational Foundation of the NRA; International Wine & Food Society; National Restaurant Association; Society of Wine Educators. Program calendar is divided into quarters. 2-year Associate degree in Culinary Arts.

Areas of Study
Beverage management; computer applications; controlling costs in food service; culinary French; culinary skill development; food preparation; food purchasing; food service communication; food service math; garde-manger; introduction to food service; kitchen management; management and human resources; meal planning; meat cutting; nutrition; nutrition and food service; restaurant opportunities; sanitation; saucier; seafood processing; soup, stock, sauce, and starch production; wines and spirits.

Facilities
2 classrooms; computer laboratory; demonstration laboratory; food production kitchen; garden; gourmet dining room; learning resource center; library; public restaurant.

Culinary Student Profile
40 full-time.

Faculty
3 full-time; 2 part-time. 4 are industry professionals; 1 is a culinary-accredited teacher. Prominent faculty: Walter N. Bronowitz.

Expenses
Tuition: $500 per quarter. Program-related fees include: $250 for textbooks; $200 for knives and tools; $250 for uniforms.

Edmonds Community College *(continued)*

Financial Aid

In 1996, 5 scholarships were awarded (average award was $500); 15 loans were granted (average loan was $2000). Employment placement assistance is available. Employment opportunities within the program are available.

Housing

Apartment-style housing available. Average on-campus housing cost per month: $300.

Application Information

Students are accepted for enrollment in January, April, and September. In 1996, 40 applied; 25 were accepted. Applicants must submit a formal application.

Contact

Walter N. Bronowitz, Department Head, Culinary Arts, 20000 68th Avenue West, Lynnwood, WA 98036-5999; 425-640-1329; Fax: 425-771-3366; E-mail: wbronowi@edcc.edu

North Seattle Community College

Culinary Arts and Hospitality

Seattle, Washington

General Information

Public, coeducational, two-year college. Urban campus. Founded in 1970. Accredited by Northwest Association of Schools and Colleges.

Program Information

Offered since 1970. Program calendar is divided into quarters. 18-month Certificate in Restaurant Cooking. 2-year Associate degree in Culinary Arts and Hospitality. 6-month Certificates in Dining Room Service; Commercial Cooking.

Areas of Study

Baking; convenience cookery; food preparation; food purchasing; kitchen management; management and human resources; meal planning; sanitation; saucier; soup, stock, sauce, and starch production.

Facilities

Bake shop; cafeteria; catering service; 4 classrooms; computer laboratory; 2 food production kitchens; gourmet dining room; laboratory; learning resource center; lecture room; library; public restaurant; teaching kitchen.

Culinary Student Profile

60 total: 50 full-time; 10 part-time. 10 are under 25 years old; 40 are between 25 and 44 years old; 10 are over 44 years old.

Faculty

3 full-time; 4 part-time. 7 are industry professionals. Prominent faculty: David Wasson and Gregg Shiosaki.

Expenses

In-state tuition: $500 per quarter full-time, $95.30 per credit part-time. Out-of-state tuition: $1600 per quarter full-time, $380.30 per credit part-time. Program-related fees include: $75 for food fees; $500 for uniforms and tools.

Financial Aid

In 1996, 3 scholarships were awarded (average award was $400). Program-specific awards include professional catering association scholarships. Employment placement assistance is available. Employment opportunities within the program are available.

Housing

Average off-campus housing cost per month: $650.

Application Information

Students are accepted for enrollment in January, April, and September. In 1996, 40 applied; 40 were accepted. Applicants must interview.

Contact

Patty Haggerty, Secretary Supervisor, Culinary Arts and Hospitality, 9600 College Way North, Seattle, WA 98103-3599; 206-527-3779; Fax: 206-527-3635.

Olympic College

Commercial Cooking/Food Service

Bremerton, Washington

General Information
Public, coeducational, two-year college. Suburban campus. Founded in 1946. Accredited by Northwest Association of Schools and Colleges.

Program Information
Offered since 1978. Member of American Culinary Federation; United States Personal Chef Association. Program calendar is divided into quarters. 18-month Associate degree in Culinary Arts. 9-month Certificate in Culinary Arts/Dining Room Service.

Areas of Study
Baking; beverage management; buffet catering; computer applications; controlling costs in food service; convenience cookery; culinary skill development; food preparation; food purchasing; food service communication; food service math; garde-manger; introduction to food service; management and human resources; meal planning; meat fabrication; menu and facilities design; nutrition and food service; sanitation; seafood processing; soup, stock, sauce, and starch production.

Facilities
Cafeteria; catering service; classroom; food production kitchen; gourmet dining room; learning resource center; lecture room; library; public restaurant; snack shop; student lounge; teaching kitchen.

Culinary Student Profile
64 total: 40 full-time; 24 part-time.

Faculty
2 full-time; 2 part-time. 2 are industry professionals; 2 are culinary-accredited teachers. Prominent faculty: Nick Giovanni and Steve Lammers.

Special Programs
5-day gourmet cooking class, specialty training in specific cuisine areas.

Expenses
Tuition: $496 per quarter full-time, $99.20 per 1-2 credits part-time. Program-related fees include: $50 for lab fees (includes lunch).

Financial Aid
In 1996, 3 scholarships were awarded (average award was $200). Employment placement assistance is available.

Housing
Average off-campus housing cost per month: $350.

Application Information
Students are accepted for enrollment in January, April, and September. In 1996, 85 applied; 64 were accepted. Applicants must interview.

Contact
Steve Lammers, Chef Instructor, Commercial Cooking/Food Service, 1600 Chester Avenue, Bremerton, WA 98337-1699; 360-478-4576; Fax: 360-478-4650; E-mail: hjgiovan@olympic.ctc.edu

Renton Technical College

Culinary Arts Department

Renton, Washington

General Information
Public, coeducational, two-year college. Urban campus. Founded in 1942. Accredited by Northwest Association of Schools and Colleges.

Program Information
Accredited by American Culinary Federation Education Institute. Member of American Culinary Federation; American Culinary Federation Educational Institute; American Institute of Baking; Educational Foundation of the NRA; National Restaurant Association; Women Chefs and Restaurateurs. Program calendar is divided into quarters. 2-year Associate degree in Culinary Arts.

Areas of Study
Baking; beverage management; buffet catering; computer applications; confectionery show pieces;

Renton Technical College *(continued)*

controlling costs in food service; convenience cookery; culinary French; culinary skill development; food preparation; food purchasing; food service communication; food service math; garde-manger; international cuisine; introduction to food service; kitchen management; management and human resources; meal planning; meat cutting; meat fabrication; menu and facilities design; nutrition; nutrition and food service; patisserie; restaurant opportunities; sanitation; saucier; seafood processing; soup, stock, sauce, and starch production; wines and spirits.

Facilities

Bake shop; bakery; cafeteria; catering service; classroom; coffee shop; computer laboratory; demonstration laboratory; food production kitchen; garden; gourmet dining room; laboratory; learning resource center; lecture room; library; snack shop; student lounge; teaching kitchen.

Culinary Student Profile

35 full-time. 10 are under 25 years old; 20 are between 25 and 44 years old; 5 are over 44 years old.

Faculty

2 full-time; 2 part-time. Prominent faculty: Erhand Volke and David Pisegna.

Expenses

Application fee: $25. Tuition: $575 per quarter. Program-related fees include: $275 for knives; $150 for uniforms; $275 for books.

Financial Aid

In 1996, 20 scholarships were awarded (average award was $1000); 25 loans were granted (average loan was $2000). Program-specific awards include Celebrity Chef Foundation scholarship. Employment placement assistance is available. Employment opportunities within the program are available.

Application Information

Students are accepted for enrollment in January, April, July, and September. In 1996, 50 applied; 35 were accepted. Applicants must submit a formal application and have a high school diploma or GED.

Contact

David Pisegna, Chef Instructor, Culinary Arts Department, 3000 Fourth Street, NE, Renton, WA 98056; 206-235-2352; Fax: 206-235-7832.

Seattle Central Community College

Hospitality and Culinary Arts

Seattle, Washington

General Information

Public, coeducational, two-year college. Urban campus. Founded in 1941. Accredited by Northwest Association of Schools and Colleges.

Program Information

Offered since 1941. Accredited by American Culinary Federation Education Institute. Member of American Culinary Federation Educational Institute; Council on Hotel, Restaurant, and Institutional Education; The Bread Bakers Guild of America; Washington Restaurant Association. Program calendar is divided into quarters. 1-year Certificate in Specialty Desserts and Breads. 18-month Certificate in Culinary Arts. 2-year Associate degrees in Hospitality Management; Culinary Arts.

Areas of Study

Baking; buffet catering; computer applications; controlling costs in food service; culinary skill development; food preparation; food service math; garde-manger; international cuisine; introduction to food service; management and human resources; nutrition; patisserie; restaurant opportunities; sanitation; saucier; soup, stock, sauce, and starch production.

Facilities

Bake shop; bakery; cafeteria; catering service; 5 classrooms; computer laboratory; 5 food production kitchens; gourmet dining room; 3

lecture rooms; library; public restaurant; snack shop; student lounge; student activity center.

Culinary Student Profile
90 full-time.

Faculty
7 full-time; 2 part-time. 1 is an industry professional; 6 are culinary-accredited teachers; 1 is a registered dietitian. Prominent faculty: Diana Dillard and Kejiro Miyata.

Special Programs
Chef-of-the-Day (students create a menu and oversee its production for the restaurant).

Expenses
In-state tuition: $484 per quarter full-time, $48.40 per credit part-time. Out-of-state tuition: $1909 per quarter full-time, $190.90 per credit part-time. Program-related fees include: $55 for lab fees.

Financial Aid
In 1996, 15 scholarships were awarded (average award was $500).

Application Information
Students are accepted for enrollment in January, April, and September. Applications are accepted continuously for fall, winter, and spring. In 1996, 90 were accepted. Applicants must submit a formal application.

Contact
Joy Gulmon-Huri, Manager, Hospitality and Culinary Arts, 1701 Broadway, Seattle, WA 98122-2400; 206-587-5425; Fax: 206-344-4323; E-mail: jgolmo@seaccc.sccd.ctc.edu; World Wide Web: http://seaccd.sccd.ctc.edu/~cculhosp/index.html

Skagit Valley College

Culinary Arts-Hospitality Management

Mount Vernon, Washington

General Information
Public, coeducational, two-year college. Rural campus. Founded in 1926. Accredited by Northwest Association of Schools and Colleges.

Program Information
Offered since 1979. Accredited by American Culinary Federation Education Institute, National Restaurant Association Educational Foundation: Pro Mngt. Member of American Culinary Federation; American Culinary Federation Educational Institute; Council on Hotel, Restaurant, and Institutional Education; Educational Foundation of the NRA; National Restaurant Association. Program calendar is divided into quarters. 1-year Certificate in Professional Cooking. 2-year Associate degree in Culinary Arts and Hospitality Management.

Areas of Study
Baking; beverage management; buffet catering; computer applications; controlling costs in food service; culinary skill development; food preparation; food purchasing; food service communication; food service math; garde-manger; international cuisine; introduction to food service; kitchen management; management and human resources; meal planning; menu and facilities design; nutrition; restaurant opportunities; sanitation; soup, stock, sauce, and starch production; wines and spirits.

Facilities
Bake shop; bakery; cafeteria; catering service; 2 classrooms; coffee shop; computer laboratory; demonstration laboratory; food production kitchen; gourmet dining room; learning resource center; lecture room; library; snack shop; student lounge; teaching kitchen.

Culinary Student Profile
35 full-time. 10 are under 25 years old; 15 are between 25 and 44 years old; 10 are over 44 years old.

Faculty
2 full-time; 1 part-time. 2 are culinary-accredited teachers. Prominent faculty: Dani Cox and Martin Hahn.

Expenses
Tuition: $486 per quarter full-time, $48.60 per credit part-time.

Financial Aid
In 1996, 5 scholarships were awarded (average award was $750). Employment placement

Skagit Valley College ***(continued)***

assistance is available. Employment opportunities within the program are available.

Housing
Apartment-style housing available. Average on-campus housing cost per month: $320.

Application Information
Students are accepted for enrollment in January, March, and September. Application deadline for fall is September 11. Application deadline for winter is January 5. Application deadline for spring is March 30. In 1996, 57 applied; 51 were accepted. Applicants must submit a formal application.

Contact
Lyle W. Hildahl, Director, Culinary Arts-Hospitality Management, 2405 College Way, Mount Vernon, WA 98273-5899; 360-416-7618; Fax: 360-416-7890.

South Puget Sound Community College

Olympia, Washington

General Information
Public, coeducational, two-year college. Suburban campus. Founded in 1970. Accredited by Northwest Association of Schools and Colleges.

Program Information
Offered since 1990. Member of American Culinary Federation; American Culinary Federation Educational Institute. Program calendar is divided into quarters. 18-month Associate degrees in Food Service Management; Food Service Technology. 9-month Certificates in Food Service Technology; Commercial Baking.

Areas of Study
Baking; buffet catering; computer applications; controlling costs in food service; convenience cookery; culinary French; culinary skill development; food preparation; food purchasing; food service communication; food service math; garde-manger; international cuisine; introduction to food service; kitchen management; management and human resources; meal planning; meat cutting; meat fabrication; menu and facilities design; nutrition; nutrition and food service; patisserie; restaurant opportunities; sanitation; saucier; seafood processing; soup, stock, sauce, and starch production; wines and spirits.

Facilities
Bake shop; bakery; cafeteria; catering service; classroom; computer laboratory; food production kitchen; gourmet dining room; laboratory; learning resource center; lecture room; library; public restaurant; student lounge; teaching kitchen; herb garden.

Culinary Student Profile
50 total: 40 full-time; 10 part-time.

Faculty
2 full-time; 4 part-time. 2 are industry professionals; 2 are master chefs; 2 are culinary-accredited teachers. Prominent faculty: Bill Wiklendt.

Special Programs
Gourmet Club.

Expenses
In-state tuition: $1800 per 3 quarters full-time, $56 per credit part-time. Out-of-state tuition: $3400 per 3 quarters full-time, $110 per credit part-time. Program-related fees include: $200 for books; $200 for assistant chef equipment; $120 for uniforms.

Financial Aid
In 1996, 10 scholarships were awarded (average award was $500); 3 loans were granted (average loan was $2000). Employment placement assistance is available. Employment opportunities within the program are available.

Housing
Average off-campus housing cost per month: $300.

Application Information
Students are accepted for enrollment in January, April, and September. In 1996, 39 applied. Applicants must submit a formal application.

Contact
Fred Durinski, Food Service Director, Food Service Technology, 2011 Mottman Road, SW, Olympia, WA 98512-6292; 360-754-7711 Ext. 347; Fax: 360-664-0780.

South Seattle Community College

Hospitality and Food Science Division

Seattle, Washington

General Information
Public, coeducational, two-year college. Suburban campus. Founded in 1970. Accredited by Northwest Association of Schools and Colleges.

Program Information
Offered since 1975. Accredited by American Culinary Federation Education Institute. Program calendar is divided into quarters. 15-month Certificate in Hospitality Production. 18-month Certificates in Pastry and Specialty Baking; Production Management. 2-year Associate degrees in Pastry and Specialty Baking; Hospitality Production Management.

Areas of Study
Baking; beverage management; buffet catering; computer applications; confectionery show pieces; controlling costs in food service; convenience cookery; culinary skill development; food preparation; food purchasing; food service communication; food service math; garde-manger; international cuisine; introduction to food service; kitchen management; meal planning; meat cutting; nutrition; nutrition and food service; patisserie; sanitation; saucier; seafood processing; soup, stock, sauce, and starch production.

Facilities
Bake shop; 2 bakeries; cafeteria; 2 catering services; 8 classrooms; coffee shop; 3 computer laboratories; 3 food production kitchens; garden; gourmet dining room; laboratory; learning resource center; 8 lecture rooms; library; 2 public restaurants; snack shop; student lounge.

Culinary Student Profile
150 full-time.

Faculty
7 full-time; 4 part-time. 7 are industry professionals; 4 are culinary-accredited teachers.

Expenses
Tuition: $484 per quarter full-time, $48.40 per credit part-time. Program-related fees include: $200 for handtools; $200 for uniforms; $400 for textbooks.

Financial Aid
In 1996, 12 scholarships were awarded (average award was $500); 8 loans were granted (average loan was $600). Employment placement assistance is available. Employment opportunities within the program are available.

Housing
Average off-campus housing cost per month: $500.

Application Information
Students are accepted for enrollment in January, March, June, and September. In 1996, 160 applied; 160 were accepted.

Contact
Joanne Craig, Administrative Supervisor, Hospitality and Food Science Division, 6000 16th Avenue, SW, Seattle, WA 98106-1499; 206-764-5344; Fax: 206-768-6728.

Spokane Community College

Culinary Arts Department

Spokane, Washington

General Information
Public, coeducational, two-year college. Suburban campus. Founded in 1963. Accredited by Northwest Association of Schools and Colleges.

Program Information
Offered since 1963. Accredited by American Culinary Federation Education Institute. Member of American Culinary Federation; Educational Foundation of the NRA; National Restaurant

Spokane Community College *(continued)*

Association; Washington State Restaurant Association. Program calendar is divided into quarters. 12-month Certificate in Baking. 18-month Associate degrees in Hotel and Restaurant Management; Culinary Arts.

Areas of Study

Baking; beverage management; buffet catering; computer applications; controlling costs in food service; culinary skill development; food preparation; food purchasing; food service communication; food service math; garde-manger; international cuisine; introduction to food service; kitchen management; management and human resources; meal planning; meat cutting; meat fabrication; menu and facilities design; nutrition; nutrition and food service; patisserie; restaurant opportunities; sanitation; saucier; seafood processing; soup, stock, sauce, and starch production; wines and spirits.

Facilities

Bake shop; bakery; cafeteria; 4 classrooms; computer laboratory; 2 food production kitchens; gourmet dining room; learning resource center; 4 lecture rooms; library; public restaurant; student lounge; 2 teaching kitchens; pastry shop.

Culinary Student Profile

70 full-time.

Faculty

5 full-time. 2 are industry professionals; 3 are culinary-accredited teachers. Prominent faculty: Douglas A. Fisher and Robert Lombardi.

Expenses

Application fee: $15. Tuition: $500 per quarter. Program-related fees include: $150 for knives; $200 for uniforms; $250 for books.

Financial Aid

Employment placement assistance is available.

Application Information

Students are accepted for enrollment in January and April. Applications are accepted continuously for fall, winter, and spring. In 1996, 80 applied; 70 were accepted. Applicants must submit a formal application and letters of reference.

Contact

Kenna Hanson, Registrar Assistant, Culinary Arts Department, North 1810 Greene Street, Spokane, WA 99207-5399; 509-533-7003; Fax: 509-533-8839.

Shepherd College

Culinary Arts Program

Shepherdstown, West Virginia

General Information

Public, coeducational, four-year college. Rural campus. Founded in 1871. Accredited by North Central Association of Colleges and Schools.

Program Information

Offered since 1993. Member of American Culinary Federation; National Restaurant Association. Program calendar is divided into semesters. 2-year Associate degree in Culinary Arts.

Areas of Study

Baking; beverage management; buffet catering; computer applications; confectionery show pieces; controlling costs in food service; convenience cookery; culinary French; culinary skill development; food preparation; food purchasing; food service communication; food service math; garde-manger; international cuisine; introduction to food service; kitchen management; management and human resources; meal planning; meat cutting; meat fabrication; menu and facilities design; nutrition; nutrition and food service; patisserie; restaurant opportunities; sanitation; saucier; seafood processing; soup, stock, sauce, and starch production; wines and spirits; dining room service; introduction to hospitality.

Facilities

Bake shop; bakery; 2 cafeterias; 2 catering services; 2 classrooms; 2 computer laboratories; demonstration laboratory; food production kitchen; garden; gourmet dining room; 2 laboratories; 2 learning resource centers; 21 lecture rooms; 2 libraries; public restaurant; snack shop; student lounge; 2 teaching kitchens; greenhouse.

Culinary Student Profile
20 total: 15 full-time; 5 part-time.

Faculty
Prominent faculty: Paul M. Saab and Judy Cole.

Expenses
Application fee: $25. In-state tuition: $1114 per year full-time, $93 per credit part-time. Out-of-state tuition: $2674 per year full-time, $223 per credit part-time. Program-related fees include: $400 for uniforms, knives, and lab fees.

Financial Aid
In 1996, 6 scholarships were awarded (average award was $2000); 1 loan was granted (loan was $2000). Employment placement assistance is available. Employment opportunities within the program are available.

Housing
Coed, apartment-style, and single-sex housing available.

Application Information
Students are accepted for enrollment in January, May, and August. In 1996, 20 applied; 15 were accepted. Applicants must submit a formal application, SAT or ACT scores, and academic transcripts.

Contact
Vicki Jenkins, Post Secondary Coordinator, Culinary Arts Program, Box 268, Route 6, Martinsburg, WV 25401; 304-754-5318; Fax: 304-754-7933.

West Virginia Northern Community College

Culinary Arts Department

Wheeling, West Virginia

General Information
Public, coeducational, two-year college. Suburban campus. Founded in 1972. Accredited by North Central Association of Colleges and Schools.

Program Information
Offered since 1974. Program calendar is divided into semesters. 1-year Certificate in Culinary Arts. 2-year Associate degree in Culinary Arts.

Areas of Study
Baking; confectionery show pieces; controlling costs in food service; culinary skill development; food preparation; food purchasing; garde-manger; international cuisine; management and human resources; meal planning; menu and facilities design; nutrition; patisserie; sanitation; saucier; seafood processing; soup, stock, sauce, and starch production; wines and spirits.

Facilities
Bake shop; 3 classrooms; 4 computer laboratories; demonstration laboratory; food production kitchen; gourmet dining room; 3 learning resource centers; 3 libraries; 3 student lounges; teaching kitchen.

Culinary Student Profile
52 total: 40 full-time; 12 part-time. 15 are under 25 years old; 25 are between 25 and 44 years old; 12 are over 44 years old.

Faculty
2 full-time. 2 are culinary-accredited teachers. Prominent faculty: Audrey Secreto and Marian Grubor.

Expenses
Tuition: $708 per semester full-time, $59 per credit hour part-time. Program-related fees include: $150 for books; $150 for uniforms.

Financial Aid
Employment placement assistance is available.

Housing
Average off-campus housing cost per month: $200.

Application Information
Students are accepted for enrollment in January and August. In 1996, 20 applied; 20 were accepted. Applicants must submit a formal application.

Contact
Bonnie Ellis, Admissions Counselor, Culinary Arts Department, 1704 Market Street, Wheeling, WV 26003; 304-233-5900 Ext. 4218.

Chippewa Valley Technical College

Culinary Arts

Eau Claire, Wisconsin

General Information

Public, coeducational, two-year college. Suburban campus. Founded in 1912. Accredited by North Central Association of Colleges and Schools.

Program Information

Offered since 1968. Member of Interntional Food Service Executives Association; Wisconsin Restaurant Association. Program calendar is divided into semesters. 2-year Associate degree in Culinary Arts.

Areas of Study

Baking; beverage management; buffet catering; computer applications; controlling costs in food service; culinary French; culinary skill development; food preparation; food purchasing; food service communication; food service math; garde-manger; introduction to food service; kitchen management; management and human resources; meat fabrication; restaurant opportunities; sanitation.

Facilities

Cafeteria; 4 classrooms; 6 computer laboratories; demonstration laboratory; 2 food production kitchens; gourmet dining room; laboratory; learning resource center; library; public restaurant; student lounge.

Culinary Student Profile

34 total: 24 full-time; 10 part-time.

Faculty

3 full-time. 3 are culinary-accredited teachers. Prominent faculty: Karen Demaree and Paul Waters.

Expenses

Application fee: $30. Tuition: $54.20 per credit. Program-related fees include: $200 for uniforms; $165 for tools.

Financial Aid

In 1996, 2 scholarships were awarded (average award was $300). Employment placement assistance is available. Employment opportunities within the program are available.

Application Information

Students are accepted for enrollment in January, June, and September. In 1996, 29 applied; 23 were accepted. Applicants must submit a formal application.

Contact

Admissions, Culinary Arts, 620 West Clairemont Avenue, Eau Claire, WI 54701-6120; 715-833-6246; Fax: 715-833-6470; E-mail: ghinrichsen@mail.chippewa.tec.wi.us

Fox Valley Technical College

Culinary Arts Department

Appleton, Wisconsin

General Information

Public, coeducational, two-year college. Urban campus. Founded in 1967. Accredited by North Central Association of Colleges and Schools.

Program Information

Offered since 1967. Accredited by American Culinary Federation Education Institute. Member of American Culinary Federation; American Culinary Federation Educational Institute; Council on Hotel, Restaurant, and Institutional Education; Educational Foundation of the NRA; Interntional Food Service Executives Association; National Restaurant Association; Wisconsin Restaurant Association. Program calendar is divided into semesters. 1-year Certificate in Food Preparation Assistant. 2-year Associate degree in Culinary Arts.

Areas of Study

Baking; buffet catering; computer applications; confectionery show pieces; controlling costs in food service; convenience cookery; culinary skill development; food preparation; food purchasing;

food service communication; food service math; garde-manger; international cuisine; introduction to food service; kitchen management; management and human resources; meal planning; meat cutting; meat fabrication; menu and facilities design; nutrition; nutrition and food service; patisserie; sanitation; saucier; seafood processing; soup, stock, sauce, and starch production.

Facilities
Bake shop; bakery; cafeteria; catering service; demonstration laboratory; 4 food production kitchens; gourmet dining room; learning resource center; library; snack shop.

Culinary Student Profile
100 total: 75 full-time; 25 part-time.

Faculty
4 full-time; 1 part-time. 3 are industry professionals; 2 are culinary-accredited teachers. Prominent faculty: Albert Exenberger and Michae Lang.

Expenses
Application fee: $25. Tuition: $1100 per semester full-time, $68 per credit part-time. Program-related fees include: $200 for knives (beginning set); $250 for books per semester; $50 for uniforms.

Financial Aid
Employment placement assistance is available. Employment opportunities within the program are available.

Housing
Average off-campus housing cost per month: $350.

Application Information
Students are accepted for enrollment in January and September. In 1996, 40 applied; 40 were accepted. Applicants must submit a formal application.

Contact
Jeffrey S. Igel, Culinary Arts Manager, Culinary Arts Department, 1825 North Bluemound, PO Box 2277, Appleton, WI 54913-2277; 920-735-5643; Fax: 920-831-5410; E-mail: igel@foxvalley.tec.wi.us

Madison Area Technical College

Culinary Trades Department

Madison, Wisconsin

General Information
Public, coeducational, two-year college. Urban campus. Founded in 1911. Accredited by North Central Association of Colleges and Schools.

Program Information
Accredited by American Culinary Federation Education Institute. Member of American Culinary Federation; American Culinary Federation Educational Institute; Interntional Food Service Executives Association; The Bread Bakers Guild of America; Retail Bakers of America. Program calendar is divided into semesters. 1-year Diplomas in Baking/Pastry Arts; Food Service Production. 2-year Associate degree in Culinary Arts.

Areas of Study
Baking; computer applications; controlling costs in food service; culinary skill development; food preparation; food purchasing; food service communication; food service math; garde-manger; introduction to food service; management and human resources; meal planning; meat cutting; menu and facilities design; nutrition; sanitation; soup, stock, sauce, and starch production.

Facilities
Bake shop; bakery; 2 cafeterias; catering service; computer laboratory; demonstration laboratory; 2 food production kitchens; gourmet dining room; learning resource center; lecture room; library; snack shop; student lounge; 2 teaching kitchens.

Culinary Student Profile
85 total: 60 full-time; 25 part-time. 55 are under 25 years old; 22 are between 25 and 44 years old; 8 are over 44 years old.

Faculty
4 full-time; 3 part-time. 1 is a master baker.

Special Programs
2-credit internship.

Madison Area Technical College *(continued)*

Expenses
Application fee: $25. Tuition: $1200 per semester full-time, $57.45 per credit part-time. Program-related fees include: $35 for knives; $70 for uniforms.

Financial Aid
In 1996, 8 scholarships were awarded (average award was $350); 35 loans were granted (average loan was $2500). Employment placement assistance is available. Employment opportunities within the program are available.

Application Information
Students are accepted for enrollment in January and August. Application deadline for fall is July 1. Application deadline for spring is November 15. In 1996, 91 applied; 90 were accepted. Applicants must submit a formal application and academic transcripts.

Contact
Linda Williams, Administrative Clerk, Culinary Trades Department, 3550 Anderson Street, Madison, WI 53704-2599; 608-246-6369; Fax: 608-246-6316.

Moraine Park Technical College

Culinary Arts/Food Service Production

Fond du Lac, Wisconsin

General Information
Public, coeducational, two-year college. Rural campus. Founded in 1967. Accredited by North Central Association of Colleges and Schools.

Program Information
Offered since 1976. Member of Council on Hotel, Restaurant, and Institutional Education; National Restaurant Association; Wisconsin Restaurant Association. Program calendar is divided into semesters. Certificate in Culinary Basics. Technical Diploma in School Food Service. 1-semester Certificates in Food Production; Bakery Deli. 1-year Technical Diplomas in Food Service Production; Dietary Manager studies. 2-year Associate degree in Culinary Arts.

Areas of Study
Baking; buffet catering; computer applications; controlling costs in food service; culinary skill development; food preparation; food purchasing; food service math; garde-manger; international cuisine; introduction to food service; meal planning; menu and facilities design; nutrition; restaurant opportunities; sanitation; saucier; soup, stock, sauce, and starch production; waiter/ waitress.

Facilities
Bake shop; bakery; cafeteria; catering service; 2 classrooms; 3 computer laboratories; demonstration laboratory; 2 food production kitchens; gourmet dining room; learning resource center; library; snack shop; student lounge; teaching kitchen.

Culinary Student Profile
40 total: 20 full-time; 20 part-time. 30 are under 25 years old; 8 are between 25 and 44 years old; 2 are over 44 years old.

Faculty
3 full-time; 1 part-time. 3 are industry professionals. Prominent faculty: David Weber and Mary Martin.

Expenses
Application fee: $20. Tuition: $5679 per 2 years full-time, $60.45 per credit part-time. Program-related fees include: $1190 for equipment, supplies, and uniform rental; $965 for book and modules.

Financial Aid
In 1996, 3 scholarships were awarded (average award was $500). Employment placement assistance is available.

Housing
Average off-campus housing cost per month: $400.

APPLICATION INFORMATION
Students are accepted for enrollment in January and August. In 1996, 29 applied; 29 were accepted. Applicants must submit a formal application.

CONTACT
Stan Arpke, Counselor, Culinary Arts/Food Service Production, 235 North National Avenue, PO Box 1940, Fond du Lac, WI 54936-1940; 920-924-3197.

NICOLET AREA TECHNICAL COLLEGE

Culinary Arts

Rhinelander, Wisconsin

GENERAL INFORMATION
Public, coeducational, two-year college. Rural campus. Founded in 1968. Accredited by North Central Association of Colleges and Schools.

PROGRAM INFORMATION
Member of American Culinary Federation; Educational Foundation of the NRA; National Restaurant Association; Wisconsin Restaurant Association. Program calendar is divided into semesters. 1-year Technical Diploma in Food Service Production. 13-credit Certificate in Catering. 14-credit Certificate in Baking. 2-year Associate degree in Culinary Arts. 8-course Certificate in Food Service Management.

AREAS OF STUDY
Baking; buffet catering; computer applications; controlling costs in food service; culinary skill development; food preparation; food purchasing; food service math; garde-manger; international cuisine; management and human resources; meal planning; nutrition and food service; sanitation; wines and spirits; food practicum; restaurant practicum.

FACILITIES
Cafeteria; classroom; demonstration laboratory; food production kitchen; gourmet dining room; learning resource center; lecture room; library; student lounge; teaching kitchen.

CULINARY STUDENT PROFILE
30 total: 20 full-time; 10 part-time.

FACULTY
2 full-time; 2 part-time. Prominent faculty: Linda Arnot and Kyle Gruening.

SPECIAL PROGRAMS
Internship in culinary arts (2 credits).

EXPENSES
Application fee: $25. Tuition: $1100 per semester full-time, $75 per credit part-time. Program-related fees include: $155 for books per semester; $145 for uniforms.

FINANCIAL AID
In 1996, 5 scholarships were awarded (average award was $500). Program-specific awards include Wisconsin Restaurant Association scholarships, NRA Education Foundation scholarships, Professional Management Program scholarships. Employment placement assistance is available. Employment opportunities within the program are available.

HOUSING
Average off-campus housing cost per month: $400.

APPLICATION INFORMATION
Students are accepted for enrollment in January and August. Application deadline for fall is August 15. Application deadline for spring is January 15. Applicants must submit a formal application, academic transcripts, and ASSET scores.

CONTACT
Susan Kordula, Admissions, Culinary Arts, Nicolet A.T.C., Box 518, Rhinelander, WI 54501-0518; 715-365-4451; Fax: 715-365-4445; E-mail: skordula@nicolet.tec.wi.usa

The Postilion School of Culinary Art

Fond du Lac, Wisconsin

General Information
Private, coeducational, culinary institute. Urban campus. Founded in 1949.

Program Information
Offered since 1949. 5-course Diploma in Classic Cuisine.

Areas of Study
Classic cuisine.

Facilities
Bake shop; classroom; 2 food production kitchens; garden; gourmet dining room; lecture room; 2 teaching kitchens.

Culinary Student Profile
24 full-time.

Faculty
1 full-time.

Expenses
Tuition: $2000 per 2 weeks (includes meals).

Financial Aid
Employment placement assistance is available.

Application Information
Applicants must submit a formal application, letters of reference, and a portfolio.

Contact
Liane Kuony, Owner, 220 Old Pioneer Road, Fond du Lac, WI 54935; 414-922-4170.

Waukesha County Technical College

Center for Hospitality Management and Culinary Arts Studies

Pewaukee, Wisconsin

General Information
Public, coeducational, two-year college. Suburban campus. Founded in 1923. Accredited by North Central Association of Colleges and Schools.

Program Information
Offered since 1971. Accredited by American Culinary Federation Education Institute. Member of American Culinary Federation; American Culinary Federation Educational Institute; Council on Hotel, Restaurant, and Institutional Education; Educational Foundation of the NRA; Interntional Food Service Executives Association; National Restaurant Association. Program calendar is divided into semesters. 1-year Diploma in Culinary Arts. 2-year Associate degrees in Hospitality and Tourism Management; Culinary Management. 3-year Certificate in ACF Culinary Apprenticeship.

Areas of Study
Baking; beverage management; computer applications; controlling costs in food service; culinary skill development; food preparation; food purchasing; food service math; garde-manger; international cuisine; introduction to food service; kitchen management; management and human resources; menu and facilities design; nutrition; restaurant opportunities; sanitation; soup, stock, sauce, and starch production.

Facilities
2 cafeterias; 3 computer laboratories; demonstration laboratory; food production kitchen; gourmet dining room; learning resource center; 2 lecture rooms; library; public restaurant; student lounge; 2 teaching kitchens.

Culinary Student Profile
105 total: 65 full-time; 40 part-time. 69 are under 25 years old; 30 are between 25 and 44 years old; 6 are over 44 years old.

Faculty
4 full-time; 6 part-time. 8 are industry professionals; 2 are culinary-accredited teachers. Prominent faculty: James Holden and Michael Leitzke.

Expenses
Tuition: $54.20 per credit. Program-related fees include: $40 for uniform service.

Financial Aid
In 1996, 3 scholarships were awarded (average award was $500). Employment placement assistance is available.

Application Information
Students are accepted for enrollment in January and August. Applications are accepted continuously for spring. Application deadline for fall is July 15. In 1996, 74 applied. Applicants must submit a formal application.

Contact
Mic Pietrykowski, Education Assistant, Center for Hospitality Management and Culinary Arts Studies, 800 Main Street, Pewaukee, WI 53072-4601; 414-691-5303; Fax: 414-691-5155; E-mail: mpietrykowski@waukesha.tec.wi.us

INTERNATIONAL PROGRAMS

Canberra Institute of Technology

School of Tourism and Hospitality

Canberra, Australia

General Information
Public, coeducational, two-year college. Suburban campus. Founded in 1976.

Program Information
Offered since 1976. Member of Tourism Council of Australia. Program calendar is divided into semesters. 1-year Diplomas in Food and Beverage Service; Culinary Skills. 30-month Advanced Diploma in Hospitality.

Areas of Study
Baking; beverage management; buffet catering; computer applications; confectionery show pieces; controlling costs in food service; convenience cookery; culinary French; culinary skill development; food preparation; food purchasing; food service communication; food service math; garde-manger; international cuisine; introduction to food service; kitchen management; management and human resources; meal planning; meat cutting; meat fabrication; menu and facilities design; nutrition; nutrition and food service; patisserie; restaurant opportunities; sanitation; saucier; seafood processing; soup, stock, sauce, and starch production; wines and spirits.

Facilities
Bake shop; bakery; cafeteria; 8 classrooms; coffee shop; 2 computer laboratories; 6 demonstration laboratories; 2 food production kitchens; garden; 4 gourmet dining rooms; laboratory; 3 learning resource centers; 2 lecture rooms; library; 4 public restaurants; snack shop; student lounge; 6 teaching kitchens.

Culinary Student Profile
2,000 total: 1,000 full-time; 1,000 part-time.

Faculty
32 full-time; 40 part-time.

Expenses
Tuition: A$7800 per year.

Financial Aid
Employment placement assistance is available. Employment opportunities within the program are available.

Application Information
Students are accepted for enrollment in February and July. Applicants must submit a formal application.

Contact
Gordon McDonald, Business Manager, School of Tourism and Hospitality, GPO Box 826, Canberra, Australia; 61-2-62073542; Fax: 61-2-62073209; E-mail: gordon.mcdonald@cit.act.edu.au

Crow's Nest College of TAFE

Sydney, New South Wales, Australia

General Information
Public, coeducational, culinary institute. Suburban campus. Founded in 1989.

Program Information
Offered since 1989. Program calendar is divided into semesters. 1-year Certificates in Commercial

Crow's Nest College of Tafe ***(continued)***

Cookery (Asian); Catering Supervision; Dietary Practices. 18-week Certificates in Catering Operations; Cookery. 18-week Statement of Attainment in Coffee Shop and Fast Food.

Areas of Study
Culinary skill development; food preparation; food purchasing; international cuisine; nutrition; nutrition and food service; soup, stock, sauce, and starch production; Asian cuisine.

Facilities
Cafeteria; classroom; coffee shop; computer laboratory; food production kitchen; garden; gourmet dining room; laboratory; lecture room; library; public restaurant; student lounge; teaching kitchen.

Culinary Student Profile
170 total: 45 full-time; 125 part-time. 50 are under 25 years old; 90 are between 25 and 44 years old; 30 are over 44 years old.

Faculty
6 full-time; 15 part-time. 5 are industry professionals; 4 are master chefs; 12 are culinary-accredited teachers.

Expenses
Tuition: A$90 per semester. Program-related fees include: A$200 for tools; A$150 for uniforms; A$40 for books.

Financial Aid
Employment placement assistance is available.

Application Information
Students are accepted for enrollment in February and July. Application deadline for summer is January 10. Application deadline for winter is June 6. In 1996, 220 applied; 170 were accepted. Applicants must submit a formal application.

Contact
Geoff Tyrrell, Senior Head Teacher, Tourism and Hospitality, 149 West Street, Crows Nest, Sydney, New South Wales, Australia; 61-2-99654433; Fax: 61-2-99654408; E-mail: geoff.tyrrell@tafensw.edu.au

Le Cordon Bleu

Sydney, New South Wales, Australia

General Information
Private, coeducational, culinary institute. Urban campus. Founded in 1995.

Program Information
Offered since 1995. Accredited by American Culinary Federation Education Institute. Program calendar is divided into quarters. 10-week Certificates in Intermediate Pastry; Basic Pastry; Intermediate Cuisine; Basic Cuisine.

Areas of Study
Baking; culinary French; culinary skill development; patisserie; Australian cuisine.

Facilities
Cafeteria; classroom; demonstration laboratory; garden; gourmet dining room; learning resource center; lecture room; library; public restaurant; snack shop; student lounge; teaching kitchen.

Culinary Student Profile
12 part-time.

Faculty
4 full-time; 4 part-time. 2 are industry professionals; 4 are culinary-accredited teachers. Prominent faculty: Geoff Montgomery and Dana Junokas.

Expenses
Application fee: A$100. Tuition: A$5000 per 10 weeks. Program-related fees include: A$100 for registration; A$140 for uniforms; A$300 for equipment.

Financial Aid
In 1996, 6 scholarships were awarded (average award was A$4750).

Housing
Average off-campus housing cost per month: A$800.

Application Information
Students are accepted for enrollment in January, April, July, and October. Application deadline for spring is continuous with a recommended date of October 16. Application deadline for summer is

continuous with a recommended date of January 20. Application deadline for fall is continuous with a recommended date of April 14. In 1996, 45 applied; 45 were accepted. Applicants must submit a formal application.

Contact

Geoff Montgomery, Manager, 4-Ryde Tafe, 250 Blaxland Road, Sydney, New South Wales, Australia; 61-2980-88307; Fax: 61-2980-76541.

See affiliated programs in London, Ontario, Paris, and Tokyo.

See display on page 264.

William Angliss Institute

Hospitality Management/Food Technology/Retail Food Studies

Melbourne, Victoria, Australia

General Information

Public, coeducational, four-year college. Urban campus. Founded in 1940.

Program Information

Offered since 1950. Member of Restaurant and Caterers Association of Victoria. Program calendar is year-round. 1-year Certificate in Food Technology. 15-week Certificate in Cookery (Patisserie). 18-month Certificate in Hospitality. 2-year Diplomas in Food Technology; Hospitality. 25-week Certificate in Baking. 26-week Certificates in Cookery; Cookery (Asian). 3-year Diploma in Hospitality Management. 4-week Certificate in Confectionery.

Areas of Study

Baking; beverage management; buffet catering; computer applications; confectionery show pieces; controlling costs in food service; convenience cookery; culinary skill development; food preparation; food purchasing; food service communication; food service math; garde-manger; international cuisine; introduction to food service; kitchen management; management and human resources; meal planning; meat cutting; meat fabrication; menu and facilities design; nutrition; nutrition and food service; patisserie; restaurant opportunities; sanitation; saucier; seafood processing; soup, stock, sauce, and starch production; wines and spirits; Asian cookery; catering consultation.

Facilities

Bake shop; 4 bakeries; cafeteria; catering service; coffee shop; 3 computer laboratories; 3 demonstration laboratories; 3 food production kitchens; garden; 2 gourmet dining rooms; 2 laboratories; 2 learning resource centers; 4 lecture rooms; library; 2 public restaurants; snack shop; student lounge; 7 teaching kitchens.

Culinary Student Profile

550 total: 420 full-time; 130 part-time.

Faculty

45 full-time.

Expenses

Tuition: A$6000 per 6 months. Program-related fees include: A$700 for knives and uniforms.

Housing

Average off-campus housing cost per month: A$450.

Application Information

Students are accepted for enrollment year-round. In 1996, 1,670 applied; 280 were accepted. Applicants must submit a formal application and letters of reference.

Contact

Vern Bruce, International Manager, Hospitality Management/Food Technology/Retail Food Studies, 555 LaTrobe Street, PO Box 4052, Melbourne, Victoria, Australia; 61-3-96701330; Fax: 61-3-96701330; World Wide Web: http://www.angliss.vic.edu.au

Canadore College of Applied Arts & Technology

School of Hospitality and Tourism

North Bay, Ontario, Canada

General Information

Public, coeducational, technical college. Rural campus. Founded in 1967.

Canadore College of Applied Arts & Technology *(continued)*

PROGRAM INFORMATION
Offered since 1967. Member of American Culinary Federation; Council on Hotel, Restaurant, and Institutional Education. Program calendar is divided into semesters. 1-year Certificate in Chef Training Pre-Employment. 2-year Diploma in Culinary Management. 3-year Diploma in Culinary Administration.

AREAS OF STUDY
Baking; beverage management; buffet catering; computer applications; confectionery show pieces; controlling costs in food service; convenience cookery; culinary French; culinary skill development; food preparation; food purchasing; food service communication; food service math; garde-manger; international cuisine; introduction to food service; kitchen management; management and human resources; meal planning; meat fabrication; menu and facilities design; nutrition; nutrition and food service; patisserie; restaurant opportunities; sanitation; saucier; seafood processing; wines and spirits.

FACILITIES
Bake shop; bakery; cafeteria; 4 classrooms; 2 computer laboratories; food production kitchen; gourmet dining room; learning resource center; lecture room; library; public restaurant; student lounge; teaching kitchen.

CULINARY STUDENT PROFILE
95 total: 90 full-time; 5 part-time. 76 are under 25 years old; 16 are between 25 and 44 years old; 3 are over 44 years old.

FACULTY
11 full-time; 2 part-time. 10 are industry professionals; 3 are master chefs. Prominent faculty: Daniel Esposito and Raymond McGuire.

SPECIAL PROGRAMS
Tour of culinary sites in the Caribbean region, provincial and international student competitions.

EXPENSES
Application fee: Can$30. Tuition: Can$1702 per year full-time, Can$130 per course part-time. Program-related fees include: Can$250 for knives; Can$200 for uniforms.

FINANCIAL AID
In 1996, 15 scholarships were awarded (average award was Can$250). Employment placement assistance is available. Employment opportunities within the program are available.

HOUSING
Single-sex housing available. Average on-campus housing cost per month: Can$350. Average off-campus housing cost per month: Can$400.

APPLICATION INFORMATION
Students are accepted for enrollment in September. Application deadline for fall is May 15. In 1996, 211 applied; 54 were accepted. Applicants must submit a formal application.

CONTACT
Daniel Esposito, Professor, School of Hospitality and Tourism, 100 College Drive, North Bay, Ontario, Canada; 705-474-7600 Ext. 5218; Fax: 705-474-2384; E-mail: espositod@cdrive.canadorec.on.ca

CULINARY INSTITUTE OF CANADA

Charlottetown, Prince Edward Island, Canada

GENERAL INFORMATION
Public, coeducational, culinary institute. Urban campus. Founded in 1984.

PROGRAM INFORMATION
Offered since 1984. Member of Council on Hotel, Restaurant, and Institutional Education. Program calendar is divided into trimesters. 1-year Certificate in Pastry Arts. 2-year Diploma in Culinary Arts.

AREAS OF STUDY
Baking; beverage management; buffet catering; computer applications; confectionery show pieces; controlling costs in food service; convenience cookery; culinary French; culinary skill development; food preparation; food purchasing; food service communication; food service math; garde-manger; international cuisine; introduction to food service; kitchen management; management

and human resources; meal planning; meat cutting; meat fabrication; menu and facilities design; nutrition; nutrition and food service; patisserie; restaurant opportunities; sanitation; saucier; seafood processing; soup, stock, sauce, and starch production; wines and spirits.

Facilities
Bake shop; bakery; cafeteria; 5 catering services; 14 classrooms; 3 computer laboratories; 3 demonstration laboratories; 6 food production kitchens; gourmet dining room; learning resource center; lecture room; 2 public restaurants; snack shop; 2 student lounges; 6 teaching kitchens.

Culinary Student Profile
190 total: 180 full-time; 10 part-time.

Faculty
10 full-time; 5 part-time. 6 are industry professionals; 9 are culinary-accredited teachers.

Expenses
Application fee: Can$20. Tuition: Can$6400 per year. Program-related fees include: Can$400 for books and knives; Can$750 for lab fees (includes one meal per day); Can$1000 for laundry and cleaning.

Financial Aid
In 1996, 15 scholarships were awarded (average award was Can$1000). Employment placement assistance is available.

Housing
Average off-campus housing cost per month: Can$500.

Application Information
Students are accepted for enrollment in May, September, and October. Application deadline for spring is continuous with a recommended date of Febuary 28. Application deadline for fall is July 1. In 1996, 125 applied; 80 were accepted. Applicants must submit a formal application.

Contact
David Harding, Culinary Programs Manager, 4 Sydney Street, Charlottetown, Prince Edward Island, Canada; 902-566-9584; Fax: 902-566-9568.

George Brown College of Applied Arts & Technology

George Brown College Hospitality Institute

Toronto, Ontario, Canada

General Information
Public, coeducational, two-year college. Urban campus. Founded in 1960.

Program Information
Offered since 1967. Member of Council on Hotel, Restaurant, and Institutional Education; Educational Foundation of the NRA; Sommelier Guild of Canada. Program calendar is divided into trimesters. 1-year Certificates in Baking and Pastry Arts; Chef Training Aboriginal; Chef Training Pre-Employment. 1-year Post Diploma in Culinary Arts (Italian). 18-week Certificates in Advanced Food Preparation; Basic Food Preparation. 2-year Diploma in Culinary Management. 24-week Certificate in Chinese Cuisine. 3-year Certificate in Apprentice Cook. 3-year Diploma in Patissier-Apprentice Baker.

Areas of Study
Baking; beverage management; computer applications; confectionery show pieces; controlling costs in food service; culinary French; culinary skill development; food preparation; food purchasing; food service communication; food service math; garde-manger; international cuisine; introduction to food service; kitchen management; management and human resources; meat cutting; menu and facilities design; nutrition; nutrition and food service; patisserie; sanitation; saucier; soup, stock, sauce, and starch production; wines and spirits; Italian cuisine; Sommelier.

Facilities
3 bake shops; bakery; cafeteria; catering service; 30 classrooms; 2 computer laboratories; 4 demonstration laboratories; 8 food production kitchens; gourmet dining room; 8 laboratories; learning resource center; 5 lecture rooms; library;

George Brown College of Applied Arts & Technology ***(continued)***

public restaurant; student lounge; 12 teaching kitchens; wine laboratory; beverage mixology classroom.

Culinary Student Profile
6,090 total: 2,090 full-time; 4,000 part-time.

Faculty
65 full-time; 10 part-time. 25 are industry professionals; 10 are master chefs; 40 are culinary-accredited teachers. Prominent faculty: Jaques Marie.

Expenses
Application fee: Can$30. Tuition: Can$1852 per 32 weeks full-time, Can$385 per hour part-time. Program-related fees include: Can$600 for materials; Can$300 for cutlery; Can$200 for uniforms.

Financial Aid
In 1996, 200 scholarships were awarded (average award was Can$500). Employment placement assistance is available. Employment opportunities within the program are available.

Application Information
Students are accepted for enrollment in January, March, April, September, and October. Applicants must submit a formal application and have a high school diploma.

Contact
Dan Borowec, Academic Chair, George Brown College Hospitality Institute, 300 Adelaide Street East, Toronto, Ontario, Canada; 416-415-2230; Fax: 416-415-2501.

See affiliated apprenticeship program.

Career-motivated men and women study in Canada's largest culinary and hospitality college, located in the centre of Canada's fastest-growing tourism city. Excellent job opportunities for students and graduates abound in the culinary field. The College participates in and organizes major international student culinary competitions. It was chosen to represent Canada as the official Canadian Junior Team for the upcoming 3 years, including the Berlin Culinary World Olympics. The Culinary and Hospitality College organizes the World Championship of Student Culinarians' *Taste of Canada* every 4 years.

Le Cordon Bleu Paris Cooking School

Ottawa, Ontario, Canada

General Information
Private, coeducational, culinary institute. Suburban campus. Founded in 1988.

Program Information
Offered since 1988. Member of Women Chefs and Restaurateurs; Canadian Federation Of Chefs and Cooks. Program calendar is divided into trimesters. 12-week Certificates in Advanced Pastry; Basic Pastry; Intermediate Cuisine; Basic Cuisine. 4-week Certificate in Introduction to Catering.

Areas of Study
Buffet catering; culinary French; meat cutting; patisserie; saucier; soup, stock, sauce, and starch production; wines and spirits; sugar work.

Facilities
Demonstration laboratory; library; student lounge; teaching kitchen.

Culinary Student Profile
150 full-time. 45 are under 25 years old; 75 are between 25 and 44 years old; 30 are over 44 years old.

Faculty
3 full-time. 3 are master chefs. Prominent faculty: Philippe Guiet and Michel Denis.

Special Programs
5-day demonstration/practical workshop in pastry with patissiers from all over North America.

Expenses
Application fee: Can$50. Tuition: Can$5000 per course.

Financial Aid
In 1996, 1 scholarship was awarded (award was Can$4200). Employment placement assistance is available.

Housing
Average off-campus housing cost per month: Can$600.

Application Information
Students are accepted for enrollment in January, April, July, and September. Application deadline for fall is continuous with a recommended date of September 22. Application deadline for winter is continuous with a recommended date of January 5. Application deadline for spring is continuous with a recommended date of April 6. In 1996, 175 applied. Applicants must submit a formal application.

Contact
Admissions, 400-1390 Prince of Wales Drive, Ottawa, Ontario, Canada; 613-224-8603; Fax: 613-224-9966; E-mail: macinnis@magmacom.com; World Wide Web: http://www.cordonbleu.net

See affiliated programs in London, Paris, Sydney, and Tokyo.

See display on page 264.

Niagara College of Applied Arts and Technology

Culinary Skills Diploma

Niagara Falls, Ontario, Canada

General Information
Public, coeducational, two-year college. Suburban campus. Founded in 1967.

Program Information
Offered since 1989. Member of The Canadian Federation of Chefs. Program calendar is divided into semesters. 2-year Diploma in Culinary Skills.

Areas of Study
Baking; beverage management; computer applications; controlling costs in food service; convenience cookery; culinary French; culinary skill development; food preparation; food purchasing; food service communication; food service math; kitchen management; meal planning; nutrition; nutrition and food service; patisserie; sanitation; soup, stock, sauce, and starch production; wines and spirits.

Facilities
Bake shop; cafeteria; 7 classrooms; 2 computer laboratories; demonstration laboratory; food production kitchen; learning resource center; lecture room; library; public restaurant; 3 teaching kitchens.

Culinary Student Profile
96 full-time. 87 are under 25 years old; 9 are between 25 and 44 years old.

Faculty
9 full-time; 4 part-time. 1 is an industry professional; 1 is a culinary-accredited teacher; 7 are certified Chef de Cuisine.

Special Programs
Students participate in community and industry special events.

Expenses
Application fee: Can$30. Tuition: Can$2000 per semester. Program-related fees include: Can$475 for knives; Can$120 for uniforms; Can$250 for food per semester.

Financial Aid
Employment placement assistance is available. Employment opportunities within the program are available.

Housing
Average off-campus housing cost per month: Can$600.

Application Information
Students are accepted for enrollment in January and September. In 1996, 300 applied; 86 were accepted. Applicants must submit a formal application.

Contact
Admissions, Culinary Skills Diploma, 300 Woodlawn Road, PO Box 1005, Welland, ON L3B

Niagara College of Applied Arts and Technology *(continued)*

5S2, Canada; 905-735-2211 Ext. 7529; Fax: 905-735-0419; E-mail: doster@niagarac.on.ca

See affiliated apprenticeship program.

Pacific Institute of Culinary Arts

Vancouver, British Columbia, Canada

General Information
Private, coeducational, culinary institute. Urban campus. Founded in 1996.

Program Information
Offered since 1996. Member of British Columbia Restaurant Association. Program calendar is divided into quarters. 6-month Diplomas in Baking and Pastry Arts; Culinary Arts.

Areas of Study
Baking; culinary French; restaurant opportunities; soup, stock, sauce, and starch production.

Facilities
Bake shop; catering service; classroom; gourmet dining room; lecture room; public restaurant; student lounge; teaching kitchen.

Culinary Student Profile
36 full-time.

Faculty
7 full-time. 7 are chefs.

Expenses
Application fee: Can$25. Tuition: Can$8975 per diploma. Program-related fees include: Can$295 for uniforms and shoes; Can$300 for textbooks; Can$400 for knives.

Financial Aid
In 1996, 3 scholarships were awarded (average award was Can$500). Employment placement assistance is available. Employment opportunities within the program are available.

Housing
Average off-campus housing cost per month: Can$500.

Application Information
Students are accepted for enrollment in January, April, July, and September. Applicants must submit a formal application and letters of reference.

Contact
Sue Singer, Director of Admissions, 1505 West 2nd Avenue, Vancouver, BC V6H 3Y4, Canada; 800-416-4040; Fax: 604-734-4408; E-mail: admissions@picularts.bc.ca; World Wide Web: http://www.picularts.bc.ca

Southern Alberta Institute of Technology

Professional Cooking

Calgary, Alberta, Canada

General Information
Public, coeducational, two-year college. Urban campus. Founded in 1916.

Program Information
Offered since 1948. Member of Confrerie de la Chaine des Rotisseurs; Council on Hotel, Restaurant, and Institutional Education; Canadian Restaurant Association. Program calendar is divided into trimesters. 12-month Diploma in Professional Cooking.

Areas of Study
Baking; computer applications; confectionery show pieces; controlling costs in food service; culinary skill development; food preparation; food purchasing; food service communication; food service math; garde-manger; international cuisine; introduction to food service; kitchen management; management and human resources; meal planning; meat cutting; meat fabrication; menu and facilities design; nutrition; nutrition and food service; patisserie; restaurant opportunities; sanitation; saucier; seafood processing; soup, stock, sauce, and starch production; wines and spirits; ice carving/fat sculpture.

Facilities
2 bakeries; 3 cafeterias; catering service; 5 classrooms; 2 coffee shops; computer laboratory; 3 demonstration laboratories; 3 food production kitchens; gourmet dining room; 2 laboratories; learning resource center; lecture room; library; 2 public restaurants; 6 snack shops; student lounge; 2 teaching kitchens; test kitchen.

Culinary Student Profile
200 total: 180 full-time; 20 part-time.

Faculty
26 full-time; 7 part-time. 30 are industry professionals; 1 is a master chef; 2 are culinary-accredited teachers. Prominent faculty: Gerd Steinmeyer and Margaret Turner.

Special Programs
5-day study tours to food processing and manufacturing sites.

Expenses
Application fee: Can$100. Tuition: Can$1200 per semester.

Financial Aid
In 1996, 46 scholarships were awarded (average award was Can$400); 45 loans were granted. Employment placement assistance is available. Employment opportunities within the program are available.

Housing
Apartment-style housing available. Average on-campus housing cost per month: Can$320. Average off-campus housing cost per month: Can$400.

Application Information
Students are accepted for enrollment in January, May, and September. In 1996, 300 applied; 180 were accepted. Applicants must submit a formal application, letters of reference, and academic transcripts and complete a questionna.

Contact
Wolfgang Stampe, Coordinator, Student Services, Professional Cooking, Hospitality Careers

Southern Alberta Institute of Technology *(continued)*

Department, 1301-16th Avenue, NW, Calgary, Alberta, Canada; 403-284-8612; Fax: 403-284-7034.

St. Clair College

Culinary Arts

Windsor, Ontario, Canada

General Information
Public, coeducational, two-year college. Urban campus. Founded in 1967.

Program Information
Program calendar is divided into semesters. 2-year Diploma in Culinary Arts.

Areas of Study
Baking; beverage management; buffet catering; computer applications; culinary skill development; food preparation; food purchasing; food service communication; food service math; garde-manger; international cuisine; introduction to food service; kitchen management; management and human resources; meal planning; menu and facilities design; nutrition; nutrition and food service; patisserie; restaurant opportunities; sanitation; saucier; soup, stock, sauce, and starch production; wines and spirits; customer service; hospitality marketing.

Facilities
Cafeteria; catering service; 2 classrooms; computer laboratory; demonstration laboratory; 2 food production kitchens; 2 gardens; learning resource center; 2 lecture rooms; library; public restaurant; 2 snack shops; student lounge; 2 teaching kitchens.

Culinary Student Profile
40 full-time.

Faculty
2 full-time; 3 part-time. 1 is an industry professional; 2 are master chefs. Prominent faculty: Michel Crovisier and Rainer Schindler.

Special Programs
Guest lecture and tours of local casinos, banquet facilities, and wineries.

Expenses
Application fee: Can$30. Tuition for residents: Can$1800 per year. Tuition for nonresidents: Can$10,000 per year.

Financial Aid
In 1996, 1 scholarship was awarded (award was Can$1000). Employment placement assistance is available.

Application Information
Students are accepted for enrollment in September. Application deadline for fall is continuous with a recommended date of March 31. Applicants must submit a formal application.

Contact
Dennis Sanson, Support Officer, Culinary Arts, Department of Hospitality and Casino Studies, Windsor, Ontario, Canada; 519-972-2727 Ext. 4403; Fax: 519-972-0801; E-mail: dsanson@stclairc.on.ca

Chopsticks Cooking Centre

Hong Kong, China

General Information
Private, coeducational, culinary institute. Urban campus. Founded in 1971.

Program Information
Offered since 1971. Member of International Association of Culinary Professionals. Program calendar is year-round. Certificate in Chinese Cuisine.

Areas of Study
Baking; culinary skill development; international cuisine; soup, stock, sauce, and starch production.

Facilities
Teaching kitchen.

Culinary Student Profile
300 part-time.

Faculty
2 full-time; 10 part-time. 1 is an industry professional; 10 are master chefs; 1 is a culinary-accredited teacher.

Special Programs
Culinary tour of local markets and restaurants.

Expenses
Application fee: $50. Tuition: $4000 per 4 weeks full-time, $60 per half day part-time.

Financial Aid
In 1996, 1 scholarship was awarded (award was $1500).

Housing
Average off-campus housing cost per month: $2500.

Application Information
Students are accepted for enrollment year-round. In 1996, 200 applied; 200 were accepted. Applicants must submit a formal application.

Contact
Caroline Au-yeung, Director, Unita, First Floor, Pine Hill Mansion, 128 Austin Road, Tsimshatsui, Kowloon, China; 852-2336-8433; Fax: 852-2338-1462.

Ecole des Arts Culinaires et de l'Hôtellerie

Ecully, Cedex, France

General Information
Private, coeducational, culinary institute. Urban campus. Founded in 1990.

Program Information
Member of Council on Hotel, Restaurant, and Institutional Education. Program calendar is divided into quarters. 2-year Diploma in Cuisine and Management. 4-month Certificate in Cuisine and Culture.

Areas of Study
Beverage management; buffet catering; computer applications; confectionery show pieces; controlling costs in food service; culinary French; culinary skill development; food preparation; food service math; garde-manger; kitchen management; management and human resources; meal planning; meat cutting; meat fabrication; menu and facilities design; nutrition and food service; patisserie; soup, stock, sauce, and starch production; wines and spirits.

Facilities
Cafeteria; 11 classrooms; computer laboratory; 6 demonstration laboratories; 2 food production kitchens; garden; gourmet dining room; library; public restaurant; vineyard.

Culinary Student Profile
60 total: 45 full-time; 15 part-time.

Faculty
8 full-time; 10 part-time. 2 are master chefs; 8 are culinary-accredited teachers. Prominent faculty: Alain Le Cossel and Alain Berne.

Special Programs
Numerous visits to area surrounding Lyons.

Expenses
Tuition: Fr90,000 per 2-year program, Fr56,000 per 4-month program (tuition includes housing, meals, and materials).

Financial Aid
Employment placement assistance is available. Employment opportunities within the program are available.

Housing
Coed and single-sex housing available.

Application Information
Students are accepted for enrollment in March, April, and October. Applicants must submit a formal application.

Contact
Eléonore Vial, Director of Studies, Château du Vivier, B.P. 25, Ecully, Cedex, France; 33 478 43 36-10; Fax: 33-478-43-33-51.

Ecole Lenotre

Plaisir, Cedex, France

General Information
Private, coeducational, culinary institute. Founded in 1970.

Ecole Lenotre *(continued)*

Program Information
Offered since 1970. Program calendar is divided into semesters. 1-week Diplomas in Decoration; Working with Sugar; Ice Creams and Frozen Desserts; Chocolate-Confectionery; Patisserie; Bakery-Viennoiserie; Cuisine-Catering.

Areas of Study
Baking; buffet catering; confectionery show pieces; culinary French; food preparation; garde-manger; patisserie; restaurant opportunities; saucier.

Contact
Daphné Cerisier, Secretary, 40 rue Pierre Curie, B.P.6, Plaisir, Cedex, France; 33-130-81-46-34; Fax: 33-130-54-73-70.

Ecole Superieure de Cuisine Francaise Groupe Ferrandi

Professional Bilingual Program

Paris, France

General Information
Public, coeducational, culinary institute. Urban campus. Founded in 1932.

Program Information
Offered since 1986. 12-month Diploma in Culinary Arts.

Areas of Study
Culinary French; French cuisine.

Facilities
Bakery; cafeteria; catering service; 10 classrooms; demonstration laboratory; 4 food production kitchens; gourmet dining room; laboratory; 4 lecture rooms; library; public restaurant; 10 teaching kitchens.

Culinary Student Profile
20 full-time.

Faculty
20 full-time. 20 are industry professionals.

Special Programs
3 excursions to French wine regions, 1 end-of-year gastronomic excursion, 3-month apprenticeship program after completion of 9-month program.

Expenses
Tuition: Fr86,000 per 9 months.

Application Information
Students are accepted for enrollment in September. Application deadline for fall is continuous with a recommended date of June 30. In 1996, 20 applied; 10 were accepted. Applicants must submit a formal application.

Contact
Stephanie Curtis, Coordinator, Professional Bilingual Program, 10 rue de Richelieu, Paris 75001, France; 33-140-15-04-57; Fax: 33-140-15-04-58.

L'Ecole de Patisserie Francaise

Uzes, France

General Information
Public, coeducational, culinary institute. Rural campus. Founded in 1994.

Program Information
Offered since 1994. Program calendar is divided into weeks.

Areas of Study
Baking; buffet catering; patisserie; ice cream preparation; pastry cooking.

Facilities
Bake shop; food production kitchen; garden; teaching kitchen.

Culinary Student Profile
1 full-time.

Faculty
1 full-time. Prominent faculty: Didier Richeux.

Expenses
Tuition: Fr5000 per week.

Application Information
Students are accepted for enrollment year-round. In 1996, 15 applied; 15 were accepted. Applicants must submit a formal application.

Contact
Didier Richeux, Director, 12 rue de la République, Uzes, France; 33-466-22-12-09; Fax: 33-466-22-26-36.

La Varenne

Villecien, France

General Information
Private, coeducational, culinary institute. Rural campus. Founded in 1975.

Program Information
Offered since 1975. Member of American Institute of Wine & Food; International Association of Culinary Professionals; James Beard Foundation, Inc.; Women Chefs and Restaurateurs. 4-week Diploma in French Culinary Studies.

Areas of Study
Culinary French; French cuisine.

Facilities
Garden; vineyard.

Culinary Student Profile
15 full-time.

Faculty
8 full-time; 5 part-time. 10 are master chefs. Prominent faculty: Anne Willan.

Special Programs
Market and vineyard tours, excursions to Beaune, Côte d'Or, and Paris.

Expenses
Tuition: $9995 per 4 weeks (includes housing and meals).

Application Information
Students are accepted for enrollment in September. Applicants must submit a formal application.

Contact
Janis McLean, U.S. Representative, PO Box 25574, Washington, DC 20007; 800-537-6486; Fax: 703-823-5438; E-mail: lavarenne@compuserve.com; World Wide Web: http://www.lavarenne.com

Le Cordon Bleu

The Grand Diplôme, The Diploma and Certificate Program

Paris, France

General Information
Private, coeducational, culinary institute. Urban campus. Founded in 1895.

Program Information
Offered since 1895. Member of American Institute of Wine & Food; Confrerie de la Chaine des Rotisseurs; Council on Hotel, Restaurant, and Institutional Education; International Association of Culinary Professionals. Program calendar is divided into terms. 10-week Certificates in Advanced Pastry Techniques; Principals of Decoration; Basic Pastry Techniques; Advanced French Culinary Technique; Regional French Influences; Basic French Culinary Technique. 30-week Diplomas in Basic, Intermediate, and Superior Cuisine; Cuisine and Pastry; Basic, Intermediate, and Superior Pastry.

Areas of Study
Baking; buffet catering; confectionery show pieces; culinary French; culinary skill development; patisserie; wines and spirits; table service; cheese.

Facilities
2 demonstration laboratories, food production kitchen; student lounge; 4 teaching kitchens; boutique.

Culinary Student Profile
104 total: 24 full-time; 80 part-time.

Faculty
9 full-time; 1 part-time. 8 are industry professionals; 2 are master chefs. Prominent faculty: Michel Cliche and Laurant Duchêne.

Le Cordon Bleu *(continued)*

Special Programs

1-2 day vineyard and cultural excursions, professional internships, exchange between Le Cordon Bleu schools worldwide.

Expenses

Tuition: Fr170,900 per year full-time, Fr30,000 per 10 weeks part-time. Program-related fees include: Fr1440 for uniform package; Fr2260 for equipment package (knives); Fr600 for internship administrative fees.

Financial Aid

In 1996, 10 scholarships were awarded (average award was Fr30,000). Employment placement assistance is available. Employment opportunities within the program are available.

Housing

Average off-campus housing cost per month: Fr4500.

Application Information

Students are accepted for enrollment in January, March, June, August, September, and November. In 1996, 280 applied; 260 were accepted. Applicants must submit a formal application, letters of reference, a personal essay, and a resume.

Contact

Sabine Bailly, Director of Admissions, The Grand Diplôme, The Diploma and Certificate Program, 8, rue Léon Delhomme, Paris, France; 33-153-68-22-50; Fax: 33-148-56-03-96; E-mail: sbailly@cordonbleu.net; World Wide Web: http://cordonbleu.net

See affiliated programs in London, Ontario, Sydney, and Tokyo.

See display on page 264.

Ritz-Escoffier Ecole de Gastronomie Francaise

Paris, France

General Information

Private, coeducational, culinary institute. Urban campus. Founded in 1988.

Program Information

Offered since 1988. Member of American Institute of Wine & Food; International Association of Culinary Professionals. Program calendar is divided into trimesters. 12-week Diplomas in Intermediate and Advanced Pastry; Advanced Cuisine and Pastry. 6-week Diploma in Intermediate Cuisine and Pastry.

Areas of Study

Baking; buffet catering; culinary French; culinary skill development; food preparation; garde-manger; meat cutting; patisserie; saucier; soup, stock, sauce, and starch production; wines and spirits.

Facilities

Cafeteria; food production kitchen; lecture room; library; public restaurant; teaching kitchen.

Culinary Student Profile

28 full-time. 10 are under 25 years old; 18 are between 25 and 44 years old.

Faculty

3 full-time. 3 are master chefs. Prominent faculty: Gilles Maisonneuue and Jacques Meunier.

Special Programs

1-day tours of various wine and food regions.

Expenses

Tuition: Fr5800 per week.

Financial Aid

Employment placement assistance is available.

Housing

Average off-campus housing cost per month: Fr4000.

Application Information

Students are accepted for enrollment in March, May, and November. Applicants must submit a formal application and a portfolio.

Contact

Sylvie Beaufils, Assistant Manager, 15 Place Vendome, Paris, France; 33-143-16-30-50; Fax: 33-143-16-31-50.

Ballymaloe Cookery School

Midleton, County Cork, Ireland

General Information

Private, coeducational, culinary institute. Rural campus. Founded in 1983.

Program Information

Offered since 1983. Member of International Association of Culinary Professionals. Program calendar is year-round. 3-month Certificate in Culinary Training.

Areas of Study

Baking; buffet catering; convenience cookery; international cuisine.

Facilities

Garden; library; teaching kitchen; demonstration kitchen.

Culinary Student Profile

44 full-time. 20 are under 25 years old; 15 are between 25 and 44 years old; 9 are over 44 years old.

Faculty

5 full-time; 5 part-time. 1 is a culinary-accredited teacher; 6 are alumni instructors. Prominent faculty: Darina Allen and Tim Allen.

Special Programs

Range of short courses on various topics.

Expenses

Tuition: 3775 Irish Punts per certificate. Program-related fees include: 100 Irish Punts for knives.

Financial Aid

Employment placement assistance is available.

Housing

42 culinary students housed on campus. Apartment-style housing available. Average on-campus housing cost per month: 140-192 Irish Punts.

Application Information

Students are accepted for enrollment in January and September. Applicants must submit a formal application and letters of reference.

Contact

Tim Allen, Owner, Shanagarry, Midleton, County Cork, Ireland; 353-21-646785; Fax: 353-21-646909; E-mail: enquiries@ballymaloe-cookery-school.ie

The International Cooking School of Italian Food and Wine

Bologna, Italy

General Information

Private, coeducational, culinary institute. Urban campus. Founded in 1987.

Program Information

Offered since 1987. Member of International Association of Culinary Professionals; New York Association of Cooking Teachers. Program calendar is divided into weeks. 1-week Certificate in Foundation of Italian Cooking.

Areas of Study

Italian cuisine.

Facilities

Food production kitchen; 3 gourmet dining rooms; 3 public restaurants; teaching kitchen; 2 vineyards.

Culinary Student Profile

12 full-time.

Faculty

1 full-time; 5 part-time. 2 are industry professionals; 3 are master chefs; 1 is a culinary-accredited teacher.

Special Programs

Private tours and tastings in Emilia-Romagna and Piedmont regions.

Expenses

Tuition: $3200–$3600 per week (includes housing).

Financial Aid

In 1996, 2 scholarships were awarded (average award was $3000).

Application Information
Students are accepted for enrollment in May, June, July, August, September, and October. Applicants must submit a formal application.

Contact
Mary Beth Clark, Owner, 201 East 28th Street Suite 15B, New York, NY 10016-8538; 212-779-1921; Fax: 212-779-3248; E-mail: marybethclark@worldnet.att.net

Scoula di Arte Culinaria "Cordon Bleu"

Florence, Italy

General Information
Private, coeducational, culinary institute. Urban campus. Founded in 1985.

Program Information
Offered since 1985. Accredited by Italian Association of Culinary Teachers. Member of International Association of Culinary Professionals; Commanderie Des Cordons Bleus de France. 1-month Certificates in Italian Cooking; Pastry. 2-month Certificate in Basic Cooking. 3-month Certificate in Advanced Cooking. 8-month Certificate in Professional Study on Italian Cooking.

Areas of Study
Baking; food preparation; meal planning; patisserie; saucier; seafood processing; soup, stock, sauce, and starch production; wines and spirits; chocolate; Italian regional and traditional cooking.

Facilities
Classroom; gourmet dining room; library; teaching kitchen.

Culinary Student Profile
320 total: 20 full-time; 300 part-time.

Faculty
2 full-time; 2 part-time. 2 are master chefs; 2 are culinary-accredited teachers. Prominent faculty: Gabriella Mari and Cristina Blasi.

Special Programs
1-week vacation program about Tuscan cooking, 1-week program on seasonal cooking.

Expenses
Tuition: 7,880,000 Lira per 8 months full-time, 750,000 Lira per 1 week part-time.

Financial Aid
Employment placement assistance is available.

Housing
Average off-campus housing cost per month: 1,000,000 Lira.

Application Information
Students are accepted for enrollment in January and September. Application deadline for fall is August 15. Application deadline for spring is November 30. In 1996, 320 applied.

Contact
Gabriella Mari, Director, Via Di Mezzo, Florence, Italy; 39-55-2345468; Fax: 39-55-2345468; E-mail: cordonbleu@aspiole.it

Le Cordon Bleu

Classic Cycle Program

Tokyo, Japan

General Information
Private, culinary institute. Urban campus. Founded in 1991.

Program Information
Offered since 1991. Member of Confrerie de la Chaine des Rotisseurs; James Beard Foundation, Inc. Program calendar is divided into trimesters. 3-month Certificates in Superior Pastry; Advanced Pastry; Basic Pastry; Superior Cuisine; Advanced Cuisine; Intermediate Cuisine; Basic Cuisine.

Areas of Study
Baking; buffet catering; culinary French; patisserie.

Facilities
5 classrooms; 8 demonstration laboratories; 9 food production kitchens; 11 gardens; 16 public restaurants; 19 teaching kitchens.

Le Cordon Bleu *(continued)*

Culinary Student Profile
1,150 total: 650 full-time; 500 part-time.

Faculty
6 full-time; 1 part-time. 7 are master chefs.

Expenses
Application fee: 50,000 Yen. Tuition: 550,000 Yen per 3 months. Program-related fees include: 75,000 Yen for cuisine set (knives, jacket, and aprons); 65,000 Yen for pastry set (knives, jacket, and aprons).

Financial Aid
Employment placement assistance is available.

Application Information
Students are accepted for enrollment in March, June, September, and December. In 1996, 630 applied; 630 were accepted. Applicants must submit a formal application.

Contact
Akiko Koyama, Chair, Classic Cycle Program, Roob-1, 28-13 Sarugaku-cho, Shibuya-ku, Tokyo, Japan; 813-54890141; Fax: 813-54890145; E-mail: tmaincent@cordonbleu.net

See affiliated programs in London, Ontario, Paris, and Sydney.

See display on page 264.

The New Zealand School of Food and Wine

Foundation Cookery Skills

Christchurch, New Zealand

General Information
Private, coeducational, culinary institute. Urban campus. Founded in 1994.

Program Information
Offered since 1995. Program calendar is divided into trimesters. 16-week Certificate in Foundation in Cookery Skills.

Areas of Study
Baking; controlling costs in food service; culinary French; culinary skill development; food preparation; garde-manger; international cuisine; meal planning; meat cutting; menu and facilities design; nutrition; patisserie; saucier; seafood processing; soup, stock, sauce, and starch production; wines and spirits.

Facilities
Classroom; coffee shop; food production kitchen; library; teaching kitchen.

Culinary Student Profile
12 full-time.

Faculty
Prominent faculty: Celia Hay.

Special Programs
Local vineyard tours, trips to relevant conferences, certification in wine.

Expenses
Application fee: NZ$56.25. Tuition: NZ$4350 per certificate.

Financial Aid
Employment placement assistance is available. Employment opportunities within the program are available.

Application Information
Students are accepted for enrollment in January, May, and September. Application deadline for summer is continuous with a recommended date of January 1. Application deadline for winter is continuous with a recommended date of May 1. Application deadline for spring is continuous with a recommended date of September 1. In 1996, 48 applied; 36 were accepted. Applicants must submit a formal application and letters of reference.

Contact
Celia Hay, Director, Foundation Cookery Skills, 63 Victoria Street, 1st Floor, Christchurch, New Zealand; 64-3-3797501; Fax: 64-3-3797501.

Silwood Kitchen Cordons Bleus Cookery School

Cape Town, Cape, South Africa

General Information
Private, coeducational, culinary institute. Suburban campus. Founded in 1964.

Program Information
Offered since 1964. Member of International Association of Culinary Professionals; International Wine & Food Society. Program calendar is divided into quarters. 10-month Diploma in In-Service Practical Training. 12-month Diploma in Practical Apprenticeships. 4-term Certificate in Basic Culinary Studies.

Areas of Study
Baking; buffet catering; confectionery show pieces; controlling costs in food service; culinary French; culinary skill development; food preparation; food purchasing; garde-manger; international cuisine; introduction to food service; kitchen management; management and human resources; meal planning; menu and facilities design; nutrition; nutrition and food service; patisserie; soup, stock, sauce, and starch production; wines and spirits; floral art; French menu.

Facilities
Catering service; 5 classrooms; 2 demonstration laboratories; garden; gourmet dining room; library.

Culinary Student Profile
44 full-time. 40 are under 25 years old; 4 are between 25 and 44 years old.

Faculty
6 full-time; 4 part-time. 4 are industry professionals; 6 are culinary-accredited teachers. Prominent faculty: Alicia Wilkinson and Peggy Loebenberg.

Special Programs
Visits to market and Cape winelands, visits to all leading hotel kitchens and restaurants in Cape Town.

Expenses
Application fee: 1000 Rands. Tuition: 21,100 Rands per certificate.

Financial Aid
Employment placement assistance is available. Employment opportunities within the program are available.

Housing
Average off-campus housing cost per month: 650 Rands.

Application Information
Students are accepted for enrollment in January. Application deadline for winter is November 30. In 1996, 58 applied; 44 were accepted. Applicants must submit a formal application and letters of reference and interview.

Contact
Alicia Wilkinson, Principal, Silwood Road, Rondebosch, Cape Town, Cape, South Africa; 27-21-686-4894; Fax: 27-21-686-5795.

Butlers Wharf Chef School

London, United Kingdom

General Information
Private, coeducational, culinary institute. Urban campus. Founded in 1995.

Program Information
Offered since 1995. Program calendar is year-round. 2-month Certificates in Restaurant Service; Culinary Arts/Chef. 6-month Diplomas in Restaurant Service; Culinary Arts/Chef.

Areas of Study
Culinary skill development; introduction to food service; restaurant opportunities.

Facilities
Catering service; classroom; food production kitchen; lecture room; public restaurant; teaching kitchen.

Culinary Student Profile
400 total: 200 full-time; 200 part-time.

Butlers Wharf Chef School ***(continued)***

Faculty
14 full-time; 6 part-time. 4 are industry professionals; 4 are master chefs; 8 are culinary-accredited teachers. Prominent faculty: John Roberts and Gary Witchalls.

Special Programs
Training experience in various restaurants in London.

Expenses
Tuition: £5000 per 6 months. Program-related fees include: £200 for uniforms and knives.

Financial Aid
In 1996, 100 scholarships were awarded (average award was £2500). Employment placement assistance is available. Employment opportunities within the program are available.

Application Information
Students are accepted for enrollment year-round. In 1996, 650 applied; 400 were accepted. Applicants must submit a formal application and a portfolio and interview.

Contact
Course Admission, Cardamon Building, 31 Shad Thames, London, United Kingdom; 44-171357-8842; Fax: 44-171403-2638.

Cookery at the Grange

Basics to Béarnaise

Frome, Somerset, United Kingdom

General Information
Private, coeducational, culinary institute. Rural campus. Founded in 1981.

Program Information
Offered since 1981. Program calendar is divided into four-week cycles. 1-month Certificate in Cookery.

Areas of Study
Baking; culinary skill development; food preparation; meal planning; meat cutting; menu and facilities design; soup, stock, sauce, and starch production; wines and spirits; general cookery.

Facilities
Garden; student lounge; teaching kitchen.

Culinary Student Profile
20 full-time.

Faculty
3 full-time; 3 part-time.

Expenses
Tuition: £1750 per certificate.

Housing
20 culinary students housed on campus.

Application Information
Students are accepted for enrollment in January, February, May, July, August, September, and November. Applicants must submit a formal application.

Contact
William and Jane Averill, Basics to Béarnaise, The Grange, Whatley, Frome, Somerset, United Kingdom; 44-1373-836579; Fax: 44-1373-836579.

Le Cordon Bleu

Le Cordon Bleu Classic Cycle Programme

London, United Kingdom

General Information
Private, coeducational, culinary institute. Urban campus. Founded in 1895.

Program Information
Member of American Institute of Wine & Food; Confrerie de la Chaine des Rotisseurs; Council on Hotel, Restaurant, and Institutional Education; International Association of Culinary Professionals. Program calendar is divided into quarters. 10-week Certificates in Advanced Technique (including sugar and chocolate); Principles of Decoration; Basic Pastry Technique; Advanced Technique; Regional Traditional Technique; Basic French Culinary Technique.

30-week Diplomas in Cuisine and Pastry; Basic, Intermediate, and Superior Pastry; Basic, Intermediate, and Superior Cuisine.

Areas of Study
Baking; buffet catering; confectionery show pieces; culinary French; culinary skill development; introduction to food service; meal planning; nutrition; nutrition and food service; patisserie; wines and spirits; table service; cheese.

Facilities
2 bakeries; 2 demonstration laboratories; food production kitchen; student lounge; 4 teaching kitchens.

Culinary Student Profile
120 total: 85 full-time; 35 part-time.

Faculty
7 full-time; 3 part-time. 7 are industry professionals; 1 is a culinary-accredited teacher; 2 are specialists in cheese and nutrition.

Special Programs
Restaurant market tours and hotel visits, exchanges between Le Cordon Blue Schools worldwide.

Expenses
Tuition: £17,296 per 9 months full-time, £2230 per 10 weeks part-time. Program-related fees include: £450 for uniforms and equipment.

Financial Aid
In 1996, 10 scholarships were awarded (average award was £2230). Employment placement assistance is available. Employment opportunities within the program are available.

Housing
Average off-campus housing cost per month: £300.

Application Information
Students are accepted for enrollment in January, March, June, August, September, and October. Applicants must submit a formal application.

Contact
Anne-Laure Trehorel, Admissions, Le Cordon Bleu Classic Cycle Programme, 114 Marylebone Lane, London, United Kingdom; 44-171-935-3503; Fax: 44-171-935-7621; E-mail: a.trehorel@cordonbleu.co.uk; World Wide Web: http://cordonbleu.net

See affiliated programs in Ontario, Paris, Sydney, and Tokyo.

See display on page 264.

Leith's School of Food and Wine

London, United Kingdom

General Information
Public, coeducational, culinary institute. Urban campus. Founded in 1975.

Program Information
Offered since 1975. Program calendar is divided into trimesters. 1-month Certificate in Practical Cookery. 1-term Certificate in Food and Wine. 1-year Diploma in Food and Wine. 2-term Diploma in Food and Wine. 5-class Certificate in Wine.

Areas of Study
Easy dinner party; fish cookery.

Facilities
Lecture room; library; student lounge; 3 teaching kitchens.

Culinary Student Profile
546 total: 96 full-time; 450 part-time.

Faculty
17 full-time; 1 part-time. 18 are industry professionals. Prominent faculty: Caroline Waldegrave and Carol Jane Jackson.

Special Programs
Excursions to Billingsgate and Smithfield markets, 5-day trip to wine area in France.

Expenses
Tuition: £9200 per year full-time, £380 per week part-time. Program-related fees include: £92 for knives; £26 for supplementary equipment to knife set; £35 for course literature.

Leith's School of Food and Wine *(continued)*

Financial Aid
Employment placement assistance is available. Employment opportunities within the program are available.

Application Information
Students are accepted for enrollment in January, March, April, July, August, September, October, and December. Applicants must submit a formal application and letters of reference and interview.

Contact
Judy Van Der Sande, Registrar, 211 St Alban's Grove, London, United Kingdom; 44-171-2290177; Fax: 44-171-9375257.

The Manor School of Fine Cuisine

Widmerpool, Nottingham, United Kingdom

General Information
Private, coeducational, culinary institute. Rural campus. Founded in 1986.

Program Information
Offered since 1986. Program calendar is year-round. 1-month Certificate in Cordon Bleu Intensive.

Areas of Study
Baking; buffet catering; culinary French; food preparation; international cuisine; kitchen management; meal planning; menu and facilities design; saucier; seafood processing; soup, stock, sauce, and starch production; wines and spirits; chocolate; vegetarian cuisine.

Facilities
2 classrooms; 2 demonstration laboratories; 2 food production kitchens; 2 gardens; gourmet dining room; lecture room; library; student lounge; 2 teaching kitchens.

Culinary Student Profile
20 total: 10 full-time; 10 part-time.

Faculty
2 full-time; 2 part-time. 2 are culinary-accredited teachers. Prominent faculty: Claire Tuttey and Amanda Lacey.

Expenses
Tuition: £1300 per month.

Financial Aid
In 1996, 12 scholarships were awarded (average award was £250). Employment placement assistance is available. Employment opportunities within the program are available.

Housing
9 culinary students housed on campus. Average on-campus housing cost per month: £200.

Application Information
Students are accepted for enrollment year-round. Applicants must submit a formal application.

Contact
Claire Tuttey, Principal, Old Melton Road, Widmerpool, Nottingham, United Kingdom; 44-194981371; Fax: 44-194981371.

Tante Marie School of Cookery

Woking, Surrey, United Kingdom

General Information
Private, coeducational, culinary institute. Suburban campus. Founded in 1954.

Program Information
Offered since 1954. Program calendar is divided into trimesters. 3-month Certificate in Cordon Bleu Cookery. 9-month Diploma in Cordon Bleu Cookery.

Areas of Study
Baking; buffet catering; confectionery show pieces; controlling costs in food service; convenience cookery; culinary French; culinary skill development; food preparation; food purchasing; garde-manger; international cuisine; introduction to food service; kitchen management; meal planning; meat cutting; meat fabrication; menu and facilities design; nutrition; nutrition and food

service; patisserie; restaurant opportunities; sanitation; saucier; seafood processing; soup, stock, sauce, and starch production; wines and spirits.

Facilities
Coffee shop; garden; gourmet dining room; lecture room; 2 libraries; student lounge; 5 teaching kitchens; vineyard; demonstration theatre.

Culinary Student Profile
84 full-time. 48 are under 25 years old; 30 are between 25 and 44 years old; 6 are over 44 years old.

Faculty
6 full-time; 4 part-time. 2 are industry professionals; 8 are culinary-accredited teachers. Prominent faculty: Mrs. Beryl A. Childs and Mrs. Susan B. Alexander.

Special Programs
4-day course on wines and spirits, 1-day course on food safety and hygiene.

Expenses
Application fee: £300. Tuition: £3000 per term. Program-related fees include: £100 for uniforms and equipment.

Financial Aid
Employment placement assistance is available. Employment opportunities within the program are available.

Housing
Average off-campus housing cost per month: £350.

Application Information
Students are accepted for enrollment in January, May, and September. Application deadline for spring is continuous with a recommended date of November 1. Application deadline for summer is continuous with a recommended date of March 1. Application deadline for fall is continuous with a recommended date of July 1. In 1996, 120 applied; 120 were accepted. Applicants must submit a formal application.

Contact
Margaret A. Stubbington, Registrar, Carlton Road, Woking, Surrey, United Kingdom; 44-1483-726957; Fax: 44-1483-724173.

Thames Valley University

School of Tourism, Hospitality, and Leisure

Slough, United Kingdom

General Information
Public, coeducational, university. Suburban campus. Founded in 1948.

Program Information
Offered since 1948. Accredited by Academie Culinaire Française. Member of Council on Hotel, Restaurant, and Institutional Education. Program calendar is divided into terms. 2-year Advanced Diploma in International Culinary Arts. 3-year Diploma in International Culinary Arts.

Areas of Study
Baking; beverage management; buffet catering; computer applications; confectionery show pieces; controlling costs in food service; convenience cookery; culinary French; culinary skill development; food preparation; food purchasing; food service communication; food service math; garde-manger; international cuisine; introduction to food service; kitchen management; management and human resources; meal planning; meat cutting; menu and facilities design; nutrition; nutrition and food service; patisserie; restaurant opportunities; sanitation; saucier; seafood processing; soup, stock, sauce, and starch production; wines and spirits.

Facilities
2 bakeries; cafeteria; catering service; 20 classrooms; 2 coffee shops; 4 computer laboratories; demonstration laboratory; 3 food production kitchens; garden; gourmet dining room; 3 laboratories; learning resource center; 7 lecture rooms; library; public restaurant; snack shop; student lounge; 4 teaching kitchens.

Culinary Student Profile
570 total: 70 full-time; 500 part-time.

Faculty
46 full-time; 6 part-time. Prominent faculty: John Huber and David Foskett.

Thames Valley University ***(continued)***

Expenses

Tuition: £1000 per term full-time, £400 per year part-time.

Financial Aid

In 1996, 10 scholarships were awarded (average award was £500). Employment placement assistance is available. Employment opportunities within the program are available.

Housing

Average off-campus housing cost per month: £50.

Application Information

Students are accepted for enrollment in January, February, March, September, October, and November. In 1996, 1,050 applied. Applicants must submit a formal application and letters of reference.

Contact

David Foskett, Associate Dean, School of Tourism, Hospitality, and Leisure, Wellington Street, Slough, United Kingdom; 44-753697604; Fax: 44-753697682.

Culinary APPRENTICESHIP Profiles

ACF Birmingham Chapter

Birmingham, Alabama

Program Information
Sponsored by Jefferson State Community College and approved by the American Culinary Federation. Degree program available through Jefferson State Community College.

Placement Information
Participants are placed in one of 20 locations, including restaurants, country clubs, private clubs, and hotels. Most popular placement locations are the Hot and Hot Fish Club, the Vestavia Country Club, and Jimmy's.

Apprentice Profile
Number of apprenticeships offered: 28. 18 apprentices were under 25 years old; 6 were between 25 and 44 years old; 4 were over 44 years old. Applicants must submit a formal application, letters of reference, and an essay.

Expenses
Basic cost of participation is $34 per credit hour for state residents, $59 per credit hour for nonresidents. Program-related fees include: $300 for knives and uniforms; $140 for annual ACF dues.

Entry-level compensation
$6.50 per hour.

Contact
George White, Director, 2601 Carson Road, Birmingham, AL 35215-3098; 205-856-7898; Fax: 205-853-0340; E-mail: tinar912@aol.com

Maricopa Skill Centers

Phoenix, Arizona

Placement Information
Participants are placed in cafeterias.

Apprentice Profile
Number of apprenticeships offered: 30. Applicants must submit a formal application.

Expenses
Application fee: $5. Basic cost of participation is $2430 per 6 months. Program-related fees include: $200 for lab fee.

Contact
Dan Bochicchio, Chef Instructor, 1245 East Buckeye Road, Phoenix, AZ 85034; 602-238-4300; Fax: 602-238-4307; E-mail: bochicchio@gwc.maricopa.edu; World Wide Web: http://www.gwc.maricopa.edu/msc/

See affiliated professional program.

Tucson Culinary Alliance

Tucson, Arizona

Program Information
Degree program available through Pima Community College and Northern Arizona University.

Placement Information
Participants are placed in one of 35 locations, including family run/chef-owned restaurants, health/elder care facilities, institutional/business and industrial kitchens, chain hotels/resorts, and catering operations.

Apprentice Profile
Number of apprenticeships offered: 10. Applicants must submit a formal application and letters of reference and participate in a 500-hour pre-apprenticeship program.

Expenses
Application fee: $50. Basic cost of participation is $750 per year for state residents, $1000 per year for nonresidents. Program-related fees include: $75 for knife set.

Entry-level compensation
$5.15 per hour.

Contact
Lorraine Adler, Training Manager, 3124 East Pima Street, Building B, Tucson, AZ 85716; 520-327-3594; Fax: 520-298-6733.

Chefs de Cuisine Association of California

Los Angeles, California

Apprentice Profile
Number of apprenticeships offered: 3. 3 apprentices were between 25 and 44 years old. Applicants must submit a formal application.

Expenses
Program-related fees include: $65 for annual ACF dues.

Contact
Mary Diltz, Office Secretary, 209 West Alameda, Suite 103, Burbank, CA 91502; 818-559-5218; Fax: 818-559-5219.

Orange Empire Chefs Association

Costa Mesa, California

Program Information
Sponsored by Orange Coast College and approved by the American Culinary Federation. Degree program available through Orange Coast College. Special apprenticeships are available in pastry.

Placement Information
Participants are placed in one of 25 locations, including hotels, clubs, and restaurants. Most popular placement location is the Ritz Carlton Hotel.

Apprentice Profile
Number of apprenticeships offered: 28. 18 apprentices were under 25 years old; 10 were between 25 and 44 years old. Applicants must submit a formal application.

Expenses
Basic cost of participation is $13 per unit for state residents, $139 per unit for nonresidents. Program-related fees include: $100 for knives and uniforms; $100 for ACFEI apprentice registration; $55 for annual ACF dues.

Contact
Bill Barber, Chef Instructor, 2701 Fairview Road, Costa Mesa, CA 92628; 714-432-5835 Ext. 2; Fax: 714-432-5609.

San Francisco Culinary/ Pastry Program

San Francisco, California

Placement Information
Participants are placed in one of 6 locations, including hotels, restaurants, and clubs.

Apprentice Profile
Number of apprenticeships offered: 10. Applicants must submit a formal application.

Contact
Joan Ortega, Director, 760 Market Street, Suite 1066, San Francisco, CA 94102; 415-989-8726; Fax: 414-989-2920.

ACF Culinarians of Colorado

Denver, Colorado

Placement Information
Most popular placement locations are the Brown Palace Hotel and the Hyatt Regency Hotel.

Apprentice Profile
Number of apprenticeships offered: 15. Applicants must submit a formal application and letters of reference and have a high school diploma and a food handlers card.

Expenses
Basic cost of participation is $650 per year. Program-related fees include: $50 for junior membership dues in ACF.

Contact
Michelle Stiers, Training Director, 820 16th Street, Suite 421, Denver, CO 80202; 303-571-5653; Fax: 303-571-4050.

Disney's Culinary Academy

Lake Buena Vista, Florida

Program Information
Degree program available through Valencia Community College.

Placement Information
Participants are placed in one of 200 locations, including resort hotels and theme parks.

Apprentice Profile
Number of apprenticeships offered: 42. Applicants must submit a formal application, letters of reference, an essay, and academic transcripts and have a high school diploma or GED.

Expenses
Application fee: $25. Basic cost of participation is $2500 per year.

Entry-level compensation
$6.20 per hour.

Contact
Cheryl Stieler, Administration Manager, PO Box 10000, Lake Buena Vista, FL 32830; 407-827-4706; Fax: 407-827-4709; E-mail: stielerc@aol.com

ACF First Coast Chapter

Jacksonville, Florida

Program Information
Degree program available through St. Augustine Technical Center. Special apprenticeships are available in baking and pastry.

Placement Information
Participants are placed in one of 10 locations, including resort hotels, city hotels, country clubs, city clubs, and cruise ships. Most popular placement locations are the Ponte Vedra Inn and Club, the Omni Hotel, and the River Club.

Apprentice Profile
Number of apprenticeships offered: 2. Applicants must submit a formal application and letters of reference.

Expenses
Basic cost of participation is $300 per year. Program-related fees include: $200 for knives; $200 for books.

Entry-level compensation
$7 per hour.

Contact
John Wright, Apprenticeship Chairman, 5437 Calloway Court, Jacksonville, FL 32209-2814; 904-765-2140.

Tampa Bay Chefs and Cooks Association

Tampa, Florida

Program Information
Degree program available through Hillsborough Community College.

Apprentice Profile
Number of apprenticeships offered: 10. 6 apprentices were under 25 years old; 4 were between 25 and 44 years old. Applicants must submit a formal application.

Expenses
Application fee: $20. Basic cost of participation is $41.50 per credit for state residents, $154 per credit for nonresidents.

Entry-level compensation
$6.00 per hour.

Contact
George Pastor, Education Director, Hillsborough Community College, Tampa, FL 33614; 813-253-7316; Fax: 813-253-7400.

Volusia County Chefs Association

Daytona Beach, Florida

Program Information
Degree program available through Daytona Beach Community College.

Volusia County Chefs Association ***(continued)***

Placement Information
Most popular placement locations are the Hilton Hotel and the Adams Mark Hotel.

Apprentice Profile
Number of apprenticeships offered: 68. Applicants must submit a formal application and two letters of interest, have a high school diploma, and interview.

Expenses
Basic cost of participation is $108 per class for state residents, $230 per class for nonresidents. Program-related fees include: $100 for uniforms and cutlery; $100 for ACF dues and textbooks.

Entry-level compensation
$6.25 per hour.

Contact
Jeff Conklin, Executive Chef, PO Box 2811, 1200 Volusia Avenue, Daytona Beach, FL 32120; 904-255-8131 Ext. 3735; Fax: 904-254-4465.

ACF Golden Isles of Georgia, Culinary Association

Sea Island, Georgia

Placement Information
Participants are placed in one of 4 locations, including restaurants, dining rooms, banquet halls, and cafeterias.

Apprentice Profile
Number of apprenticeships offered: 9. 6 apprentices were under 25 years old; 3 were between 25 and 44 years old. Applicants must submit a formal application, letters of reference, and a portfolio; take math and English tests; and interview.

Entry-level compensation
$325 per week.

Contact
Franz J. Buck, Executive Chef, The Cloister Hotel, Sea Island, GA 31561; 912-638-3611 Ext. 5724; Fax: 912-638-5159.

ACF Maui Chefs and Cooks Association

Kahului, Hawaii

Program Information
Degree program available through Maui Community College.

Placement Information
Participants are placed in one of 2 locations, including hotels. Most popular placement locations are the Kealani Hotel and the Westin Maui Hotel.

Apprentice Profile
Number of apprenticeships offered: 6. 6 apprentices were under 25 years old. Applicants must submit a formal application and have a high school diploma.

Expenses
Basic cost of participation is $477 per semester for state residents, $2856 per semester for nonresidents.

Entry-level compensation
$9.00 per hour.

Contact
Robert Santos, Apprenticeship Chair, 310 Kaahumanu Avenue, Kahului, HI 96732; 808-984-3225; Fax: 808-984-3314; E-mail: santos@mccada.mauicc.hawaii.edu

ACF Chef de Cuisine/Quad Cities

Bettendorf, Iowa

Program Information
Sponsored by Scott Community College and approved by the American Culinary Federation. Degree program available through Scott Community College.

Placement Information
Participants are placed in one of 30 locations, including country clubs, hotels, restaurants, and casinos. Most popular placement locations are the Radisson Hotel chain and the President Riverboat Casino.

Apprentice Profile
Number of apprenticeships offered: 26. 12 apprentices were under 25 years old; 13 were between 25 and 44 years old; 1 was over 44 years old. Applicants must submit a formal application.

Expenses
Application fee: $20. Basic cost of participation is $51 per credit hour for state residents, $71 per credit hour for nonresidents. Program-related fees include: $100 for uniforms; $150 for tool kit; $100 for training log and textbooks.

Entry-level compensation
$5.50 per hour.

Contact
Bradley Scott, Director of Culinary Arts, 500 Belmont Road, Bettendorf, IA 52722; 319-441-4246.

ACF Greater Kansas City Chefs Association

Overland Park, Kansas

Program Information
Sponsored by Johnson County Community College and approved by the American Culinary Federation. Degree program available through Johnson County Community College.

Apprentice Profile
Number of apprenticeships offered: 140. Applicants must submit a formal application.

Expenses
Application fee: $10. Basic cost of participation is $46 per credit hour for state residents, $122 per credit hour for nonresidents.

Entry-level compensation
$6.00 per hour.

Contact
Jerry Vincent, Academic Director, c/o Johnson County Community College, 12345 College Boulevard, Overland Park, KS 66210; 913-469-8500 Ext. 3250; Fax: 913-469-2560; E-mail: jvincent@johnco.cc.ks.us

ACF New Orleans Chapter

New Orleans, Louisiana

Program Information
Sponsored by Delgado Community College and approved by the American Culinary Federation. Degree program available through Delgado Community College.

Placement Information
Participants are placed in one of 80 locations, including restaurants, hotels, hospitals, and convention centers.

Apprentice Profile
Number of apprenticeships offered: 150. Applicants must submit a formal application and letters of reference.

Expenses
Application fee: $25. Basic cost of participation is $600 per semester for state residents, $1700 per semester for nonresidents.

Entry-level compensation
$5.25 per hour.

Contact
Iva Bergeron, Director, 615 City Park Avenue, New Orleans, LA 70119; 504-483-4208; Fax: 504-483-4893; E-mail: iberge@dcc.edu

ACF Michigan Chefs de Cuisine Association

Farmington Hills, Michigan

Program Information
Sponsored by Oakland Community College and approved by the American Culinary Federation. Degree program available through Oakland Community College.

ACF Michigan Chefs de Cuisine Association *(continued)*

Placement Information
Participants are placed in one of 60 locations, including restaurants, clubs, and hotels. Most popular placement locations are the Detroit Athletic Club, the Golden Mushroom Restaurant, and the Westin Hotel.

Apprentice Profile
Number of apprenticeships offered: 30. Applicants must submit a formal application and letters of reference.

Expenses
Basic cost of participation is $46 per credit hour. Program-related fees include: $150 for books; $120 per semester for lab fees.

Entry-level compensation
$10 per hour.

Contact
Kevin Enright, Apprentice Coordinator, 27055 Orchard Lake Road, Farmington Hills, MI 48324; 248-471-7786; Fax: 248-471-7553.

ACF Chefs & Cooks of Springfield/Ozark

Springfield, Missouri

Program Information
Sponsored by Ozarks Technical Community College and approved by the American Culinary Federation. Degree program available through Ozarks Technical Community College.

Placement Information
Participants are placed in one of 24 locations, including restaurants, country clubs, nursing homes, hotels, and resorts. Most popular placement location is Big Cedar Resort.

Apprentice Profile
Number of apprenticeships offered: 16. Applicants must submit a formal application.

Expenses
Application fee: $30. Basic cost of participation is $43 per credit hour for state residents, $63 per credit hour for nonresidents.

Entry-level compensation
$11 per hour.

Contact
James Lekander, Instructor, PO Box 5958, Springfield, MO 65801; 417-895-7282; Fax: 417-895-7161.

ACF Professional Chefs of Omaha

Omaha, Nebraska

Program Information
Sponsored by Metropolitan Community College and approved by the American Culinary Federation. Degree program available through Metropolitan Community College.

Placement Information
Participants are placed in one of 15 locations, including country clubs, hotels, and restaurants. Most popular placement location is the Happy Hollow Country Club.

Apprentice Profile
Number of apprenticeships offered: 25. 15 apprentices were under 25 years old; 10 were between 25 and 44 years old. Applicants must submit a formal application.

Expenses
Basic cost of participation is $27.50 per credit hour. Program-related fees include: $100 for logbook; $50 for annual ACF dues.

Entry-level compensation
$5.25 per hour.

Contact
James E. Trebbien, Division Representative, PO Box 3777, Omaha, NE 68103; 402-457-2510; Fax: 402-457-2515; E-mail: jtrebbien@metropo.mccneb.edu

ACF Greater North New Hampshire Chapter

Dixville Notch, New Hampshire

Program Information
Sponsored by New Hampshire Community Technical College and approved by the American Culinary Federation. Degree program available through New Hampshire Community Technical College.

Placement Information
Participants are placed in one of 20 locations, including hotels and restaurants. Most popular placement locations are the Balsams Grand Hotel, the American Club, and Mohonk Mountain House.

Apprentice Profile
Number of apprenticeships offered: 15. Applicants must submit a formal application and an essay.

Expenses
Basic cost of participation is $110 per credit for state residents, $253 per credit for nonresidents. Program-related fees include: $119 for knives.

Entry-level compensation
$5.15 per hour.

Contact
Phil Learned, Food Director/Managing Partner, Balsams Grand Hotel, Route 26, Dixville Notch, NH 03576; 603-255-3861; Fax: 603-255-4670.

ACF Chefs of Santa Fe

Santa Fe, New Mexico

Program Information
Sponsored by Santa Fe Community College and approved by the American Culinary Federation. Degree program available through Santa Fe Community College.

Placement Information
Participants are placed in one of 10 locations, including hotels, restaurants, and hospitals. Most popular placement location is the LaFonda Hotel.

Apprentice Profile
Number of apprenticeships offered: 20. Applicants must submit a formal application.

Expenses
Basic cost of participation is $20 per credit hour for state residents, $45 per credit hour for nonresidents. Program-related fees include: $150 for ACF apprentice registration.

Entry-level compensation
$6.00 per hour.

Contact
Bill Weiland, Culinary Director, 6401 Richards Avenue, Santa Fe, NM 87505; 505-438-1600; Fax: 505-438-1237; E-mail: bweiland@santa-fe.cc.nm.us

ACF Paso Del Norte Chapter

Santa Teresa, New Mexico

Placement Information
Participants are placed in one of 2 locations, including restaurants and hotels.

Apprentice Profile
Applicants must submit a formal application.

Entry-level compensation
$4.75 per hour.

Contact
Frankie Herrera, Chef/Instructor, David L. Carrasco Job Corps Center, 11155 Gateway West, El Paso, TX 79935; 915-594-0022; Fax: 915-591-0166.

High Sierra Chefs Association

Carson City, Nevada

Placement Information
Participants are placed in one of 2 locations, including hotels and casinos. Most popular placement location is Harrah's Lake Tahoe Casino.

High Sierra Chefs Association *(continued)*

Apprentice Profile
Number of apprenticeships offered: 8. Applicants must have previous employment experience.

Expenses
Basic cost of participation is $216 per year for state residents, $624 per year for nonresidents. Program-related fees include: $100 for ACF log book and books; $35 for annual ACF dues; $10 for local organization dues.

Entry-level compensation
$6.54 per hour.

Contact
Paul J. Lee, Executive Sous Chef, Highway 50, PO Box 8, Lake Tahoe, NV 89449; 702-588-6611 Ext. 2205; Fax: 702-586-6643; E-mail: plee@laketahoe.harrahs.com

Capital District-Central New York Chapter of the ACF

Albany, New York

Program Information
Sponsored by Schenectady County Community College and approved by the American Culinary Federation. Degree program available through Schenectady County Community College. Special apprenticeships are available in culinary and pastry arts.

Placement Information
Participants are placed in one of 5 locations, including country clubs, restaurants, and hotels.

Apprentice Profile
Number of apprenticeships offered: 10. 10 apprentices were under 25 years old. Applicants must submit a formal application, letters of reference, and an essay.

Expenses
Basic cost of participation is $87 per credit hour for state residents, $135 per credit hour for nonresidents. Program-related fees include: $125 for knives; $100 for uniforms.

Entry-level compensation
$6.00 per hour.

Contact
Scott A. Vadney, Apprenticeship Chair, PO Box 128, Rensselaerville, NY 12147-0128; 518-797-3222; Fax: 518-797-3692; E-mail: confcntr@aol.com

New School Culinary Arts

New York, New York

Program Information
Sponsored by New School for Social Research.

Placement Information
Participants are placed in one of 50 locations, including restaurants, catering operations, pastry shops/bakeries, gourmet shops, and test kitchens. Most popular placement locations are Zoë, Bouley, and the Tea Box Cafe.

Apprentice Profile
Number of apprenticeships offered: 180. Applicants must interview.

Expenses
Basic cost of participation is $2275 per 5 weeks. Program-related fees include: $450 for materials.

Contact
Gary A. Goldberg, Executive Director, 100 Greenwich Avenue, New York, NY 10011; 212-255-4141; Fax: 212-807-0406; E-mail: admissions@newschool.edu

ACF Cleveland Chapter

Willowick, Ohio

Program Information
Degree program available through Cuyahoga Community College.

Placement Information
Participants are placed in one of 30 locations, including country clubs, hotels, hospitals, and restaurants.

Apprentice Profile
Number of apprenticeships offered: 12. Applicants must submit a formal application and interview.

Expenses
Basic cost of participation is $39.80 per credit hour for state residents, $105.50 per credit hour for nonresidents. Program-related fees include: $100 for log book and text book; $40 for annual local ACF chapter dues.

Contact
Richard Fulchiron, Chef Instructor, Cuyahoga Community College, 2900 Community College Avenue, Cleveland, OH 44115; 216-987-4087; Fax: 216-987-4086.

ACF Columbus Chapter

Columbus, Ohio

Program Information
Sponsored by Columbus State Community College and approved by the American Culinary Federation. Degree program available through Columbus State Community College.

Placement Information
Participants are placed in one of 45 locations, including hotels, catering firms, clubs, and restaurants.

Apprentice Profile
Number of apprenticeships offered: 80. Applicants must submit a formal application, letters of reference, an essay, and academic transcripts.

Expenses
Application fee: $10. Basic cost of participation is $59 per credit hour. Program-related fees include: $125 for lab fees.

Entry-level compensation
$6.00 per hour.

Contact
Carol Kizer, Apprenticeship Coordinator, 550 East Spring Street, Columbus, OH 43215; 614-227-5126; Fax: 614-227-5146.

Bucks County Community College

Newtown, Pennsylvania

Program Information
Sponsored by Bucks County Community College. Degree program available through Bucks County Community College.

Placement Information
Participants are placed in one of 80 locations, including hotels, restaurants, contract food services, extended-care facilities, and supermarkets.

Apprentice Profile
Number of apprenticeships offered: 55. Applicants must submit a formal application.

Expenses
Application fee: $30. Basic cost of participation is $66 per credit hour for state residents, $198 per credit hour for nonresidents. Program-related fees include: $160 for knives and pastry kits; $100 for uniform.

Contact
Admissions Department, Swamp Road, Newtown, PA 18940; 215-968-8100; World Wide Web: http://www.bucks.edu

See affiliated professional program.

ACF Delaware Valley Chefs Association

Plymouth Meeting, Pennsylvania

Program Information
Degree program available through Community College of Philadelphia. Special apprenticeships are available in pastry.

ACF Delaware Valley Chefs Association *(continued)*

Placement Information
Participants are placed in one of 300 locations, including restaurants, university dining halls, country clubs, catering houses, and hotels.

Apprentice Profile
Number of apprenticeships offered: 22. Applicants must submit a formal application and letters of reference and interview.

Expenses
Basic cost of participation is $800 per year. Program-related fees include: $75 for ACF dues.

Entry-level compensation
$6.50 per hour.

Contact
William Tillinghast, Associate Director, PO Box 504, Plymouth Meeting, PA 19462; 215-895-1143.

ACF Laurel Highlands Chapter

Youngwood, Pennsylvania

Program Information
Sponsored by Westmoreland County Community College and approved by the American Culinary Federation. Degree program available through Westmoreland County Community College.

Apprentice Profile
Number of apprenticeships offered: 71. 47 apprentices were under 25 years old; 22 were between 25 and 44 years old; 2 were over 44 years old. Applicants must submit a formal application and take a physical exam.

Expenses
Application fee: $10. Basic cost of participation is $1104 per semester for state residents, $1656 per semester for nonresidents. Program-related fees include: $150 for uniforms; $125 for knives; $200 for lab fees.

Entry-level compensation
$5.15 per hour.

Contact
Susan Kuhn, Admissions Coordinator, 400 Armbrust Road, Youngwood, PA 15697-1895; 412-925-4064; Fax: 412-925-1150; E-mail: kuhnsl@wccc.westmoreland.cc.pa.us

Opryland Hotel Culinary Institute

Nashville, Tennessee

Program Information
Sponsored by Opryland Hotel Culinary Institute and approved by the American Culinary Federation. Degree program available through Volunteer State Community College.

Placement Information
Most popular placement location is the Opryland Hotel.

Apprentice Profile
Number of apprenticeships offered: 25. Applicants must submit a formal application, letters of reference, and academic transcripts.

Expenses
Basic cost of participation is covered by the Opryland Hotel.

Entry-level compensation
$5.25 per hour.

Contact
Dina Starks, Culinary Apprenticeship Coordinator, 2800 Opryland Drive, Nashville, TN 37214; 615-871-7765; Fax: 615-871-7872.

See affiliated professional program.

Texas Chefs Association-Dallas Chapter

San Antonio, Texas

Program Information
Sponsored by El Centro College and approved by the American Culinary Federation. Special apprenticeships are available in pastry.

Placement Information
Participants are placed in one of 30 locations, including hotels and clubs. Most popular placement locations are the Hyatt Regency Hotel, the Windham Anatole Hotel, and Harvey Hotels.

Apprentice Profile
Number of apprenticeships offered: 60. 20 apprentices were under 25 years old; 35 were between 25 and 44 years old; 5 were over 44 years old. Applicants must submit a formal application and a biographical essay.

Expenses
Basic cost of participation is $374 per semester for state residents, $644 per semester for nonresidents. Program-related fees include: $212 for ACF membership application.

Entry-level compensation
$5.00 per hour.

Contact
Gus Katsigris, Director, El Centro College, Main and Lamar Streets, Dallas, TX 75202; 214-860-2213; Fax: 214-860-2335.

Virginia Chefs Association

Richmond, Virginia

Program Information
Sponsored by J. Sargent Reynolds Community College and approved by the American Culinary Federation. Degree program available through J. Sargeant Reynolds Community College.

Placement Information
Most popular placement locations are Colonial Williamsburg, the Country Club of Virginia, and the Hermitage County Club.

Apprentice Profile
Number of apprenticeships offered: 25. Applicants must submit a formal application, letters of reference, an essay, and academic transcripts.

Expenses
Basic cost of participation is $3335 per 3 years for state residents, $10,771 per 3 years for nonresidents. Program-related fees include: $600 for textbooks; $140 for ACF fees, manuals, and membership dues; $25 for practical exam application fee.

Entry-level compensation
$6.50 per hour.

Contact
D. Bruce Clarke, Apprenticeship Chairman, 3204 Old Gun Road East, Midlothian, VA 23113; 804-272-0522; Fax: 804-786-5465; E-mail: srbarrd@jsr.cc.va.us; World Wide Web: http://www.jsr.cc.va.us/dtcbusdiv/hospitality

Washington State Chefs Association

Seattle, Washington

Placement Information
Participants are placed in one of 6 locations, including hotels, clubs, and restaurants. Most popular placement locations are the Ranier Club and Westin Hotels.

Apprentice Profile
Number of apprenticeships offered: 15. 7 apprentices were under 25 years old; 8 were between 25 and 44 years old. Applicants must submit a formal application and an essay, have previous employment experience, and interview.

Expenses
Basic cost of participation is $25 per quarter. Program-related fees include: $50 for annual ACF fees; $100 for books and materials; $300 for nutrition management and sanitation course.

Entry-level compensation
$7.00 per hour.

Contact
Jamie Callison, Chef, 8720 231st Street, SW, Edmonds, WA 98026; 425-776-2310.

ACF Chefs of Milwaukee, Inc.

West Allis, Wisconsin

Program Information
Degree program available through Milwaukee Area Technical College.

Placement Information
Participants are placed in one of 60 locations,

ACF Chefs of Milwaukee, Inc. *(continued)*

including restaurants, hotels, private clubs, and catering operations. Most popular placement locations are the Sanford Restaurant, the American Club, and Grenadier's Restaurant.

Apprentice Profile
Number of apprenticeships offered: 30. Applicants must submit a formal application.

Expenses
Basic cost of participation is $350 per semester.

Entry-level compensation
$5.50 per hour.

Contact
John Reiss, Chef/Instructor, 700 West State Street, Milwaukee, WI 53233; 414-297-6861; Fax: 414-297-7990; E-mail: reissj@milwaukee.tec.wi.us

ACF Fox Valley Chapter

Appleton, Wisconsin

Program Information
Sponsored by Fox Valley Technical College and approved by the American Culinary Federation. Degree program available through Fox Valley Technical College.

Placement Information
Participants are placed in one of 12 locations, including hotels, country clubs, and restaurants. Most popular placement location is the Bergstrom Corporation-Appleton.

Apprentice Profile
Number of apprenticeships offered: 24. 12 apprentices were under 25 years old; 12 were between 25 and 44 years old. Applicants must submit a formal application.

Expenses
Basic cost of participation is $180 per semester for state residents, $746 per semester for nonresidents.

Entry-level compensation
$6.00 per hour.

Contact
Albert Exenberger, Apprenticeship Chair, Fox Valley Technical College, 825 North Mound Drive, Appleton, WI 54914; 920-831-5491; Fax: 920-735-2582.

George Brown College of Applied Arts & Technology

Toronto, Ontario, Canada

Program Information
Special apprenticeships are available in patissier.

Placement Information
Participants are placed in restaurants, hotel kitchens, bakeries, hospitals, and schools. Most popular placement locations are the Royal York Hotel and the King Edward Hotel.

Apprentice Profile
Number of apprenticeships offered: 197. 118 apprentices were under 25 years old; 69 were between 25 and 44 years old; 10 were over 44 years old. Applicants must be sponsored by the government and a resident of Ontario.

Contact
Mr. D. Borowec, Chair, 300 Adelaide Street East, Toronto, Ontario, Canada; 416-415-2231; Fax: 416-415-2501; E-mail: dborowec@gbrowne.on.ca

See affiliated professional program.

Niagara College of Applied Arts and Technology

Niagara Falls, Ontario, Canada

Program Information
Sponsored by Niagara College of Applied Arts and Technology.

Apprentice Profile
Number of apprenticeships offered: 2. Applicants must be employed and sponsored by the government.

Expenses
Program-related fees include: Can$450 for knives; Can$120 for uniforms; Can$300 for books.

Contact
Terry Reid, Training Consultant, 5881 Dunn Street, Niagara Falls, Ontario, Canada; 905-988-5528; Fax: 905-988-9250.

See affiliated professional program.

Index

Certificates or Degrees Offered

Programs Offering Certificates or Diplomas

United States

Alabama

Alaska

Arizona

Arkansas

California

Colorado

Connecticut

Delaware

Florida

Georgia

Hawaii

Idaho

Illinois

Iowa

Kansas

Kentucky

Louisiana

Maine

Maryland

Massachusetts

Michigan

Minnesota

Mississippi

Missouri

Montana

Nevada

New Hampshire

New Jersey

New Mexico

New York

Programs Offering Associate Degrees

Programs Offering Bachelor's Degrees

Programs Offering Master's Degrees

Programs offering Doctoral Degrees

INDEX

Alphabetical Index of Schools and Programs

(A) = Apprentice Programs (P) = Professional Programs

D

E

F

G

H

I

J

K

L

(A) = Apprentice Programs (P) = Professional Programs

M

N

O

P

R

S

(A) = Apprentice Programs (P) = Professional Programs

T

U

V

W

(A) = Apprentice Programs (P) = Professional Programs